Einstein

T0396017

Albert Einstein (1879–1955) was the most influential physicist of the 20th century. Less well known is that fundamental philosophical problems, such as concept formation, the role of epistemology in developing and explaining the character of physical theories, and the debate between positivism and realism, played a central role in his thought as a whole.

Thomas Ryckman shows that already at the beginning of his career – at a time when the twin pillars of classical physics, Newtonian mechanics and Maxwell's electromagnetism, were known to have but limited validity – Einstein sought to advance physical theory by positing certain physical principles as secure footholds. That philosophy produced his greatest triumph, the general theory of relativity, and his greatest failure, an unwillingness to accept quantum mechanics. This book shows that Einstein's philosophy grew from a lifelong aspiration for a unified theoretical representation encompassing all physical phenomena. It also considers how Einstein's theories of relativity and criticisms of quantum theory shaped the course of 20th-century philosophy of science.

Including a chronology, glossary, chapter summaries, and suggestions for further reading, Einstein is an ideal introduction to this iconic figure in 20th-century science and philosophy. It is essential reading for students of philosophy of science, and is also suitable for those working in related areas such as physics, history of science, or intellectual history.

Thomas Ryckman is Professor of Philosophy at Stanford University, USA. He is author of The Reign of Relativity: Philosophy in Physics 1915–1925 (2005).

Routledge Philosophers

Edited by Brian Leiter

University of Chicago

Routledge Philosophers is a major series of introductions to the great Western philosophers. Each book places a major philosopher or thinker in historical context, explains and assesses their key arguments, and considers their legacy. Additional features include a chronology of major dates and events, chapter summaries, annotated suggestions for further reading, and a glossary of technical terms.

An ideal starting point for those new to philosophy, they are also essential reading for those interested in the subject at any level.

Recently published:

Kant, second edition
Paul Guyer

Hume
Don Garrett

Dewey
Steven Fesmire

Freud, second edition
Jonathan Lear

Habermas
Kenneth Baynes

Peirce
Albert Atkin

Plato
Constance Meinwald

Plotinus
Eyjólfur Emilsson

Forthcoming:

Merleau-Ponty, second edition
Taylor Carman

Levinas
Michael Morgan

Cassirer
Samantha Matherne

Kierkegaard
Paul Muench

Anscombe
Candace Vogler

Mill
Daniel Jacobson

Berkeley
Lisa Downing and David Hilbert

Nietzsche
Maudemarie Clark

Du Bois
Chike Jeffers

Marx
Jaime Edwards and Brian Leiter

Sartre
Kenneth Williford

Bergson
Mark Sinclair

Arendt
Dana Villa

Adam Smith
Samuel Fleischacker

Thomas Ryckman

Einstein

Routledge
Taylor & Francis Group

LONDON AND NEW YORK

First published 2017
by Routledge
2 Park Square, Milton Park, Abingdon, Oxon OX14 4RN

and by Routledge
711 Third Avenue, New York, NY 10017

Routledge is an imprint of the Taylor & Francis Group, an informa business

British Library Cataloguing-in-Publication Data
A catalogue record for this book is available from the British Library

Library of Congress Cataloging-in-Publication Data
Names: Ryckman, Thomas, author.
Title: Einstein / by Thomas Ryckman.
Description: 1 [edition]. | New York : Routledge, 2017. |
    Series: Routledge philosophers | Includes bibliographical
    references and index.
Identifiers: LCCN 2016056482 | ISBN 9780415773263 (hardback:
    alk. paper) | ISBN 9780415773270 (pbk.: alk. paper) |
    ISBN 9781315175829 (e-book)
Subjects: LCSH: Einstein, Albert, 1879–1955. | Physicists—Biography. |
    Relativity (Physics) | Special relativity (Physics) | Quantum theory.
Classification: LCC QC16.E5 R93 2017 | DDC 530.092 [B]—dc23
LC record available at https://lccn.loc.gov/2016056482

ISBN: 978-0-415-77326-3 (hbk)
ISBN: 978-0-415-77327-0 (pbk)
ISBN: 978-1-315-17582-9 (ebk)

Typeset in Joanna MT and Din
by Apex CoVantage, LLC

for Arthur

Contents

Acknowledgements

I am grateful to the community of excellent Einstein scholars, from whom I have learned just how difficult it is to write sensibly and informatively about someone literally larger than life. Among them, I recognize in particular Olivier Darrigol, Joroen van Dongen, Peter Galison, Herbert Goenner, Hanoch Gutfreund, Gerald Holton, Don Howard, Michel Janssen, Daniel Kennefick, Diana Kormos-Buchwald, Anne J. Kox, Dennis Lehmkuhl, Arthur I. Miller, John Norton, Hans C. Ohanian, Jurgen Renn, Tilman Sauer, and John Stachel. I am grateful to my colleagues, past and present, in the Department of Philosophy at Stanford for their continuous encouragement, and to Michael Friedman, Helen Longino, and Paolo Mancosu for intellectual sustenance. Heartfelt thanks to Olivier Darrigol and Arthur Fine for extremely helpful criticism and comments on drafts of various chapters. I benefitted from comments of Patrick Austin, Harald Wiltsche and Rasmus Winther while Marlene Griffiths Bagdikian assisted in turning much of a teutonically ponderous written text into passable English prose. Above all, I owe the inspiration, or audacity, to undertake this project, almost a common labor of love, to Arthur Fine: mentor, colleague, and friend.

Chronology

1879   Born March 14 in Ulm, Germany, first child of Hermann and Pauline (*née* Koch), bourgeois, largely non-observant, Jews.

1880   Family moves to Munich where Hermann and brother Jakob form electrical engineering firm *Electrotechnische Fabrik J. Einstein & Cie.*

1881   Sister Maria (Maja) born on November 18; lifelong confidant, she joins Albert in Princeton in 1939.

1888   Enters the newly established Luitpold Gymnasium (today, Albert Einstein Gymnasium) in Munich on October 1, only Jew among some 70 classmates.

1889   21-year-old medical student Max Talmud, befriended by Einstein family, introduces Albert to Kant's *Critique of Pure Reason* and to other philosophical and scientific writings. Begins study of differential and integral calculus.

1894   *J. Einstein & Cie* fails, going into liquidation in June; family moves to northern Italy where brothers reopen electrical business. Albert remains in Munich under care of relatives; in December withdraws from Gymnasium without graduating, joining parents in Milan, then Pavia.

1895   Fails entrance exam for Swiss Federal Polytechnical School in Zurich in October; enters cantonal school at Aarau, Switzerland.

1896   Renounces German citizenship in January to avoid military service; affirming of "no religious denomination", remains stateless until acquiring Swiss citizenship in 1901. Passes entrance exam and enters Zuirch Polytechnical School in October; among classmates are Marcel Grossmann and Mileva Marić.

1900   Graduates with diploma qualification to teach mathematics and physics at secondary schools. In December, submits first scientific paper to *Annalen der Physik*, leading German physics journal.

1901   Works as substitute teacher and tutor at small towns outside Zurich. Dissertation on molecular forces in gases submitted to University of Zurich in December, withdrawn in March 1902.

1902   At end of January, Mileva Marić gives birth to Einstein's daughter Lieserl at parent's home in Novi Sad; Einstein never sees child, presumably given for adoption. Around Easter, together with Maurice Solovine and Conrad Habicht, forms "Olympia Academy" in Bern, informal reading and discussion group of philosophical, physical, and literary texts. With Grossmann's assistance, obtains temporary employment in June as Technical Expert, Third Class, at Swiss patent office in Bern. In September, first of trilogy of papers on statistical thermodynamics published in *Annalen der Physik*.

1903   Despite objections of both families, marries Mileva in January; second paper of statistical trilogy published in April.

1904   Son Hans Albert born in Bern in May. Final trilogy paper published in June, presenting theory of energy fluctuations in equilibrium systems. Probationary employment at Bern patent office becomes permanent in September.

1905   *Annus mirabilis* with papers on light-quantum hypothesis (March), Brownian motion (April), special relativity (June), and equivalence of energy and mass (September), all appear in *Annalen der Physik*.

1906   Receives doctorate in January from University of Zurich; thesis "A New Determination of Molecular Dimensions" published in *Annalen der Physik*. "On the Theory of Light Production and Light Absorption" appears in April, the paper "announcing birth of the quantum theory" (Kuhn). Promoted that month to Technical Expert, Second Class. In November completes paper that extends quantum theory of radiation to electrons moving in solid materials, providing explanation of anomalous specific heats; published on December 28.

1907   On January 7, the "most fortunate thought of my life" leads to principle of equivalence. In September, commissioned to write report on relativity theory and its consequences for Johannes Stark's *Jahrbuch der Radioaktivität*; appearing in December, it contains first explicit statement that the principle of equivalence extends principle of relativity to uniformly accelerating systems.

1909   On May 7, appointed associate (*Ausserordentlicher*) professor of physics at University of Zurich. Two significant papers: the first, "On the Present Standpoint of the Radiation Problem", sent in January to *Physikalische Zeitschrift*, applies energy fluctuation theory to blackbody radiation, a new argument for existence of light quanta. The second, invited address in September to assembly of German scientists in Salzburg, argues theory of light to become a kind of "merger" (*Verschmelzung*) of wave and particle theories. In July, awarded honorary doctorate by the University of Geneva, first of twenty-three such honors.

1910   In July, second son Eduard is born in Zurich. Berlin physical chemist Walter Nernst (Nobel Prize, 1920) pays visit to discuss problem of specific heats.

1911   Nernst reports "brilliant confirmation of the quantum 'theory' of Planck and Einstein" in March paper providing experimentally determined values for specific heats of a number of metals and diamond at low temperatures. Appointed Professor of Theoretical Physics at the German University of Prague on April 1. In October, attends prestigious First Solvay Conference in Brussels devoted to "The Theory of Radiation and the Quanta". Youngest invitee at age 32, gives report on current state of problem of specific heats.

1912   Returns to Zurich in October as Professor of Theoretical Physics at renamed Swiss Federal Institute of Technology (ETH). Seeks instruction on Riemannian geometry and "absolute differential" (tensor) calculus from Marcel Grossmann, professor of mathematics at ETH.

1913   In June, publishes monographic *Entwurf* (Outline) of relativistic theory of gravity with Grossmann, who wrote mathematical part. In November, elected to Prussian (Berlin) Academy of Sciences. In December accepts July offer, tendered personally by Max Planck and Walther Nernst, of position as research professor in Berlin with no teaching obligations, and as founding director of future Kaiser Wilhelm Institute of Physics. Becomes Prussian citizen yet retains Swiss passport.

1914   Moves to Berlin at end of March; Mileva and two sons arrive in April, return to Zurich at end of July, separation ending in 1919 divorce. WW I begins in August; one of few signatories of immediately censored pacifist "Manifesto to the Europeans" denouncing nationalism and war, first public political declaration. In October systematic exposé of 1913 theory of gravitation in Proceedings of Berlin Academy with several new amendments; "*Lochbetrachtung*" in §12 invoked as proof of necessary restriction of covariance of gravitational field equations.

1915   At invitation of mathematician David Hilbert, gives six two-hour lectures in Göttingen from June 28 to July 5 on 1913–1914 relativistic theory of gravitation. Hilbert begins work on relativistic gravitational theory coupled to electromagnetic theory of matter. In competition with Hilbert in Göttingen, each Thursday in November Einstein presents paper to Berlin Academy. In the third, on November 18, derives exact observed value of the anomalous precession of the perihelion of Mercury. In last, on November 25, the gravitational field equations in generally covariant form.

1916   Canonical presentation of general theory of relativity (*Die Grundlagen der allgemeinen Relativitätstheorie*) published in March in *Annalen der Physik*. Elected head of the German Physical Society (*Deutsche Physikalische Gesellschaft*) in May; in June, solution of field equations in linearized approximation yield plane gravitational waves that "always propagate with the velocity of light"; the paper contains a serious error. In July, publishes new derivation of Planck's blackbody radiation law, distinguishing spontaneous and stimulated emission, basis of maser and laser technologies (1950s). In December, completes popular book on two theories of relativity (*Über die spezielle und die allgemeine Relativitätstheorie, gemeinverständlich*), appears in early 1917 with many subsequent editions.

1917   On February 8, presents "Cosmological Considerations in the General Theory of Relativity" to Berlin Academy, introducing "cosmological term" into gravitational field equations. Prompts correspondence, and further developments by, Dutch astronomer Willem de Sitter. Kaiser Wilhelm Institute of Physics finally established after three years delay due to war; becomes head on October 1.

1918   In January, first paper on gravitational radiation, deriving correct (quadrupole i.e., "spinning dumbbell") formula expressing amount of energy radiated away by massive source. In March, paper stating general theory of relativity rests upon three principles (equivalence, general covariance, and "Mach's"). In May, arranges publication of Zurich mathematician Hermann Weyl's theory of gravitation and electromagnetism in *Proceedings* of Berlin Academy, followed by appended objection. On November 9, Germany surrenders, Kaiser Wilhelm abdicates, Social Democrats in *Reichstag* proclaim a republic and form a provisional cabinet; armistice with allies signed November 11. Fighting breaks out in Germany between revolutionary socialists and nationalists, including reactionary paramilitary *Freikorps*. Together with physicist Max Born and psychologist Max Wertheimer, obtains release of rector and deans of the University of Berlin held hostage in December by a revolutionary student group.

1919   Communist Spartacus League revolt in Berlin crushed by army in January. Divorce from Mileva finalized on February 14; settlement includes monetary award from any future Nobel Prize. Marries cousin Elsa Löwenthal on June 2. Weimar constitution goes into effect on August 11, creating parliamentary democracy in Germany. On November 6 at joint meeting in London of Royal Society and Royal Astronomical Society, results of British observations of solar eclipse previous May 29 announced. Key prediction of general theory of relativity confirmed; light rays from distant stars "bent" on passing through solar corona on way to Earth. Becomes media sensation and world-recognized figure.

1920 Target of anti-Semitic nationalists and opponents of relativity theory; responds in both professional and popular venues. Speaks out against oppression of Jews from Eastern Europe.

1921 January 27, lecture "Geometry and Experience" at Berlin Academy Founder's Day celebration. Departing Berlin in April, accompanies Chaim Weizmann on fundraising tour in USA on behalf of Hebrew University; in June interview states intellectuals use anti-Semitism for political purposes, and claims recent awareness of need for Jewish cultural and spiritual home in Palestine. On April 25 meets President Warren Harding at White House. In May four lectures on the theory of relativity at Princeton University; published in German and English (as The Meaning of Relativity) in 1922.

1922 In January with assistant Jakob Grommer, completes first paper on unified field theory; shows equations of five-dimensional 1919 theory of Theodor Kaluza unifying gravitation and electromagnetism lack solutions corresponding to singularity-free representations of electrons; appears in January 1923 in publication of Hebrew University. Invited by French scholars, visits France in late March and early April; tours WW I battlefields, lectures in Paris to scientists at Collége de France and philosophers at Sorbonne. In May, joins International Committee on Intellectual Cooperation (ICIC) affiliated with League of Nations. Following June 24 assassination of friend Walther Rathenau, German Foreign Minister and a Jew, withdraws temporarily from public life. In September, submits note to Zeitschrift für Physik claiming A. Friedmann's solution of the gravitational field equations showing a non-static universe rest on a mathematical mistake. Embarks on six-month journey to Far East at the end of October, visiting Japan, Ceylon, Singapore, Hong Kong, Shanghai; on return, Palestine and Spain. On November 9, awarded 1921 Nobel Prize in Physics for discovery of "law of the photoelectric effect"; German ambassador to Sweden accepts at Nobel ceremony in Stockholm on December 10.

1923   Returns to Berlin in February. Withdraws from ICIC in spring, protesting League's impotence in the face of French-Belgian occupation of the Ruhr. First papers on "affine field theory", including one translated into English in *Nature*. Idea, initiated by Arthur Eddington in 1921, pursued until 1926. On July 11, gives "Nobel Lecture" in Gothenburg, Sweden.

1924   In May, rejoins ICIC, urging League to permit German membership. Receives June letter from S.N. Bose (Calcutta), together with manuscript in English re-deriving Planck's radiation law. Translates Bose's paper into German, arranging publication in *Zeitschrift für Physik*. In July publishes analogy between Bose's statistical treatment of radiation and material gases; i.e., ideal gases of single atoms at very low temperature, showing deviations from classical gas law in this regime.

1925   Two further papers show non-classical quantum behavior of ideal gases at low temperature, using analogy to Bose's statistical method treating photons as indistinguishable particles. The first, in early January, calls attention to L. de Broglie's 1924 Paris thesis on wave nature of particles of matter and predicts existence of new state of matter. ("Bose-Einstein condensate" first created in lab in 1995.) Spends month in Argentina (March 24 to April 21), followed by visits to Uruguay and Brazil, returning to Berlin in early June.

1926   Closely follows new quantum mechanics of Heisenberg (with Born and Jordan), then of Schrödinger. In August, among notable signatories of anti-conscription manifesto, urging League of Nations to propose abolition of compulsory military service among member states.

1927   In January and February two-part paper to Berlin Academy exploring Kaluza's five-dimensional theory connecting gravitation and electromagnetism without restriction to weak gravitational fields; unknowingly duplicates 1926 results of Swedish physicist Oscar Klein. Also in January, with assistant Grommer, publishes derivation of geodesic equation (law of motion for freely falling material point) from field equations of gravitation, results previously obtained by

Hermann Weyl in 1923. Fifth Solvay Conference in Brussels in last week of October; begins discussion of foundations of quantum mechanics with Niels Bohr.

1928   Collapses in February in Davos, Switzerland; after diagnosis of heart inflammation, spends four months in bed in Berlin; while recuperating, secretary Helen Dukas hired in April. In June, two papers outline new geometrical framework for unified field theory, a four-dimensional space-time with "distant parallelism" (*Fernparallelismus*).

1929   Exposition of 1928 theory becomes January public sensation, first printing in Berlin Academy *Proceedings* (1,000 copies) sells out instantly, three additional printings, also by newspapers in London and New York; displayed in window of Selfridges department store in London. January 17 *Proceedings of the National Academy of Sciences* publishes Edwin Hubble's observational evidence showing expansion of universe consistent with Friedmann's solutions in 1922.

1930   Attends Sixth Solvay Conference in Brussels in October; further discussions with Bohr on foundations of quantum mechanics. November 9 article on "Religion and Science" in the *New York Times* affirms faith in "cosmic religion" revealed by the "miraculous order that manifests itself in all of nature as well as in the world of ideas" but without a personal God. On December 2, leaves for New York, Cuba, then Pasadena (Caltech), remaining until early March. "Two-Percent Speech" in New York on December 14 states that just 2 percent of men resisting military conscription would be too many to jail, forcing governments to settle disputes peaceably. After lengthy correspondence with French geometer Élie Cartan, publishes outline of geometry of 1928–1929 distant parallelism theory in December issue of *Mathematische Annalen*, accompanied by Cartan's historical note showing new geometry to be a special case of more general and earlier results.

1931   Responding to 1929 observations of Hubble, accepts non-static (expanding) character of universe, formally renounces the cosmological constant in April. In October, paper with new assistant Walther Mayer introduces novel approach to unified field theory: to each event of four-dimensional space-time is associated a vector space with five spatial dimensions, permitting mixed tensors linking the two manifolds. Achieves formal unification of gravitation and electricity, but authors conclude no light shed on particulate structure of matter nor on known quantum facts. Publication in Leipzig of *Hundert Autoren gegen Einstein.*

1932 Resigns from ICIC due to lack of effectiveness, yet accepts ICIC invitation to dialogue with another thinker on war and politics; choosing Sigmund Freud, correspondence published in Paris in summer 1933 as pacifist pamphlet *Warum Krieg (Why War?)*. Advocates surrender of national sovereignty to supranational organization in order to eliminate war. Second paper with Mayer appears in April generalizing 1931 theory, formally reproducing field equations of gravitation and electromagnetism, but no solutions correspond to elementary particles; "distant parallelism" given up in late March. Institute of Advanced Study (IAS) in Princeton, to begin operations in autumn 1933, extends offer in June to become one of six initial IAS professors; accepts, provided Mayer comes as assistant; arrangement approved in October. In German elections in July, Nazis win 230 of 607 *Reichstag* seats becoming largest party. In November first of four papers with Mayer on yet another approach to unified field theory, introducing concept of "semivector". Leaves in December with Mayer and wife Elsa for ostensible three-month visit to Caltech; depart Germany with thirty-three pieces of luggage.

1933 Adolf Hitler appointed Chancellor of Germany on January 30 by *Reichspräsident* Hindenburg; *Reichstag* fire in March used as pretext for dictatorial decrees violating Weimar constitution. Einstein apartment in Berlin (5 *Haberland-strasse*) raided and ransacked; most documents preserved, previously taken to French Embassy. On return to Europe (Antwerp) on March 28, promptly renounces German citizenship and resigns from Berlin Academy. Albert and Elsa reside temporarily in Belgium, Albert leaving for UK in early June to give Herbert Spencer Lecture in Oxford on June 10. On October 3rd speaks at benefit for refugee

German scholars before 9,000 in London's Royal Albert Hall; on October 7, Einsteins, Mayer and Dukas depart Southhampton, arriving in New York on October 17 and Princeton that evening.

1935   January article (in *Polity*) states pacifists must adopt means other than passive resistance to oppose "the warlike programs of political adventurers". In March receives Franklin Medal in Philadelphia. On May 15, *Physical Review* publishes EPR (Einstein-Podolsky-Rosen), and July 1 paper with Nathan Rosen proposing topological solution to particle problem in general relativity. In August, purchases single family house in Princeton at 112 Mercer Street.

1936   "Physics and Reality" published in *The Journal of the Franklin Institute* in March. Short note in December 4 *Science* derives lens-like action on light from one star traversing the gravitational field of another star. Elsa dies after long illness on December 20.

1937   A paper ("Do Gravitational Waves Exist?") co-authored with Rosen, denies existence of gravitational waves; withdrawn after submission to *Physical Review* following a critical referee report, published with altered conclusion in January issue of *The Journal of the Franklin Institute*. Never again submits to *Physical Review*. In Princeton for April 16 recital, contralto Marian Anderson accepts Einstein's offer of lodging after being refused room at whites-only Nassau Inn; they remain friends until Einstein's death in 1955.

1938   Further work with assistants Leopold Infeld and Banesh Hoffman on geodetic law in general relativity; with assistant Peter Bergmann on five-dimensional Kaluza-Klein theory considered purely as classical field theory (without Planck's constant). Bestselling popular book (*The Evolution of Physics*), co-written with Infeld, supports latter's stay at Institute. Experiments of physical chemists Otto Hahn and Fritz Strassmann at Kaiser Wilhelm Institute interpreted in December, by Lise Meitner and Otto Frisch in Sweden, as showing fission of uranium nuclei.

1939   Vacationing at Peconic on eastern Long Island in July, Einstein dictates famous letter to President Roosevelt warning of military uses of atomic fission. WW II begins in Europe on 1 September with Germany's invasion of Poland.

1940 March 19 statement in *New York Times* supports Bertrand Russell ("Great spirits have always found violent opposition from mediocrities"); nonetheless, Russell's contract to teach at the College of the City of New York rescinded at court trial on March 30. In ceremony at Trenton, New Jersey, becomes naturalized US citizen retaining Swiss citizenship on October 1.

1943 January paper with Wolfgang Pauli Jr. proves non-existence of stationary, non-singular particle solutions in a five-dimensional generalization of relativistic theory of gravity. Begins *per diem* consulting arrangement on high explosives with US Navy Bureau of Ordinance in May.

1944 Freshly rewritten manuscript of 1905 relativity paper auctioned in War Bond drive at Kansas City on February 3 raising $6.5 million (approx. $90 million in 2016); manuscript subsequently presented to Library of Congress. Two papers, the first with assistant Valentin Bargmann, on "bi-vector fields"; bi-vectors proposed as new mathematical concept to replace metric tensor in four-dimensional space-time continuum.

1945 Germany surrenders on May 7; WW II ends on September 2, following use of atomic weapons against Japan. In October adopts final approach to unified field theory, four-dimensional space-time with asymmetric metric tensor and complex components; papers on this theory co-authored with last research assistant Bruria Kaufmann in 1954 and 1955.

1946 On May 23 named chair of Emergency Committee of Atomic Scientists (ECAS), group promoting international control of atomic energy and warning of danger of proliferation of atomic weapons. Remains head until organization disbanded in 1951. Writes "Autobiographical Notes" for forthcoming *Einstein: Philosopher-Scientist* volume edited by philosopher Paul Schilpp.

1948 Essay "Quantum Mechanics and Reality" (*Quanten-Mechanik und Wirklichkeit*) sent in April to Swiss journal *Dialectica*, and separately to Max Born. After abdominal surgery, diagnosed in December with aneurysm of major abdominal aorta.

1949 Recuperating in Florida, in February completes responses ("Remarks Concerning the Essays Brought Together in this Co-operative Volume") to essays in Schilpp volume.

1950 Signs last will and testament in March, leaving papers to Hebrew University. April *Scientific American* contains "On the Generalized Theory of Gravitation", popular exposition of developments leading to current approach to unified field theory; contains self-description as "tamed metaphysicist".

1951 Death of sister Maja in Princeton.

1952 Declines offer of Presidency of Israel in November, following death of Chaim Weitzmann. Essay "Relativity and the Problem of Space", added to fifteenth English edition (1954) of popular relativity book from 1917, concludes "Spacetime does not claim an existence on its own, but only as a structural quality of the field".

1953 Paper in Max Born *Festschrift* argues quantum mechanical descriptions cannot be extended to classical limit. Letter to Brooklyn schoolteacher in May, urging non-compliance with Joseph McCarthy's Senate Internal Security Subcommittee, elicits denunciations from major US newspapers, including *The New York Times*. Further attacks follow support in December given to electrical engineer union member refusing to answer questions posed by the committee.

1954 In April, advises Robert Oppenheimer, Director of IAS, not to cooperate with Atomic Energy Commission hearing on suspension of Oppenheimer's security clearance; Oppenheimer chooses to testify, losing clearance at the end of June.

1955 Signs last letter, to Bertrand Russell on April 11, agreeing to lend name for "Einstein-Russell Manifesto" on nuclear disarmament. Dies April 18 in Princeton hospital from ruptured aneurysm of abdominal aorta. Helen Dukas inhabits 112 Mercer Street until her death in 1982.

1979 Celebration of centennial of Einstein's birth with conferences in Jerusalem, Berlin and Princeton (The Institute of

Advanced Study); commemorative proclamations by Jimmy Carter, 39th President of the USA, and Pope John Paul II.

1987 Princeton University Press begins publication of *Collected Papers* and correspondence.

2005 Centennial of *annus mirabilis* named "World Year of Physics" by International Union of Pure and Applied Physics (IUPAP).

2015 Centennial of general theory of relativity celebrated by Physical Review journals, American Institute of Physics, NASA, IUPAP, and conferences around the world; on September 14 gravitational waves, predicted in 1916, are first observed.

Introduction

"I do not feel comfortable and at home in any of the 'isms'. It always seems to me as though such an *ism* is strong only so long as it nourishes itself on the weakness of its counter-*ism*. But if the latter is struck dead and it is alone on an open field, then it proves to be wobbly on its legs. Therefore, *away with the grousing* (**los mit der Stänkerei**)!"
Albert Einstein, letter to mathematician Edward Study, September 25, 1918[1]

Albert Einstein was a theoretical physicist, not a philosopher in any customary sense. His contributions to, and influence upon, physical theory are rivaled, if at all, only by those of Isaac Newton at the beginning of the modern age. A rapid succession of publications, appearing within a few months of each other in the *annus mirabilus* 1905 brought Einstein the attention of Europe's most prominent physicists. At the time obscurely employed as a full-time patent clerk in Bern, Switzerland, any one of three of these would suffice to ensure a glowing reputation in the annals of 20th-century physics. Best known is the paper outlining the special theory of relativity, showing that the recognized incompatibility between Maxwell's theory of electromagnetism and classical mechanics required modifying Newtonian mechanics. Einstein himself favored the one stating the light-quantum hypothesis; of the three, only this he considered "very revolutionary". It contains the first argument that Planck's quantum of energy is a physical entity. And then the paper outlining the theory of Brownian motion within a few years led to experiments resulting in universal acceptance of the reality of atoms. His greatest achievement came not in 1905 but in 1915, with the relativistic theory of gravity, or "general relativity", completed after eight years of struggle in the middle of World War I. Conceived as a theory to remove the last vestiges of fixed background structure from the physical doctrines of space and time, by 1917 Einstein had applied it to the entire universe, in so doing creating modern relativistic cosmology.

The theories of relativity, special and general, are two of three props of contemporary fundamental physical theory, extending from elementary particle theory, where special relativity is essential and gravitation is ignored, to inflationary cosmology and the origin and fate of the universe. Even Einstein's failures are spectacular. The theory of light quanta and his call for a dual wave-particle theory of light in 1909 effectively set the table for the other revolution of 20th-century physics, the indeterministic quantum theory he would never accept. After the arrival of quantum mechanics in 1925–1926, Einstein remained a skeptic, yet called attention to phenomena (most notably, "entanglement") nearly all the founders of quantum mechanics either dismissed as an imperfectly posed "paradox" or chose to ignore. More than any other scientist, Einstein created the basis for the current understanding of nature and indeed the universe. *Time Magazine*'s recognition as "Person of the Century" (December 26, 1999) was both deserved and not at all surprising.

In what sense is Einstein to be considered a *philosopher*? To Abraham Pais, particle physicist and author of what remains the sole scientific biography, "calling Einstein a philosopher sheds as much light on him as calling him a musician".[2] Though perhaps too harshly put, the observation is largely correct. Einstein was at most a "philosopher-scientist" (per the subtitle of Schlipp 1949). Whereas philosophers typically develop a characteristic viewpoint in dialogue with, or response to, other philosophers or philosophical traditions, as did Kant with respect to Hume (empiricism) and Leibniz (rationalism), there is very little evidence of this kind of exchange or transmission to be found in Einstein's writings. Letters occasionally mention philosophical readings, briefly reporting praise or disagreement. There is a late reminiscence crediting texts of Hume and Ernst Mach as assisting the realization, in 1905, of the unphysical nature of the prevailing, and largely implicit, presupposition of absolute time (absolute simultaneity). Yet as the 1918 epigraph indicates, Einstein did "not feel comfortable and at home in any of the -*isms*", and he shared the modern physicist's disdain for the "squabbling" among them. Disregard for philosophical pigeon holes is evidenced in his one summary self-characterization as "a type of unscrupulous opportunist", appearing alternately "as *realist* . . . as *idealist* . . . as *positivist* . . . as *Platonist* or *Pythagorean*".[3]

To deem Einstein a philosopher even in the attenuated sense of "philosopher-scientist" would suggest that there is in fact an Einsteinian *philosophy*. Is there an overall narrative to be fashioned to encompass all or most of the observations of a philosophical character scattered in his writings? A looming difficulty is the largely occasional nature of many of Einstein's philosophical remarks. Trying to find consistency presents a challenge to any synthetic portrayal of Einstein's philosophy. This is not to say that selective attention to certain comments or to different periods of Einstein's career hasn't produced various attempts to assimilate him to one or another philosophical heading. Logical empiricist philosophers took inspiration from various remarks pre-1925 that appear little more than broadcasts of textbook positivism. But positivism of any variety cannot abide later notable expressions of faith in mathematical speculation, or "logical simplicity", as a method capable of revealing nature's innermost secrets.

The apparently extreme swings of direction are frustrating to anyone looking for a convincing portrait of Einstein's philosophical views with broad textual support. A natural accommodating strategy is then to identify a pivotal change in philosophical affinity or allegiance. A well-known attempt of this kind is that of physicist and historian of science Gerald Holton who viewed the completion of the general theory of relativity as the key episode.[4] Holton, the leading Einstein scholar of his generation, portrayed Einstein's career as a philosophical odyssey, a "pilgrimage . . . starting on the historic ground of positivism", heavily under the influence of Mach. Over time, "apostasy" from Mach became more and more apparent, particularly following the triumph of general relativity, a success Einstein subsequently attributed to his belated adoption of a mathematical "speculative-constructive" method. Einstein's philosophical terminus is then a "rationalistic realism" that essentially turned Mach's phenomenalistic positivism "on its head" (see Chapter 8). Holton's narrative is instructive and has been highly influential.

Yet even while accurate in its broad contours, the Holton template doesn't quite fit the facts; the young Einstein was at least as much a follower of Boltzmann (and proponent of the reality of atoms) as of Mach (who deemed atomism to be a "metaphysical" hypothesis). On the other hand, the older Einstein, Holton's "rationalistic realist", can still appear almost a pragmatist in the mold of Dewey, declaring "the whole of science" to be "nothing more than a refinement of everyday thinking".[5] The challenge of interpolating between ostensibly conflicting philosophical pronouncements still remains.

A clue lies in the Herbert Spencer lecture, given at Oxford on June 10, 1933. Lecturing for the first time in English, Einstein spoke "On the Method of Theoretical Physics", a topic with definite, if somewhat constrained, philosophical import. Holton and subsequently others emphasize that the lecture can be considered something of a philosophical watershed, a public "coming out", proclaiming a faith in mathematical simplicity as trustworthy guide to theoretical comprehension of the most fundamental underlying processes of nature. For present purposes, that of identifying a philosophy that can be warranted as *Einstein's*, another message in the lecture stands out, issued in a provocative challenge near the beginning:

> If you wish to learn from the theoretical physicist anything about the methods which he uses, I would give you the following piece of advice: Don't listen to his words, examine his achievements.[6]

This book heeds Einstein's advice, identifying a philosophy that is indeed his, though "philosophy" is here to be understood in the first instance in the indicated sense, namely, as illustrated by his main achievements in physical theory. It is a philosophy manifested over his entire career, consisting in the advance or critique of theory through the use, and presumed general validity, of certain principles; physical, formal, methodological, and metaphysical. While Einstein's various remarks do reveal a philosophical viewpoint informed by decades of reading, it is a "philosophy of principles" guiding theoretical practice and shaping philosophical pronouncements rather than the other way round; as he once put it, "the (physicist) himself knows best . . . where the shoe pinches him".[7]

Philosophy of principles

Historians of physics have remarked that Einstein's strivings and accomplishments in physical theory have a distinctive character: they reveal the quest of the *natural philosopher*, of one who believes that genuine understanding of physical reality is attainable only from the standpoint of a unified theoretical foundation for physics. The pursuit of such a foundation can be seen across the more than fifty-year span of his life as a physicist.

In the 17th century the idea there might be a unified foundation for all of physics originated with Descartes, for whom extension was the primary attribute of matter, motion was caused exclusively by contact, and physical explanation was necessarily causal. Classical mechanics, largely created by Newton, departed from the Cartesian model by introducing the notion of force, a program adequate for all of physical theory for more than two centuries. All physical phenomena were to be represented in terms of the concepts of space, time, material point, and force. Shortly after 1900, Einstein recognized that classical mechanics could no longer serve as the foundation it had been essentially since 1687. In his self-described "obituary" in 1946, Einstein recalled that in the wake of "Planck's trailblazing work" in 1900, it had become clear "that neither mechanics nor thermodynamics could claim exact validity, except in limiting cases".[8] Following Maxwell and Boltzmann, most theorists regarded thermodynamics as resting upon the "molecular-kinetic theory of heat", i.e., the assumption that physical systems are composed of molecules obeying Newtonian laws of motion and collision, and that macroscopic thermodynamic properties of systems (temperature, pressure, density, entropy, etc.) are brought about by motions and collisions of vast numbers of molecules.

From a completely different direction, Maxwell's field theory of electromagnetism successfully unified physical optics and the theories of electricity and magnetism, showing that light itself was an electromagnetic phenomenon. This was a revelation to the young Einstein. But in 1900 Planck had also shown what he himself was reluctant to admit, that thermal radiation – the energy emitted and absorbed by matter – had a particulate and non-Maxwellian character. At least by 1905, Einstein understood that Maxwell's theory was not fully compatible with Planck's law successfully describing the spectrum of blackbody radiation. If – Einstein argued from two directions in 1905 – the point particle mechanics of Newton failed for very rapid periodic processes (as in matter-radiation interactions) as well as for systems moving at or near the speed of light, what might replace Newtonian mechanics, assumed the foundation of all of physics, including thermodynamics? It was "as if the ground had been pulled out from under one's feet, with no firm foundation on which to build anywhere" (1946, p. 45). Progress toward a new unified foundation, if possible at all, might be made only by building from a secure basis. Where, amidst the turmoil and uncertainty of turn-of-the century physical theory, might this be found?

By 1905 Einstein had placed epistemic bets on three prescriptive injunctions: the relativity principle, the light principle, and what he termed "Boltzmann's principle", the definitional connection between thermodynamic entropy and probability on which Boltzmann anchored the probabilistic interpretation of the second law of thermodynamics. The first two, both postulates of special relativity, affirmed just those features of mechanics and electromagnetism deemed essential and indispensable to each. Any future advance in mechanics or electromagnetic theory would have to admit their empirical validity. But this was a lesson Einstein learned from the principle named after Boltzmann. By at least 1904, Einstein realized theoretical progress or unification can be attained by according *certain physical principles wider empirical validity than a particular theory based upon on that principle* (see Chapter 3).

The cited "obituary" obliquely refers to this realization. In explaining how he came to introduce the postulate that the laws of physics must satisfy the principle of relativity, Einstein relates that he was inspired by the example of thermodynamics. Thermodynamics impressed him as a particular *type* of physical theory, a *theory of principle*: in essence, one whose empirical validity in most general forms, extending beyond particular applications or formulations, could be summarized by confirmed facts of projected universal scope, the prohibition of two kinds of perpetual motion, corresponding to the first two laws of thermodynamics. In Boltzmann's principle and in the energy principle (conservation of energy) Einstein recognized two core postulates underlying all thermodynamic phenomena but of broader validity than recognized in the existing statistical treatment of thermodynamics. He would then use this understanding to highlight the significance of energy fluctuation phenomena, natural processes lying outside the purview of classical thermodynamic treatment, pointing to the non-classical and quantum character of the physical micro-world.

Principles also guided Einstein to the general theory of relativity. The path was opened in 1907 by an insight that led to the "principle of equivalence". Loosely construed (see Chapter 6 for a careful statement), it affirms that in certain ideal situations, an observer inside a closed cabinet would not be able to observationally distinguish between the effects of gravity and those of uniform acceleration in a gravity-free region. The insight proved the crucial heuristic for extending the principle of relativity from the inertial frames of special relativity to the non-inertial frames of accelerating systems, and so to gravitational phenomena. The motivation for such an extension was deeply philosophical, prompted by Mach's earlier critique of Newton's explanation of inertial effects, a treatment that implicitly relied on the metaphysical notion of absolute space. On the eight-year path to general relativity, Einstein repeatedly referred to Mach's attempt to "relativize inertia", and on completion of the theory, Einstein would state (wrongly, as is now known) that global (cosmological) solutions of the theory's field equations rest upon "Mach's principle". That principle, as will be seen, was intimately intertwined in Einstein's thought with another, a "principle of general relativity" that turns out to be a merely formal principle, the "principle of general covariance", pertaining to the invariant form equations take in any permissible coordinate system. General relativity, resting on the three above principles, also prompted Einstein's statement of still another principle in 1931, the so-called "cosmological principle", affirming "all locations in the universe are equivalent; in particular the locally averaged density of stellar matter should therefore be the same everywhere".[9] The cosmological principle remains today a widely held methodological assumption of many cosmological models (see Chapter 11).

The philosophy of prescriptive principles, comprised of those above, and others to be discussed, informed not only Einstein's achievements as a theoretical physicist but also his failures. These are essentially two: an unwillingness to accept the irreducibly probabilistic character of quantum theory, and correspondingly the ill-fated effort to replace it, a three-decades-long attempt to produce a "unified field theory". This was to be an extension of general relativity in which particles were mathematically characterized through particular solutions of the field equations, and the quantum behavior of matter and radiation indeed could be deemed probabilistic, but only in the sense of classical statistical mechanics where probabilities and averages are proxies for knowledge of exact positions and velocities of enormous numbers of particles. In effect, both failures are a consequence of Einstein's conception of what a fundamental physical theory is, and what it should aim to do. These aspects of his philosophy are alluded to in a letter to Schrödinger of June 17, 1935:

> From the point of view of principles, I absolutely do not believe in a statistical basis for physics in the sense of quantum mechanics, despite the singular success of the formalism of which I am well aware. I do not believe that such a theory can be made general relativistic. Aside from that, I consider the renunciation of a spatio-temporal setting for real events to be idealistic-spiritualistic. This epistemology-soaked orgy ought to burn itself out.[10]

And again in 1949:

> What does not satisfy me in [quantum mechanics], from the standpoint of principle [*Was mich an der Theorie vom prinzipiellen Standpunkt aus nicht befriedigt*], is its attitude towards that which appears to me to be the programmatic aim of all physics: the complete description of any (individual) real situation (as it supposedly exists irrespective of any act of observation or substantiation).[11]

On the other hand, since Einstein's death the tide of positivism in theoretical quantum physics has waned. Programs today within the foundations of quantum mechanics seek to implement something of a realist vision of quantum theory, though that vision is not Einstein's. Just what that vision amounts to is taken up in this book's latter chapters; there it will be shown to fall rather short of the contemporary doctrine termed "scientific realism". Still another reason from his divergence from realism appears in a distinction he drew between "theories of principle" and "constructive theories".

Principle theories vs. constructive theories

Responding to a query posed by an editor of the Times of London in late November 1919, the suddenly famous Albert Einstein wrote a brief account of relativity theory in German, sending it to the Times' Berlin correspondent. Appearing in the Times on November 28 as "Einstein On His Theory"[12] Einstein affirmed that relativity theory is a two-story building, the theory of special relativity on the ground floor, supporting the general theory one flight up. Both, Einstein noted, were examples of a particular type of physical theory termed "principle theories". Employing a distinction similar to those previously used by Henri Poincaré and H.A. Lorentz, Einstein observed that "principle theories" differ in kind from most other fundamental physical theories that are "constructive" in character.

Constructive theories "attempt to build up a picture of the more complex phenomena out of the materials of a relatively simple formal scheme from which they start out". "Building up" is naturally a synthetic process; the "more complex phenomena" are constructed from "materials of a relatively simple formal scheme" – that is, from elementary processes deemed capable of bringing about or producing the phenomena of interest. The phrase "relativity simple formal scheme" suggests that the materials of construction have some latitude to entertain a wide variety of underlying elementary processes, as long as they, and their hypothesized actions, are readily conceived and mathematically tractable. As had Lorentz, Einstein pointed to the kinetic theory as an exemplary constructive theory. The "molecular-kinetic theory" of heat (see Chapter 2) explains the thermodynamic properties of gases when the system is in thermal equilibrium under the assumption that the gas is a vast number of molecules in rapid motion. Macroscopic thermodynamic properties – pressure, volume, temperature, entropy – are explained by positing that molecular motions obey the laws of classical mechanics and by introducing probability considerations, such as the relative numbers of molecules in each distinct state (e.g., possessing a certain velocity) at a given time. The kinetic theory, and its extension by statistical mechanics to non-equilibrium systems approaching equilibrium, are canonical examples of the explanatory paradigm of reduction of one theory to another, wherein the laws of the theory pertaining to observable phenomena and relations are regarded as deriving from laws of a theory pertaining to unobservable microphysical objects and processes.[13] Einstein's 1905 paper on Brownian motion employed the "molecular-kinetic theory" to explain the observable (through a microscope) random motions of grains of pollen suspended in a liquid as due to chaotic thermal motions of the liquid's unobservable constituent molecules. In the annals of science, Einstein thus played a key role in demonstrating the reality of atoms, a triumph of realism (see Chapter 3).

Principle theories, on the other hand, have a top-down axiomatic character. Accounts of the phenomena of interest are derived on the basis of "empirically discovered" postulates or principles, i.e., "general characteristics of natural processes" which all physical events are required to satisfy. Principle theories "employ the analytic not the synthetic method"; that is, from a postulational starting point, they license logical rather than causal-dynamical inferences. Einstein identified thermodynamics as a paradigm of a principle theory. Here, the "universally experienced fact" of the impossibility of a *perpetuum mobile* stands at the pinnacle of the theory, from which a large number of empirical regularities in the theory of heat may be inferred. The theories of relativity, both special and general, he explained, are principle theories in this sense.

Einstein's intent, however, is not to rehearse an already recognized difference between types of physical theory so much as it is to point to the explanatory weaknesses of both theories of relativity. For the two types of theory differ not only methodologically and structurally, but also have distinct epistemic virtues:

> The advantages of the constructive theory are completeness, adaptability, and clearness, those of the principle theory are logical perfection and security of the foundations.[14]

The virtue of "completeness" is associated with constructive theories, a point to which return is made when considering Einstein's objection that quantum mechanics is "incomplete". Remarkably, Einstein then insists that constructive theories *alone* yield understanding:

> When we say that we have succeeded in understanding a group of natural phenomena, we always (*immer*) mean that we have found a constructive theory that embraces them.[15]

This is an astonishing admission, since the two theories of relativity are not constructive theories but principle theories, as Einstein readily admits. Yet reserving the accolade of understanding to constructive theories is not altogether surprising. It is fully in accord with the reductionist scientific maxim that explanations of phenomena remain incomplete until a causal mechanism or elementary process can be identified and can be shown to be capable of producing the observed effects. A principle theory, without an underlying component describing elementary processes, does not do this.

One can well ask what might have been Einstein's intent in choosing to introduce his theories of relativity through the prism of the principle theory/constructive theory distinction. By the end of November 1919, Einstein's relativistic theory of gravity was already four years old, though it had only just come to the wider world's attention in the London announcement on November 6 of the British solar eclipse observations of the previous May. He was well aware that the general theory of relativity was essentially incomplete; its treatment of matter-energy was nothing more than a skeleton that needed to be filled out in a field theory of matter. He had not rested on his laurels during these years but – prompted by the efforts of others to provide such a theory – was already engaged in attempting such a completion. This would become known as the program of unified field theory. It would require bringing all fundamental interactions into a geometrical broadening of the space-time setting in which he had placed gravitation. These efforts would continue for more than three decades, and they were ultimately all unsuccessful. Once before, in 1910–1911, Einstein embarked on such a constructive path when he sought to construct the quanta of radiation from a nonlinear generalization of Maxwell's theory of the electromagnetic field; during this time, he even expressed skepticism about the existence of quanta; perhaps they were not, after all, fundamental if they could be constructed, i.e., derived from more fundamental field processes. These efforts had come to naught, and were rather quickly given up. But the broader idea of constructing particles, with their discrete size and charge, within a field theory returned with the completion of general relativity. Now one had to begin from an unknown theory of the total field that must satisfy certain principles, above all, the "principle of general relativity".

While at various points in his career he would proffer theoretical proposals that may be regarded as "constructive" in the indicated sense, Einstein was predominately and preeminently a theorist who built upon principles regarded as universally applicable. Both theories of relativity were erected upon a respective "principle of relativity". It seems ironic and paradoxical that Einstein would extol the explanatory virtues furnished only by constructive theories even as he became the icon of scientific genius on the basis of principle theories. There is another way to understand this. The *de facto* choice in the matter stemmed not from any modesty of ambition. Rather it tells a great deal about the epistemology and methodology of Einstein, both young and old, for whom half a loaf was better than none. Just as constructive theories remained the explanatory gold standard of physics on account of their capacity to provide comprehensive understanding, the virtues of principle theories, namely, "logical perfection and security of the foundations", were assets that could be acquired within the existing state of physical knowledge. To Einstein, the way forward lay in attempting to build empirically confirmed theories upon prescriptive injunctions, principles stipulated to be universally valid that accordingly placed restrictions upon the possible laws of nature. Einstein's principles are not explanations, but postulates. This path admittedly gives something of a rationalist *a priori* character to the two theories of relativity. Late in life, Einstein portrayed the fateful decision to pursue the route of a principle theory, the path that led him to special relativity, as his one real option, a kind of Hobson's choice, writing of his "despair" at the time of any prospect of finding any true constructive laws underlying the pertinent electrodynamical phenomena.[16]

Still, the example set with both special and general relativity – the use of principles of invariance (for that is what both principles of relativity are, however named) to constrain the class of possible laws of nature – proved a salutary lesson to theoretical physics in the

20th century and beyond. Appropriately, Princeton physicist Eugene Wigner, winner of the 1963 Nobel Prize for the pioneering use of symmetry principles in nuclear and molecular physics, described this as "Einstein's greatest discovery".

> One can say, in fact, that Einstein's greatest discovery was the establishment of the importance of invariance principles.[17]

Together with a lifelong passion for theoretical unification, one lasting bequest to theoretical physics is the innovative use of principles of symmetry to constrain the dynamics of theories, viewing these principles not as corollaries to the dynamical laws but as axioms restricting the allowable dynamics.

Because of the close interconnection between what is here understood as Einstein's philosophy and his achievements and endeavors in physical theory, this book has a considerably different format than might be expected in a book about a particular philosopher. Physics and the history of physics occupy much of the center stage. An effort has been made to present this material in a way that is both accessible to non-specialists and is reasonably self-contained regarding Einstein's innovations and interventions. Very little higher mathematics has been used, and where it has, for example, with the metric tensor $g_{\mu\nu}$ of general relativity, the symbols occur only for expository purposes. This is not to say that this book will not make demands on the reader. But it is the author's belief that this mode of presentation is needed to avoid the sin of anachronism, of assimilating Einstein to other, and later, philosophical currents.

Notes

1 Einstein, letter to mathematician Edward Study, September 25, 1918. *CPAE* 8 (1998) Part B, Doc. 624.

2 Pais, Abraham, *"Subtle Is the Lord ..."The Science and the Life of Albert Einstein*. New York: Oxford University Press, 1982, p. 318.

3 Einstein, "Reply to Criticisms" (1949), p. 684. "(The scientist) must appear to the systematic epistemologist as a type of unscrupulous opportunist: he appears as *realist* insofar as he seeks to portray a world independent of acts of perception; as *idealist* insofar as he looks upon concepts and theories as free inventions of the human mind (not logically derivable from what is empirically given); as *positivist* insofar as he regards concepts and theories as justified only to the extent to which they afford a logical representation of relations between sensory

experiences. He may even appear as *Platonist* or *Pythagorean* insofar as he considers the viewpoint of logical simplicity as an indispensable and effective tool of his research" (translation slightly modified).

4 Holton, Gerald, "Mach, Einstein, and the Search for Reality", *Daedalus* v. 97 (1968), pp. 636–73. Page references to the reprint in Holton, *Thematic Origins of Scientific Thought: Kepler to Einstein*. Cambridge, MA: Harvard University Press, 1988, pp. 237–77.

5 Einstein, "Physik und Realität"; "Physics and Reality", *Journal of the Franklin Institute* v. 221, no. 3 (March, 1936), pp. 313–47; pp. 349–82. The translation by J. Picard is reprinted in Albert Einstein, *Ideas and Opinions*. New York: Crown Publishers, 1954, pp. 290–323; p. 290.

6 Einstein, "On the Method of Theoretical Physics", *Philosophy of Science* v. 1, no. 2 (April 1934), pp. 163–9; p. 163.

7 "On the Method of Theoretical Physics" (1936), p. 313; p. 290.

8 "Autobiographical Notes" (1946), p. 52; p. 53.

9 "*Zum kosmologischen Problem der allgemeinen Relativitätstheorie*", *Sitzungsberichte der Preußischen Akademie der Wissenschaften, Phys-Math. Klasse* (1931), pp. 235–7; p. 235. As reprinted in Dieter Simon (ed.), *Albert Einstein: Akademie-Vorträge: Sitzungsberichte der Preußischen Akademie der Wissenschaften 1914–1932*. Weinheim: Wiley-VCH Verlag GmbH & Co. KGaA. Published online, 9 August 2006, pp. 361–4; p. 361.

10 As translated in Arthur Fine, *The Shaky Game: Einstein Realism and the Quantum Theory*. Second edition. Chicago: University of Chicago Press, 1996, p. 68.

11 "Reply to Criticisms", p. 667.

12 Einstein, *CPAE* 7 (2002), Doc. 25; pp. 206–13. Translation as "What Is the Theory of Relativity?" in *Ideas and Opinions*, 1954, pp. 227–32.

13 The classic philosophical account of the reduction of thermodynamics to statistical mechanics is given in Nagel, Ernst, *The Structure of Science*. New York: Harcourt, Brace and World, Inc., 1960, chapter 11. A critical, and more sophisticated, discussion is Sklar, Lawrence, *Physics and Chance*. Cambridge: Cambridge University Press, 1993; chapter 9.

14 *Ibid.*, p. 228.

15 *Ibid.*

16 "Autobiographical Remarks" (1946), p. 52; p. 53.

17 "Symmetry in Nature" (1972), as reprinted in J. Mehra and A.S. Wightman (eds.), *Eugene Paul Wigner: Philosophical Reflections and Syntheses* with annotations by G. Emch. Berlin, Heidelberg, New York: Springer, 1995, pp. 382–99; p. 390.

Further reading

Six invaluable resources are:

1. *The Collected Papers of Albert Einstein*. Princeton: Princeton University Press. Abbreviated in this book as *CPAE*. Beginning with volume 1 in 1987 (*The Early Years 1879–1902*), additional volumes continue to appear up to volume 13 in 2012 (*Writings and Letters 1922–23*). Certain texts within each volume appear in

English translation in a supplementary separate volume. Remarkably, the volumes and supplements are freely available online, http://einsteinpapers.press. princeton.edu/

2. Schilpp, Paul A. (ed.), *Albert Einstein: Philosopher-Scientist*. Evanston, IL: Northwestern University Press, 1949. Besides containing interesting essays on aspects of Einstein's work and thought by both physicists and philosophers, the Schilpp volume has Einstein's "Autobiographical Notes", completed in 1946, pp. 2–95, in the original German with facing page translation; and "Reply to Criticisms", completed in 1949, in Schilpp's translation. For those who read German, the translation should be compared with Einstein's German text, "*Bemerkungen zu den diesem Bände Vereinigten Arbiten*", in the German edition of the Schilpp volume, *Albert Einstein als Philosoph und Naturforscher*. Stuttgart: W. Kohlhammer Verlag, 1955, pp. 493–511.

3. Pais, Abraham, "*Subtle Is the Lord . . ."The Science and the Life of Albert Einstein*. New York: Oxford University Press, 1982. The scientific biography, with technical details and much else.

4. *Albert Einstein: Ideas and Opinions*. New York: Crown Publishers, 1954. An English translation of a collection of popular and philosophical texts that first appeared in German in 1934.

5. Janssen, Michel, and Christoph Lehner (eds.), *The Cambridge Companion to Einstein*. New York: Cambridge University Press, 2014. Essays on various aspects of Einstein's physics and philosophy by leading Einstein scholars.

6. Calaprice, Alice, Daniel Kennefick, and Robert Schulmann (eds.), *An Einstein Encyclopedia*. Princeton, NJ and Oxford, UK: Princeton University Press, 2015. Both a good place to begin and a useful reference work.

While the number of books and other materials on Einstein's work is truly enormous, among my favorites are:

Bernstein, Jeremy, *Einstein*. New York: The Viking Press, 1973 (Modern Masters series, edited by Frank Kermode).

Fine, Arthur, *The Shaky Game: Einstein, Realism and the Quantum Theory*. Second edition. Chicago: University of Chicago Press, 1996.

Ohanian, Hans C., *Einstein's Mistakes: The Human Failings of Genius*. New York and London: W.W. Norton and Co., 2008.

One
Life and works

"I was always interested in philosophy but only as a sideline. My interest in natural science was always essentially limited to the study of principles, which best explains my conduct in its entirety. That I have published so little is attributable to the same circumstance, for the burning desire to grasp principles has caused me to spend most of my time on fruitless endeavors".

Albert Einstein letter to Maurice Solovine, October 30, 1924[1]

Albert Einstein was born in 1879, the eighth year after Otto von Bismarck's unification of Germany by "blood and iron". He died in Princeton, New Jersey, in April 1955, two and a half years after a successful test of the world's first thermonuclear weapon. His life spanned the two revolutions of 20th century physics, in both of which he was intimately involved, and two World Wars. His principal contributions to physical theory are the theories of relativity, the special theory (1905) and the general theory (1915). He was awarded the Nobel Prize in Physics in 1922 not for the theories of relativity but for "his services to physics, especially for the law of the photoelectric effect", a tribute leaving oddly unmentioned his revolutionary – and in 1922 still controversial – "light-quantum hypothesis" of 1905 on which said law relies (see Chapter 3). His many contributions to the development of the so-called "old quantum theory" (1900–1925) show that he must be regarded as a father, though a stern and critical one, of quantum mechanics. A 1935 paper with Podolsky and Rosen (two assistants at Princeton, see Chapter 7) is recognized today as discovering quantum entanglement, a phenomenon endemic to quantum mechanics.

Einstein reached intellectual maturity around the turn of the century and can be treated legitimately as a 20th-century figure. By summarizing his life through his scientific contributions the account below follows his self-description, as one increasingly disengaged from the "merely personal" while striving for a "mental grasp of things".[2] There are roughly three periods following his childhood.

Youth

Einstein's family moved to Munich shortly after he was born in the Swabian city of Ulm, on March 14, 1879. A Jewish medical student, Max Talmud (Talmey), befriended by the Einstein family, became something of a tutor to Einstein in philosophy and elementary natural science. After seven years at the Luitpold Gymnasium in Munich, where he was increasingly unhappy, Einstein withdrew at age 15 before graduating, joining his family then living in Milan. To avoid compulsory military service, he renounced his German citizenship in 1896 and remained stateless until 1901, when he was granted Swiss citizenship.

The Swiss years (1896–1914)

Though excelling in the math and physics parts, Einstein did not do well in the literature, social sciences, and zoology section, and so failed the entrance exam at the Polytechnic School in Zurich (renamed in 1910 as Federal Swiss Technical College, Eidgenössiche Technishe Hochschule or ETH); he then enrolled in secondary school at Aarau, near Zurich, where he passed the Matura exam (qualification for university entrance) in October 1896. He immediately enrolled as a student of physics and mathematics at the ETH. Among his fellow students were Marcel Grossmann, a lifelong friend who became a mathematician, and Mileva Marić, a Serbian fellow student of physics. Unable to find a secondary school teaching position upon graduating in 1900 with a diploma for teaching mathematics and physics, he initially found work as a private tutor. His first scientific paper was published in 1901, the year he received Swiss citizenship.

In early 1902 he moved to Bern, where from June 1902 until October 1909 he worked as a patent clerk, a job he acquired with the assistance of Grossmann. These were immensely productive years. Three early papers on statistical thermodynamics in 1902–1904 were published in the leading German physics journal, the *Annalen der Physik*, edited then by Max Planck and Wilhelm Wien. Although Einstein later downplayed their significance, these are exceptional papers, generalizing the statistical methods of the kinetic theory of gases independently but along the same lines as the celebrated contemporaneous work of Ludwig Boltzmann and J. Willard Gibbs. In the last of them, in 1904, Einstein employed what he termed "Boltzmann's Principle" (the connection of entropy and probability) to develop a theory of energy fluctuation phenomena. He would use this theory to probe quantum behavior in the theory of radiation (light) and the theory of matter (specific heats of substances, critical opalescence).

Together with friends Conrad Habicht and Maurice Solovine, for several years beginning April 1903 Einstein participated in a discussion group playfully named the "Olympia Academy". Among their readings were books of philosophers (Spinoza, Hume) and philosophically themed works of scientists, such as Hermann von Helmholtz, Heinrich Hertz, Ernst Mach, and Henri Poincaré. Much later, Solovine recalled that the latter's 1902 book *La science et l'hypothèse*, "held us spellbound for weeks on end". Against the objections of both families, in January 1903 he married Mileva Maric; a year earlier, she had given birth to their first child, a daughter, at her parents' home in Serbia. The child, whose existence was unknown to Einstein biographers until the 1980s, died of scarlet fever before her second birthday. A son, Hans Albert, was born in 1904; he would become a professor of engineering at the University of California, Berkeley. That year with the assistance of Grossmann, Einstein found a permanent job at the patent office in Bern.

At age 26, Einstein completed his doctoral dissertation in 1905 at the University of Zurich with a thesis "A New Determination of Molecular Dimensions". He also published four groundbreaking papers. The first, in March, is a revival of the corpuscular theory of light, largely abandoned by physicists early in the 19th century. In an investigation related to Max Planck's 1900 work on blackbody radiation, Einstein proposed that in certain circumstances radiation (light) behaved "as if" it had a particulate structure, like the molecules of gas. The so-called "light-quantum" hypothesis was indeed "very revolutionary" as Einstein characterized this paper in a letter to his friend Habicht in May;[3] the idea was not widely accepted until the early 1920s (the term "photon" was coined by Berkeley physical chemist G.N. Lewis in 1926). In May followed a first paper on Brownian motion, the visible phenomena of irregular zigzagging motions of tiny particles suspended in a liquid or, as motes of dust, in the air of a closed up room. Though the phenomenon was recognized since the 1820s, Einstein provided the first theoretical explanation: the observable motions are due to statistical fluctuations in the motions of the invisible molecules of the containing liquid or air. Polish physicist Marian von Smoluchowski advanced a similar theory in 1906. Experiments conducted in Paris in 1908 confirmed the theories, after which the existence of atoms became universally accepted. On June 30, the *Annalen der Physik* received Einstein's "On the Electrodynamics of Moving Bodies" initiating the special theory of relativity overturning Newtonian ideas of space and time. A follow-up paper in September attempted the first general proof of the famous equivalence between energy and mass, $E = mc^2$. In the annals of science, perhaps only the "plague years" 1665–1666, when Newton discovered the theory of binomial series, the method of fluxions (differential and integral calculus), advanced the theory of colors and idea of universal gravitation, can compare to Einstein's *annus mirabilis* of 1905.

The next year, 1906, Einstein published a critical examination of Max Planck's derivation of his law of blackbody radiation. Beginning with classical electromagnetism and statistical physics, in 1900 and in years subsequent, Planck derived the experimentally determined graph of an energy density function relating the frequency of thermal radiation to its temperature. To fit the known experimental data, Planck had to introduce what he considered a stopgap assumption, effectively discretizing the energy spectrum of thermal radiation. Einstein pointed out the inconsistency between Planck's starting point and the non-classical nature of his assumption, remarking that Planck had only succeeded by introducing the idea that there are discrete "atoms" of energy. Historians of physics, such as Thomas Kuhn, regard Einstein's 1906 paper as marking the actual beginning of the quantum revolution.

Prompted by Johannes Stark, an experimentalist and Nobel Prize winner, to write a survey of special relativity, Einstein began thinking about the extension of the relativity principle from inertial to accelerating frames. He published the article in 1907 in the journal edited by Stark, *Jahrbuch der Radioaktivitaät und Elektronik*. An irony, since the paper is Einstein's first step towards general relativity, the relativistic theory of gravitation that Stark, who would became a leading Nazi physicist, later attacked as exemplifying non-Aryan "Jewish physics". Also in 1907, as he later recalled, occurred "the most fortunate thought of my life", that a man falling freely (say, from a roof) would not feel his own weight.[4] Following up this insight led to the principle of equivalence, the crucial heuristic step towards general relativity. Still, most of Einstein's publications in 1907–1911 were devoted to the problems of understanding the increasing evidence of the quantum structure of matter that showed up in anomalies for classical theory. He proposed in 1907 a quantum modification that accounted for a well-known anomaly in the theory of specific heats. In 1908 he used the fluctuation theory developed in 1904 to advance a theory of critical opalescence, explaining the anomalous strong increase in the scattering (deviation in all directions) of light passing through a dense gas or liquid by the incidence of light on density fluctuations in the gas or fluid; he also showed that the scattering had a maximum near the material's "critical" point, the temperature of a phase change to another state of matter. In a lecture in Salzburg in September 1909, Einstein presciently predicted that a future theory of light would be a kind of amalgam of the wave and particle theories, the first statement of the wave-particle duality that is a cornerstone of quantum theory.

Einstein acquired his first academic job when he was made associate professor of theoretical physics at the University of Zurich in October 1909, having received his first honorary doctorate from the University of Geneva in July. For a brief period, in 1911–1912, he became professor of theoretical physics at the German University in Prague, while in October 1911, he was the youngest physicist invited to participate in the first world physics (Solvay) conference in Brussels, joining luminaries such as Lorentz, Planck, Poincaré, and Marie Curie. He returned in 1912 to Zurich as professor at the ETH, where he remained until accepting a position in Berlin in the summer of 1914. From 1907 when he first formulated the equivalence principle until finally succeeding in November 1915, his principal theoretical activity lay in attempting to generalize the special theory of relativity to apply to accelerating systems, and so to gravitational fields. In Prague in 1911 he first calculated that the principle of equivalence implied an observable deflection of light rays passing near the surface of the Sun. Fortunately, since without the correct theory of gravity, he got the wrong value, the expedition to Russia to test this prediction was interrupted by WW I.

The Berlin years (1914–1933)

Einstein's call to the Prussian Academy of Sciences and Berlin University in 1914 as a research professor without teaching duties placed him at the pinnacle of the scientific establishment in Germany. His greatest achievement, the general theory of relativity, was completed in outline in November 1915 (in something of a race with the mathematician David Hilbert). The final result is summed up in an apparently elegant, but highly complex, tensor equation. That equation stands for ten nonlinear field equations of gravity, the Einstein field equations, that must be given simultaneous solutions. For many years, very few exact solutions of this set of equations were known. The theory is inherently philosophical: It arose from a Machian-inspired (see Chapter 5) motivation to "relativize inertia". Two of the three principles Einstein placed at the basis of the theory, the "principle of general relativity" or general covariance, and what he termed "Mach's principle", gave rise to a good deal of controversy and misunderstanding, some of it on Einstein's part. A critical understanding of Einstein's adherence to the principle of general covariance only emerged in the 1980s with the work of the historians and philosophers of physics, John Stachel and John Norton.[5]

In 1917, seeking to eliminate "un-Machian" solutions to his field equations of gravitation, Einstein created relativistic cosmology, today a central part of fundamental physical theory. Einstein essentially "abolished infinity".[6] In the belief that that a "consistent theory of relativity" required the inertial mass of a body to drop to zero in the absence of all other masses, Einstein projected the first relativistic cosmological model, a spatially finite, temporally infinite, static universe with a uniform distribution of stars. In order to insure the model's stability against gravitational attraction between cosmic masses, he inserted a new "cosmological term" in his field equations that had the effect of a repulsive force to exactly balance against gravitational collapse. Almost immediately, Dutch astronomer Willem de Sitter produced an exact solution of the now-modified field equations that contained no matter at all, a universe possessing just the energy of empty space, as represented by the cosmological term. De Sitter's model universe turned out to be expanding, though this was not clear for some time. In 1922 Russian mathematician Alexander Friedmann showed that the most natural solution of Einstein's unmodified field equations resulted in a dynamic, not static, universe. Without knowledge of Friedmann's work, Belgian Jesuit priest Georges Lemaître in 1927 created a general relativistic model with the cosmological term. The theoretical developments dovetailed with astronomical observations of American Edwin Hubble reported in 1929, agreeing that the geometrical structure of the universe is described by general relativity and is expanding. When convinced of Hubble's results in the early 1930s, Einstein is reported to have called the introduction of the cosmological term his "biggest blunder". Today, however, the so-called cosmological constant is a fundamental link between quantum field theory and general relativity while in Big Bang cosmology, de Sitter's model for the matter-free early universe is widely adopted.

In the period immediately following completion of the general theory of relativity, Einstein showed considerable interest in attempts by philosophers to illuminate the theory's philosophical significance. His approval of Moritz Schlick's treatment was enthusiastic but short-lived. Schlick would go on to become the head of the Vienna Circle, the Olympus of logical positivism. Einstein's own philosophical views about his theory were rather fluid, changing in tandem with his own pursuit of a generalization of that theory to incorporate the other known field at the time, that of electromagnetism, into space-time geometry. In the 1920s he increasingly came to emphasize, with perhaps questionable accuracy of memory, the predominate role that considerations of mathematical elegance played in guiding him to the correct field equations of gravity during the period 1913–1915. These considerations became a core, and controversial, methodological theme after his Herbert Spencer lecture at Oxford in 1933.

Less widely known, but worthy of emphasis, are contributions Einstein made to the development of the quantum theory during the Berlin period. These include the first quantum derivation of Planck's radiation law in late 1916 that introduced the idea of "stimulated emission" of radiation, the theoretical basis of the laser and maser in the 1950s. In 1924–1925, spurred by works from two unknown physicists, by Louis de Broglie's Paris doctoral thesis extending wave-particle duality to electrons, and by a paper by Satyendra Nath Bose in Calcutta that derived Planck's radiation law by using an implicit assumption that light quanta are indistinguishable particles, Einstein initiated quantum statistics for gases. He also predicted the existence of a new state of matter, a so-called Bose-Einstein condensate in dilute gases, first produced in a laboratory seventy years later. These contributions built upon the "old quantum theory" but also signficantly influenced Schrödinger in the development of his wave equation in 1926. Nonetheless by the end of 1927 Einstein had turned against the quantum theory (linking its statistical character with its irrealism), proposed (and rejected) a "hidden variables" alternative, and initiated a decades-long debate with atomic physicist and founder of "complementarity", Niels Bohr.

The Princeton years (1933–1955)

Conveniently visiting at Caltech in Pasadena, California when Hitler became Chancellor of Germany in January 1933, Einstein never returned; it was rumored that Nazi newspapers secretly placed a bounty of $5,000 (approximately $90,000 in 2016 dollars) on his head.[7] World famous, he quickly became the most visible refugee from Hitler's Germany. One of six initial faculty members of the Institute of Advanced Study, he worked tirelessly from 1933 to help émigré scientists and researchers. In August 1939, he signed the well-known letter to President Franklin D. Roosevelt regarding the possibility of an atomic bomb. Becoming a naturalized US citizen in 1940, he was a persistent vocal advocate for racial justice in the USA, for nuclear disarmament, and for universal human rights. In the McCarthy period of the early 1950s, he strongly and publicly supported those who refused to testify regarding their political views at government hearings. An offer from the new state of Israel to become its first president was turned down in 1952. His last signed letter, in April 1955, was to Bertrand Russell, lending his name to a joint manifesto urging the complete renunciation of nuclear weapons by all nations.

Philosophically, these years are of interest on account of Einstein's sharpened critique of the quantum theory centering on several arguments that the theory is incomplete. Although that argument is imperfectly expressed in the Einstein-Podolsky-Rosen (EPR) paper of 1935, Einstein's correspondence of the time (especially with Schrödinger) reveals its source in the twin philosophical principles of separability and locality, deployed in the quantum context of what Schrödinger dubbed "entanglement". More than a generation of quantum physicists treated Einstein's concerns as the philosophical prejudices of a young revolutionary turned old reactionary (if not the signs of scientific senility). Nevertheless, after Northern Irish physicist John S. Bell revived interest in the EPR paper in 1964, the ensuing study of entanglement continues to produce a rich harvest of significant results in the foundations of physics and currently underwrites the newly emerging field of quantum information theory. Einstein's "other argument" that quantum mechanics is incomplete began with an example prior to, and analogous to, that of Schrödinger's famous cat in 1935. That argument concerns quantum mechanical universality and so its characterization of the transition from the quantum realm, where the superposition principle of quantum mechanics is fundamental, to the macroscopic realm of classical objects, where it is irrelevant. The problem of the existence of stable macroscopic objects in a fundamentally quantum world remains today; it is the subject ("decoherence") of an active branch of quantum research dealing with the coupling between macroscopic objects and their surrounding environment.

Einstein's philosophical engagement with, and critique of, the quantum theory in these years should be considered in conjunction with his field-theoretic attempts to frame a comprehensive theory of what he termed the "total field", from which quantum mechanics would emerge as a statistical approximation. That enterprise failed and in his stubborn pursuit of it, Einstein patiently suffered the ridicule of a younger generation of theorists eager to extend the quantum theory to the new physics of the nucleus and to quantum field theory. Though he defended himself by stating that he "had earned the right to make mistakes", a more generous appraisal from the vantage point of more than sixty years is possible. The quest for unity in the foundations of physics that remained his lifelong passion was rekindled on a different, quantum, basis by superstring theory in the 1980s and continues to thrive. Will it, or any other such attempt, ultimately succeed? Einstein's response, as fitting now as it was when given in 1950, is to answer that question with a smile.[8]

Notes

1 *Albert Einstein: Letters to Solovine.* New York: Philosophical Library, 1987, pp. 62–3.

2 "Autobiographical Remarks", pp. 6–7.

3 *CPAE* 5 (1993), Doc. 27.

4 "der glücklichste Gedanke meins Lebens", in *Grundgedanken und Methoden der Relativitätstheorie in ihrer Entwicklung dargestellt* ("Fundamental Ideas and Methods of the Theory of Relativity, Presented in Their Development"), draft of what is probably an article intended to appear in English translation in the British journal *Nature* but withdrawn by Einstein and never published. *CPAE* 7 (2002), Doc. 31.

5 Stachel, John, "Einstein's Search for General Covariance, 1912–1915", in D. Howard and J. Stachel (eds.), *Einstein and the History of General Relativity* (Einstein Studies v. 1), Basel, Boston, Berlin: Birkhäuser, 1989, pp. 63–100. This paper is based on the written version of a talk circulated since 1980. Also Norton, John, "How Einstein Found His Field Equations, 1912–1915", pp. 101–59 in the same volume.

6 Eddington, Arthur, *The Expanding Universe.* Cambridge, UK: Cambridge University Press, 1933, p. 21.

7 Vallentin, Antonina, *The Drama of Albert Einstein.* New York: Doubleday, 1954, p. 231. Vallentin was a friend of Einstein's second wife, Elsa Löwenthal.

8 "It seems to me a smile is the best answer". "Physics, Philosophy, and Scientific Progress", a speech delivered in English to the International Congress of Surgeons in Cleveland, Ohio in 1950. Text reprinted in *Physics Today* (June 2005), pp. 46–8; p. 48.

Further reading

Clark, Ronald W., *Einstein: The Life and Times.* New York and Cleveland: World Publishing Company, 1971. Although superseded in many respects by more recent biographies, Clark's is both well-written and full of interesting details later authors omit.

Fölsing, Albrecht, *Albert Einstein: A Biography*. Translated by Ewald Osers. New York: Viking Penguin, 1997. Unfortunately, the English edition of this biography omits a chapter on Einstein and philosophy in the 1993 German edition, entitled "Written in Honey" ("in Honig geschrieben").

Hoffmann, Dieter, *Einstein's Berlin: In the Footsteps of Genius*. Baltimore, MD: Johns Hopkins University Press, 2013.

Isaccson, Walter, *Einstein: His Life and Universe*. New York: Simon and Schuster, 2007.

Part I

Quantum theory

Two

On the road to Planck 1900

"We consider, however – this is the most essential point of the whole calculation – [the energy] E to be composed of a very definite number of equal parts and use thereto the constant of nature $h = 6.55 \times 10^{-27}$ erg-sec. This constant multiplied by the frequency v . . . gives us the energy element, ε".

Max Planck, lecture of December 14, 1900[1]

Introduction

At least by 1850, nearly all physicists had abandoned the theory of Lavoisier, Laplace, and Gay-Lussac that heat was a special type of substance *caloric*, an imponderable fluid passing from one body to another. In its place came a revival of the ancient atomist idea that heat is attributable to the motions of very small bodies. Though the doctrine that all matter is composed of tiny indivisible material ("ponderable") particles remained controversial, in the late 1850s the concept of *kinetic atom* gained acceptance. These invisible "spheres of action" were regarded in various ways: as mental fictions or heuristic devices, as centers of repulsive forces or as physically real tiny indivisible particles. Possessing minimal material properties of size and motion, the kinetic atom became the central concept of the kinetic theory of heat, that a body's temperature is a measure of the kinetic energy, or energy of motion, of its hypothetical microconstituents. Emil Wiechert's and J.J. Thomson's 1894–1897 discovery of electrons, "atoms of charge", led by the turn of the century to the first models of an electrically neutral atom, a composite structure whose stability was explained by the attraction of oppositely charged constituent particles. Maxwell's electromagnetic theory, spectacularly confirmed by Hertz's wireless experiments in 1887, implied that charged moving bodies emit electromagnetic radiation. Lying at the intersection of the kinetic theory of heat and Maxwell's theory are the phenomena of *thermal radiation*, electromagnetic radiation generated by the vibrational motions of charged particles in bulk matter. Investigations of thermal radiation in the last decades of the 19th century led to the quantum theory.

On the joint basis of Maxwell's theory and the kinetic theory, in the autumn of 1900 Max Planck attempted to derive the recent empirically established frequency distribution law of thermal radiation for a so-called blackbody. In so doing, he found he was compelled to introduce the notion of a discrete minimal unit of energy. It is customary to say that Planck thereby "discovered" the quantum. Planck later on modestly described his brainchild as "an act of desperation", a mathematical stratagem needed to obtain what he knew to be the correct answer. He proved a highly conservative revolutionary – so much so that Thomas Kuhn's 1978 study of Planck's reluctant embrace of the physical reality of energy quanta credits Einstein, not Planck, for the birth of the quantum theory. Kuhn argued that the quantum revolution really began with Einstein's paper of March 1906, pointing out that Planck's derivation, based on classical theory yet incorporating the above stratagem, was strictly speaking inconsistent. Nonetheless, Einstein observed, the derivation succeeded through implicit use of the non-classical light-quantum hypothesis he himself introduced in 1905. While historians of physics can disagree about the respective merits of Planck or Einstein as father of the quantum revolution (in no small measure due to genuine obscurity concerning what Planck actually believed), there is widespread recognition and general agreement that Einstein played a unique and revolutionary role in advancing and extending the quantum hypothesis, first in the theory of radiation and then pursuing it into the theory of matter.

To appreciate Einstein's role in creating the "old quantum theory" in the next chapter, it will be useful to first sketch the situation of late-19th-century physical theory that Einstein inherited. This chapter briefly surveys the rise of thermodynamics, the kinetic theory, its application to gases and statistical generalization by Boltzmann, and the developments leading up to Planck's radiation law. The next concerns how Einstein formulated then wielded what he termed "Boltzmann's principle" in demonstrating the quantum hypothesis was here to stay. The story here begins with the rise of thermodynamics in the 19th century and subsequently its statistical foundation by Boltzmann.

Historical overview of the concept of energy

The concept of energy has a long history; precursors appear in Greek antiquity and again in the Middle Ages. But it was Leibniz who foreshadowed the modern concept of energy in 1686 with his notion of "living force" (*vis viva*). In the process he gave rise to a century-long dispute about whether *vis viva* (mv^2) or linear momentum (mv) is the appropriate conserved quantity in mechanics.[2] The concept of energy is present in all but name in the 18th-century dynamics of d'Alembert and Lagrange, although to them it is not a generally conserved quantity. The general doctrine of energy conservation then emerged in the early 19th century from a combination of factors, among the most important being French engineers' interest in the theory of machines, the idea that all physical phenomena might be reduced to conservative mechanical processes, and the discovery of the interconvertibility of various forms of energy.

The advent of steam engines in the Industrial Revolution brought a generalization of the energy concept from mechanics to the phenomena of heat and the definition of energy as "the capacity of a physical system to do work". At the same time heat came to be regarded as a form of energy convertible into work. In a notable 1959 paper on simultaneous discovery in science, Thomas Kuhn pointed to the principle of energy conservation as a prime example.[3] Between 1842–1847, a group of researchers, each working largely in ignorance of the others, proposed that within an isolated system different forms of energy might be transformed from one kind to another, but total energy is conserved by these processes. Synthesizing these results in 1847, Hermann von Helmholtz formulated a law of conservation of force ("*Kraft*"), a term not yet with a univocal meaning though soon to be understood as "energy". Shortly afterwards, William Thomson, the future Lord Kelvin, coined the term "thermo-dynamics" in 1851 for the new science of the study of the effects of work, heat, and energy on a physical system.

The idea of transformational invariance appears central in establishing the concept of energy; that heat is a form of energy and that the internal energy of a system cannot be created or destroyed but only transformed from one form to another led to the first law of thermodynamics. On the other hand, Ernst Mach, in an influential 1872 monograph on the history of the principle of conservation of energy,[4] argued that the actual root of the energy concept lay, as Helmholtz had maintained, in the historically attested record of failures to produce a *perpetuum mobile*, a machine that once started, would continue to run indefinitely without any external input. While this apparently restricts the concept of energy to mechanics, Mach insisted that the prohibition of perpetual motion arose in the course of human experience long antedating mechanics and hence logically distinct from it. Einstein followed Mach in similarly regarding the impossibility of perpetual motion as a "principle", the indisputable empirical fact underlying thermodynamics.[5]

Kinetic theory

Thermodynamics, a science of the macroscopic properties of matter, developed rapidly in the first half of the 19th century. Its results are largely independent of hypotheses about underlying dynamics and microstructural composition, among which the most influential was physical atomism, a speculative hypothesis pertaining to invisible, indivisible corpuscles.[6] But from the early 19th-century atomism of another type, the "chemical atomism" (1808–1810) of John Dalton (1766–1844), was quickly accepted as it proved successful in explaining the respective weights of chemical elements in compounds. If chemistry suggested the existence of atoms of the chemical elements, the experimental data represented in the gas laws due to Boyle (1627–1684) and Mariotte (1620–1684) showing that the pressure of air is proportional to its density, suggested that the elasticity ("spring") of air is due to the motions of tiny bodies, although Boyle himself was critical of Greek atomism and Mariotte disliked atomism altogether. Molecules of air were assumed in the prototype kinetic theory of Daniel Bernoulli (1700–1782) in 1738. Bernoulli correctly understood that air pressure within a closed cylinder supporting a weighted piston is due to the repeated impacts of vast numbers of rapidly moving "very minute corpuscles". However, despite Bernoulli's derivation of Boyle's law relating the pressure and volume of a gas, there were competing theories of gas pressure and for a considerable time, no agreement on the nature of heat and temperature. Only in 1854 did Kelvin propose a generally accepted absolute temperature scale.[7] The kinetic-molecular theory remained incomplete. For almost a century, further progress in developing the concept of heat as molecular motion was delayed.

Bonn physicist Rudolf Clausius (1822–1888) formulated the two initial laws of thermodynamics. The first is the principle of conservation of energy that Clausius applied to heat and thermodynamic processes. Expressed in the form of a prohibition, the first law of thermodynamics states the impossibility of a machine that can perform work without modifying the environment, a so-called *perpetuum mobile* of the first kind. Clausius's initial version of the second law affirmed that heat cannot pass from a colder to a warmer body without compensation in the environment, as for example by a refrigerator's compressor. Another formulation, deriving from Sadi Carnot's 1824 theory of the limits of efficiency of heat engines, is that there is a fixed upper limit to the amount of work obtainable from a given amount of heat. This effectively states the impossibility of a *perpetuum mobile* of the *second* kind; namely, the impossibility of building a machine that produces work in a cycle of operation by borrowing heat from a single source (in Kelvin's formulation). Clausius subsequently gave the second law quantitative expression, in the process introducing the notion of *entropy*, the most characteristically *thermodynamic* concept, a more abstract notion lying further away from sense experience than volume or pressure or temperature. Designating entropy by the letter S, Clausius's quantitative statement of the second law is an inequality holding that the incremental change in entropy in a closed system cannot be negative.[8]

The development of the concept of energy in the 1830s and 1840s and the new thermodynamics around 1850 set the stage for the kinetic theory of gases by the end of the 1850s. In the famous characterization of Clausius, molecular motion is "the kind of motion that we call heat". Between 1855–1865, Clausius and Maxwell were able to derive the law of perfect gases, assuming molecules freely flying around within a containing vessel except for occasional elastic collisions among themselves and with the walls of the container. Temperature was thereby interpreted as the average or mean kinetic energy (or velocity) of the particles of the gas. In this way the kinetic theory explained the established empirical regularities between the macroscopic thermodynamic quantities in terms of the random motions of molecules.

The next step was to similarly account in kinetic terms for the two laws of thermodynamics, both regarded as exceptionless absolute laws by Clausius. Kinetically, entropy is a measure of the state of molecular disorder in a macroscopic system: e.g., the molecules of a gas within a container may be mostly bunched in one corner (low entropy) or distributed more or less uniformly throughout the container's volume (high entropy). But the statement that entropy never decreases posed a difficult philosophical problem for the kinetic theory according to which matter is composed of atoms in continual motion and subject to the laws of Newtonian mechanics. The basic equations of mechanics are both exceptionless and time-reversible; they do not change their form in the least if the sign of the time variable is reversed (making the substitution t $\rightarrow$ $-$ t). As Newton's second law,

$$\vec{F} = m\vec{a} = md^2x / dt^2,$$

contains the second derivative of the time variable, any solution of Newtonian equations can be transformed into another solution with the time variable reversed and the system represented as going "backwards in time" (as in heat flowing from a cold body to a hot one). How then can time-reversible laws governing the motions of the particles of a gas be reconciled with the apparently irreversible macroscopic increase of entropy posited by the second law? Could any reconciliation be made? Initially, only the far-sighted Maxwell recognized the fundamentally statistical aspect of the second law;[9] in a letter to Peter Guthrie Tait of 1867, Maxwell explained how the second law could be hypothetically violated by a "demon", a being with finite attributes that without expending any work at all could selectively segregate the "sufficiently fast" molecules of a gas from all others, so that a hot region of a gas grew hotter and the cold region colder, in violation of the second law. A less fanciful illustration of Maxwell's understanding was given in an 1871 letter to John Strutt, the future Lord Rayleigh, in which Maxwell remarked that the second law had "the same degree of truth" as the statement that a glass full of water thrown into the sea cannot be recovered. However, it was the Austrian Ludwig Boltzmann (1844–1906) who formulated the statistical explanation of the second law that remains today. Boltzmann's interpretation of the second law will be essential to Einstein's contributions to the early quantum theory.

Enter Boltzmann

Clausius, then Maxwell, introduced statistics into the study of the properties of a multi-particle system, in particular, a dilute ideal gas at thermal equilibrium in a container. From statistical arguments, Clausius deduced the probability that a gas molecule could travel on average a certain distance, a "mean free path", without collision. To do this, he assumed for simplicity that the molecules of the gas moved in random directions but with the same average velocity.

In his first paper on kinetic theory in 1860, Maxwell introduced the idea that molecular velocities are distributed according to a well-defined law, today known as the Maxwell-Boltzmann distribution. Maxwell treated velocity as a variable whose values are distributed randomly, attaching probability not to "mean free path" but to every molecular state of motion. The probability that a molecule of the gas possessed a definite kinetic property, e.g., a velocity between v and $v + dv$ or energy between E and $E + dE$, is then a relative probability, expressed as the ratio of the average number of particles of the gas with this property at a given time to the total number of particles. This information is summarized by a distribution function, indicating the velocity spread of the molecules of a gas and their most probable speeds at various equilibrium temperatures. The distribution is identical in form with the "normal distribution" bell curve of the Laplacian theory of errors and is known in statistics as a Gaussian. It is the basis of the kinetic theory and the first statistical law in physics.

Strictly speaking, Maxwell's distribution applies to an ideal gas of fictional particles moving freely within a closed container, and without interactions other than brief collisions that elastically exchange energy and momentum. More generally, the Maxwell distribution function is applicable wherever a system of classical particles has established thermal equilibrium at absolute temperature T as the result of collisions among themselves and with the walls of the container. Boltzmann subsequently sought to understand how the gas approached this equilibrium distribution, and in 1872 used kinetic theory together with Maxwell's implicit assumption that each pair of colliding molecules is uncorrelated prior to collision (the so-called "Stoßzahlansatz"), to prove that the Maxwellian velocity distribution is the unique stationary (equilibrium) distribution to which the molecules of a more realistic model gas (polyatomic molecules, subject to external force fields like gravity) would tend over time due to the laws of mechanics. As the Maxwell/Boltzmann distribution function occupies an important place in Einstein's contributions to statistical physics, note that it contains the exponential function, $\exp(-E/kT)$ where E is the energy of a molecule, T the temperature, and k a bridging constant between microphysics (molecular mechanics) and macrophysics (thermodynamics) now named for Boltzmann.[10] A key contribution of Einstein to the emerging quantum theory will be to show the physical significance of the new constant.

From the year he received his PhD (1866), Boltzmann set out to interpret the second law in the context of the kinetic theory. In 1872 he showed that the Newtonian laws of motion applied to the particles of a dilute gas, plus the Stoßzahlansatz, suffice to show that in finite time every non-equilibrium distribution of molecules will approach the Maxwell-Boltzmann equilibrium distribution where entropy is a maximum. To do this, Boltzmann constructed a mechanical function H of the velocity distribution, to be interpreted as the negative of the thermodynamic entropy function S. Boltzmann's "H theorem" is then the inequality $dH/dt \leq 0$. It is allegedly a demonstration that a gas beginning in any non-equilibrium velocity distribution monotonically moves closer and closer to equilibrium and the state of highest entropy: an irreversible process obtained from the time-reversible laws of mechanics. Objections to the H theorem from Boltzmann's former teacher and Graz colleague Josef Loschmidt (1821–1895) and from the British kinetic theorists over the course of the ensuing decade pushed Boltzmann to increasingly appreciate the important role that statistical concepts play in the theory of gases, as Maxwell had emphasized.

The turning point came in 1876 when Loschmidt pointed to an insurmountable difficulty: from fully specified initial conditions, an increase of entropy could be derived according to Newton's laws (applicable to the motions and collisions of molecules of a gas), but equally a decrease of entropy could be derived by those same laws from the final state by simply reversing each molecule's velocity. An irreversible approach to equilibrium and associated monotonic increase of entropy cannot be derived from the time-reversible laws of mechanics without introducing additional probabilistic considerations. Boltzmann subsequently sought to show how macroscopic irreversibility could be understood statistically in a gas comprised of an enormous number of molecules, even though the motions of individual molecules follow the time-reversible laws of mechanics. He interpreted the empirical fact that thermodynamic systems evolve towards thermal equilibrium as the overwhelming *tendency* of such systems to evolve from less probable low entropy states towards the most probable state, the state of maximum entropy, i.e., thermal equilibrium. As applied to the universe as a whole, this account required the assumption of a very improbable universal initial condition, a problem that still remains in cosmology. For particular isolated systems, however, Boltzmann's understanding of the second law became a cornerstone of statistical mechanics.

Statistical mechanics

In everything but name, statistical mechanics began in the late 1860s with a series of memoirs in which Boltzmann introduced various relations between entropy and probability. The most famous of these is from 1877, entitled "On the Relation Between the Second Law of Thermodynamics and Probability Calculus". Published, as were many of Boltzmann's papers, as a memoir in the *Proceedings of the Viennese Academy of Sciences*, it radically departs from Boltzmann's earlier kinetic approach; it opens with the statement that the probability of a molecular distribution could be determined in a way "completely independent of whether or how that distribution came about". Without recourse to kinetic hypotheses or assumptions about molecular

collisions, the new method derived a probability distribution for the
different thermodynamic states of a gas by simply counting the distinct number of ways each such macrostate might be realized by the
particles of the gas. Each distinct way is called a "complexion" and a
new complexion (a microstate specifying at a given instant the positions and velocities of each of the N particles of the gas) arises by as
simple a process as merely permuting the speeds of two of the gas's
N molecules (e.g., give B's velocity to particle A, and vice versa), leaving the overall thermodynamic state unchanged. The paper marked
a considerable departure from Maxwell's understanding of the use
of probability considerations in physics. Whereas Maxwell justified
the introduction of probability methods into physics by the human
(i.e., non-demonic) impossibility of specifying the exact state of
each molecule of a gas, Boltzmann essentially regarded such detailed
information as unnecessary for the purpose of providing a kinetic
foundation for thermodynamics. What mattered in accounting for
the particular values of thermodynamic variables is simply the sheer
number of distinct "complexions" associated with a given value of a
thermodynamic variable: that vast number is directly proportional to
the probability that the system is in that thermodynamic state.

Although later on in his 1877 paper Boltzmann introduced more
realistic systems (again, polyatomic molecules possessing more
degrees of freedom and subject to external forces), he used a simple model, an "unrealizable fiction", to bring out the fundamental
idea. Consider a system consisting of a large but finite number N
of identical and distinguishable "rigid absolutely elastic spherical molecules trapped in a container with perfectly elastic walls". Taking into
account only the total energy E of this "gas" of fictional molecules
isolated in a container, the next step was to distribute or partition E,
a thermodynamic observable, over the N molecules. Energy is classically a continuous variable, so this meant distributing a continuous
quantity over an enormous yet finite number of objects. Boltzmann
accordingly stipulated that the kinetic energy of each particle had to
be one of the discrete energy values: $0, \varepsilon, 2\varepsilon, 3\varepsilon, \ldots, p\varepsilon$. The total
energy E of the gas is then an integral multiple of ε distributed over
the N particles by an energy-partition (Energievertheilung), i.e., numbers
$w_0, w_1, \ldots, w_r$, such that w_0 particles have energy 0, w_1 particles have

energy ε, w_2 particles have energy 2ε, and so on, where $\Sigma\, w_i = N$. An energy-partition is thus a particular state description, a distribution of molecules according to the coarse scale of discrete energies. The energy element ε was given no physical but merely a computational meaning (Boltzmann chose it small enough so that the end result would not depend on its value) in determining the number of complexions corresponding to the given state description. Max Planck will borrow this method of discretizing a continuous variable in 1900 when using Boltzmann's statistical method in deriving the radiation law that bears Planck's name.

A given thermodynamic state (that is, for fixed E) then corresponds to a vast number of distinct complexions. Since two distinct complexions arise merely by permuting molecules, Boltzmann's method of counting complexions implicitly assumes that the molecules, though identical in all respects, nonetheless can be distinguished from one another. It may well be asked – and this will prove crucial later – how particles can be identical yet distinct from one another. Suppose one goes into a room where there is only a table on which sit two boxes, each containing a small black ball. The balls and boxes are identical in every respect but can be given different names since their placement allows them to be perceptually distinguished: A is in the box to the observer's left, B in the box to the right. Suppose one leaves the room and returns at a later time. Again two boxes, each containing a black ball. But does one really know that A remains in the box on the left, or might it be B? Unless one has supplementary information, say, by continuously monitoring the interlude with a CATV monitor, the latter has to be considered as not impossible. Hence two balls in two boxes at two different times give rise to two situations, identical yet distinct. It is not one situation encountered twice. *Identical* but *distinguishable* particles is a key implicit assumption of Boltzmann's 1877 statistical method. Later on, in his 1897 lectures on mechanics (underlying the theory of gases), Boltzmann made this assumption explicit, calling it the "first assumption of mechanics", a consequence of "the law of continuity of motion", that a mass point is identifiable as the same point at two different places at two distinct times by its continuous trajectory connecting them. As we shall see, according to standard quantum mechanics, quantum particles do not have determinate trajectories, and in 1924 Einstein will show that the statistics of one species of quantum particles (bosons) accordingly require a different method of counting that corresponds to the fact that bosons are *identical* but *indistinguishable* particles.

On the assumption of identical and distinguishable particles, a huge yet definite number of distinct complexions correspond to each thermodynamic state of the gas. The total number P of complexions associated with that state description is computed using combinatorics. As P is also the number of permutations of N molecules associated with that description, Boltzmann termed P the "permutability", computing it by the factorial formula,

$$P = N!/w_0! \, w_1! \, w_2! \ldots$$

On the assumption that each distinct complexion is equally probable, Boltzmann set the probability W (*Wahrscheinlichkeit*) of finding the gas in a given thermodynamic state as proportional to P. The most probable distribution for fixed E and fixed N will be those $w_0, w_1, w_2, \ldots$ values for which P is a maximum. Probability here means relative probability, the ratio of the number of distinct possible complexions corresponding to a definite state description to the vast total number of all complexions possible within the energy constraints on the system. As the numbers involved are so enormous, Boltzmann could use a well-known mathematical approximation (Stirling's formula) for the factorials involved, allowing the most probable distribution to be represented by an exponential function.[11]

The relationship between the entropy S of a system and its probability (W) is then as follows. Entropy is an additive quantity: the entropy S of two separate systems A and B that are brought into contact is the sum of their entropies, $S_{AB} = S_A + S_B$. But probability is not additive: two systems that have not yet interacted must be considered independent, and therefore the probability W of occurrence of the joint AB system is the product of the probabilities of the two systems separately, $W_{AB} = W_A W_B$. The mismatch requires the proportionality between entropy and probability to be logarithmic.[12] First written down in this form by Planck, who named the constant k of proportionality *Boltzmann's constant*, the statistical form of the second law ornaments Boltzmann's tomb-stone in Vienna,

$$S = k \log W.$$

The expression essentially states that a system with relatively fewer equivalent ways to arrange its components has relatively less probability W (and so entropy); alternately the one with relatively many equivalent ways of doing so (larger W) has relatively more entropy.

In defining the entropy of a thermodynamic state by the measure of its probability, Boltzmann extended the second law beyond the domain of pure thermodynamics into statistical mechanics. Boltzmann's "statistical mechanics" (a term coined by Yale physicist Willard Gibbs in 1902) thereby created the modern understanding that the second law is not an absolute law but is merely statistical: for a given time lapse, entropy increase in closed systems occurs in the vast majority of initial conditions, but not all of them. The use of the logarithm of P rather than P had another significant consequence. It enabled Boltzmann to show that in the simple case of a gas in equilibrium satisfying the Maxwell distribution, the entropy of the gas is essentially that calculated directly using only thermodynamic assumptions. That the Maxwell-Boltzmann velocity distribution for the equilibrium state could be derived from this new point of view showed not only that it is the unique stationary distribution, but that it is also the most probable.[13] Still, in the first decade of the 20th century, the statistical meaning Boltzmann accorded the second law was highly controversial. Planck, then a rising young theorist who had already built an international reputation on the basis of his lectures on thermodynamics, would not accept Boltzmann's statistical interpretation until 1914.

Blackbody radiation

According to Maxwell's dynamical theory of the electromagnetic field (1865), electromagnetic radiation is the emission of energy in the form of waves; up until the special theory of relativity in 1905, it was widely believed that these waves propagated through space in a material medium known since early in the 19th century as the luminiferous ether, a medium that the wave theory of light regarded necessary for the propagation of light (see Chapter 5). Many physicists on the continent, however, were skeptical of a theory that predicted effects of electromagnetic waves years before they were demonstrated in experiment. Only after Heinrich Hertz's experiments at Karlsruhe beginning in 1885 did Maxwell's theory gain widespread acceptance. Hertz demonstrated propagation of electric waves through space by fabricating an electrical oscillator and analyzing the surrounding field by means of small resonant circuits. In his original device, Hertz measured the wavelength (from crest to crest) as 66 cm; these are radio waves or what is now called the microwave spectrum. It was the first detection of electromagnetic waves with properties of light: frequency, refrangibility, interference, polarization, etc. Further experiments in England, Italy, and India were successful in generating waves of shorter wavelengths. By the end of the 19th century, the known spectrum of electromagnetic radiation had greatly expanded, including wavelengths of invisible infrared and ultraviolet rays adjacent to visible light but also much longer radio waves and much shorter X-rays. By 1900, the theory of radiation was fairly well encompassed by Maxwell's theory, though doubts remained about the nature of X-rays.

As noted above, quantum theory originated through the investigation of thermal radiation, radiation emitted by a hot object. When heated, a body's constituent particles vibrate more rapidly; if electrically charged, they emit radiation, electromagnetic waves carrying energy away from the body at a rate that increases with its temperature: the hotter the object, the brighter it appears. A heated iron bar, for example, will initially glow with a dull reddish light; as it becomes hotter, it will shine more and more brightly, emitting yellowish and eventually bluish-white light, spanning the visible part of the electromagnetic spectrum. All matter (above absolute zero on the Kelvin scale [−273.15 °C]) continually emits and absorbs thermal radiation. Beginning around 1860 the new science of thermodynamics was employed to investigate how the character of emitted thermal radiation depends on a body's temperature, the motions of charged particles of the body's constituent atoms and molecules.

Two initial theoretical results are important for what follows, the first by Gustav Kirchhoff and the second by Boltzmann. In Heidelberg, Gustav Kirchhoff (1824–1887), and his colleague

Robert Bunsen (1811–1899) pioneered the spectroscopic method of chemical analysis. By refracting visible light emitted by a glowing object into its spectrum, the chemical composition of the object could be inferred. In 1859, Kirchhoff discovered the presence of sodium in the sun and within a few years had shown that the solar spectrum reveals the possible presence of at least nine more chemical elements. Those who, like Charles S. Peirce, scorn the erection of "barriers to inquiry" can take comfort in the fact the Kirchhoff's results belied the 1835 prediction of positivist philosopher August Comte that the chemical composition of the stars could never be known to science.

More significantly for what follows, in 1860 Kirchhoff stated a remarkable theorem despite lacking adequate experimental evidence to back it up. Now known as Kirchhoff's law of thermal radiation, it states that for any body, the ratio of its emissive power (E_v) to its dimensionless coefficient of absorption (A_v) is characterized by a function J of the mode frequency v of the emitted or absorbed radiation and the body's temperature,

$$E_v \: / \: A_v = J(v, T).$$

Such a function is "universal" as it is independent of every other property of the body including shape or composition. A corollary holds that at constant temperature (i.e., in thermal equilibrium with its surroundings) a body will emit and absorb radiation at the same rate. Kirchhoff furthermore conceived of a theoretically useful fiction, a so-called *blackbody* that could in principle with perfect efficiency emit and absorb radiation of all possible wavelengths. A body of this kind is ideal for studying thermal radiation, the energy emerging from matter itself, as it absorbs all light of any wavelength incident upon it yet at thermal equilibrium emits radiation whose spectrum, or distribution of frequencies, is characteristic only of its temperature (in this case $A_v = 1$).

By the end of the 19th century physicists believed that thermal radiation was entirely electromagnetic in character. For this reason any putatively successful theoretical account of blackbody radiation would be fundamental, as it would draw upon two hitherto distinct domains of physics, thermal physics (thermodynamics and the kinetic theory of gases) and Maxwell's theory of electromagnetism.

A perfect blackbody does not exist in nature (although the spectrum of the cosmic microwave background (CMB), discovered by Penzias and Wilson in 1964, is very close to that of a perfect blackbody). Still, Kirchhoff proposed that radiation emitted from a blackbody at a fixed temperature could be closely approximated by equilibrium thermal radiation within a cavity whose walls were kept at uniform temperature. The emitted radiation could be studied as it exited through a small hole in the cavity. Because blackbody radiation emitted by any substance has the same frequency signature at a given temperature, it is distinct from the line spectrum of the discrete band of frequencies characteristic of particular substances studied in spectroscopy. It is ideally simple – homogeneous, isotropic and unpolarized – and it can be characterized by the function $\rho(v, T)$ giving the spectral or energy radiation density, i.e., the energy per unit volume, and per unit frequency interval $(v + dv)$ of blackbody radiation of frequency v at temperature T. Determining the graph of this function led to the first quantum concepts.[14]

Conceived as a theoretical ideal in the 1860s, by the 1890s experimenters in Berlin had devised objects whose emitted radiation approximated the theoretical blackbody spectrum to within a few percent. This required, among other challenges, the development of radiation detectors of great sensitivity and of ways of extending measurements over higher and higher frequency domains and temperatures. The first successful design and actual fabrication of a physical body closely approximating a blackbody was made by two German physicists, Wilhelm Wien (1864–1928) and Otto Lummer (1860–1925), who worked at the newly established (1887) Physical-Technological Imperial Institute (*Physikalisch-Technische Reichsanstalt*, or PTR) in Charlottenburg (after 1920, a district of Berlin). Founded by the German government at the initiative of inventor and industrialist Werner Siemens (1816–1892), the illustrious Hermann von Helmholtz (1821–1894) was the PTR's first director; later on, from 1917–1933, Einstein was a member of the PTR's board of directors. Tasked with setting calibration and metrological standards for German industry, the PTR quickly became the world center for the development and study of high-temperature technologies. Since measurement of high temperatures requires precise measurements of thermal radiation, it also became a natural place for investigation of the theoretical problem of blackbody radiation.

Theorists could build upon several attested experimental laws. First was Austrian physicist Josef Stefan's 1879 determination that the total energy emitted by a hot body, i.e., the density of thermal radiation of all frequencies at a given temperature $E(T)$, was proportional to the fourth power of the temperature. In 1884 his former student, Boltzmann, then professor of experimental physics in Graz, showed that Stefan's experimental law is valid only for blackbodies and succeeded in deriving it from thermodynamic and electromagnetic assumptions. The so-called Stefan-Boltzmann law states that the total energy $E(T)$ emitted at all frequencies by a blackbody is proportional to the fourth power of the absolute temperature (in degrees Kelvin),

$$E(T) = aT^4. \text{ (Stefan-Boltzmann)}.$$

Here a is a new constant, the Stefan-Boltzmann constant that, it would turn out, could be expressed in terms of two even newer ones, the one Planck termed Boltzmann's constant k and the constant h that became associated with Planck's name.

A similar mixture of thermodynamic and electromagnetic reasoning enabled Wien to derive his *displacement law* in 1894 showing that as temperature rises, the blackbody spectrum is simply rescaled according to the rule $\rho(v, T) = v^3 f(v/T)$, with f as a universal function of the ratio v/T only. In conjunction with the first experimental measures of the spectral distribution of blackbody radiation conducted at the PTR in the early 1890s, and by analogy with Maxwell's exponential law for the distribution of velocities in a gas, the displacement law allowed Wien to theoretically propose a form for $\rho(v, T)$ that fit the then-known experimental data quite well:

$$\rho(v, T) = \alpha v^3 e^{-\beta v/T}, \text{ (Wien)}$$

wherein α and β are two constants. Wien's "discoveries regarding the laws governing the radiation of heat" at the PTR merited the Nobel Physics prize in 1911. As will be seen, however, in autumn 1900 two teams of experimentalists also working at the PTR demonstrated that even though Wien's law correctly described blackbody radiation of short wavelength and thus high frequencies, it broke down in the low-frequency (infrared) range of the radiation spectrum.

Another proposal for determining the function $\rho(v, T)$ describing the frequency distribution of blackbody radiation came from Britain in 1900; it is associated with the names of John Strutt, the third Lord Rayleigh (1842–1919) and James Jeans (1877–1946). Rayleigh, then Cavendish professor at Cambridge, was an expert in optics, having explained the blue color of the daytime sky with his eponymous formula for the scattering of sunlight by molecules of the constituent gases of the Earth's atmosphere.[15] He now considered the problem of determining $\rho(v, T)$ analogous to that of determining the thermal equilibrium of all modes of vibration of the ether, considered an elastic solid. The analogy appeared to call for a straightforward use of "the Boltzmann-Maxwell doctrine of the partition of energy". A result of statistical mechanics, the equipartition theorem states that in any system of a large number of particles at equilibrium (constant temperature), the available energy is distributed equally (on average) among all modes of motion or degrees of freedom (say, translational motion along the x-axis, or rotational motion about the z-axis); the energy of each degree of freedom is ½ kT where T is the absolute temperature.[16] But Rayleigh presumably recognized that in blackbody radiation there are infinitely many degrees of freedom of high frequency, so unrestricted use of the equipartition theorem would lead to an infinite concentration of energy at the high frequencies. He accordingly restricted its application, proposing a result that applied only to "the graver modes", i.e., low frequencies, while keeping the total energy finite. The following expression (though with an erroneous factor, corrected by Jeans in 1905) is then valid in the limit of low radiation frequency and high temperature,

$$\rho(v, T) = \left(\frac{8\pi v^2}{c^3} \right) kT, \text{ (Rayleigh-Jeans)}$$

where the coupling term is determined by Maxwell's theory, $kT = \bar{E}$ is the mean total energy, and k is again Boltzmann's constant relating energy (at the individual particle level) to temperature.

It is worth lingering a moment on the form of this expression: the first factor on the right (containing c) is a consequence of Maxwell's electromagnetism, the second factor (with k) of the kinetic theory of heat. When integrated over all frequencies, this formula evidently leads to an absurd result, infinite energy of the radiation. In Rayleigh's eyes, this circumstance was not even worth noting, for he expected the equipartition theorem to fail at high frequencies and at low temperatures, analogous to similar violations observed in gases. Of primary interest is the form of the two laws, of Wien and of Rayleigh-Jeans. Wien's law broke down in the region of low frequencies, Rayleigh-Jeans in the region of high frequencies. Planck's correct formulation of the radiation law will combine them. Einstein will show in 1909 that Wien's law has a "particle" character while Rayleigh-Jeans's has a "wave" character.

Enter Planck

Kirchhoff had moved from Heidelberg to the University of Berlin in 1875 to take up the newly established chair in theoretical physics, the first in Germany. Upon his death in 1879, the prestigious chair was first offered to Boltzmann who declined, and then to Heinrich Hertz who also declined. Only in 1889, after a vacancy of nearly ten years, was the chair next offered to Max Planck (1858–1947), who accepted. The advanced experiments on thermal radiation at the PTR in Charlottenburg made Berlin a fertile location for Planck.

Planck has been aptly caricaturized as "believing in God, in Germany, and in the absolute validity of the two principles of thermodynamics".[17] Through self-study, Planck was a disciple of Clausius and like Clausius, Planck regarded both laws of thermodynamics as absolute and exceptionless. Indeed, the first law, the principle of conservation of energy, is still generally regarded as absolute today, though its exceptionless status was famously challenged in a 1924 paper of Bohr, Kramers, and Slater (to which Einstein responded critically, see Chapter 4). Already in his doctoral dissertation of 1879, Planck displayed a preoccupation with the second law of thermodynamics. Planck viewed the status of the two laws of thermodynamics quite differently. Whereas the initial and final states of a process in nature were equivalent from the standpoint of the first law, this was not true in the second law, a fact signaling the direction in which processes must take place. Within a few years Planck deliberately reformulated Clausius's version to bring out the strong presumption of irreversibility: *all isolated systems move irreversibly from states of lower to states of higher entropy*. Whereas Maxwell and Boltzmann deemed the second law to be merely a statistical law pertaining to unimaginably large numbers of particles and so subject to exceptions, Planck could not accept that the law of entropy increase was not immutably valid. For this reason he opposed the attempt to base the laws of thermodynamic evolution and approach to equilibrium on the time-reversible propositions of mechanics. For the same reason he was a critic of Boltzmann's atomism and its role in the chain of reasoning leading to the unsavory conclusion that the second law is not absolute.

In 1895, Planck embarked on a highly ambitious program to tackle the riddle of blackbody radiation, the puzzle of why *any* object regardless of composition, maintained at constant temperature, emits the same exact spectrum of radiation. He began with thermodynamics, and in particular the second law, hoping to show that irreversibility could be theoretically derived using only the resources of electrodynamics, i.e., Maxwell's theory augmented to include interactions between field quantities (electric, magnetic, current) and their material sources. Thermal radiation was to be understood exclusively as an electromagnetic phenomenon. Planck sought to tackle two problems at once: appeal to electromagnetic interaction between matter and radiation to explain both thermodynamic irreversibility and the growing experimental data on blackbody radiation. His initial idea, roughly stated, was that the particular mechanism producing irreversibility is the transition of an electromagnetic wave from an incident plane wave in the process of absorption to a spherical one in the process of emission. Boltzmann, recalling his lesson from Loschmidt, at once pointed out that Planck's strategy could not by itself succeed in explaining irreversibility since the laws of electrodynamics also permitted the time reverse of an emitted spherical wave. Like the laws of Newtonian mechanics, those of electrodynamics are completely time-reversible.

Chastened by this exchange with Boltzmann, Planck over the next decade or so adopted various versions of a hypothesis of "elementary disorder" to account for thermodynamic irreversibility. He sought to show that the statistical form of the second law had to be a *mathematical consequence* of an elementary randomness, i.e., of an *a priori* initial independence of the individual elements when considered statistically. In this way the second law retained its absolute status as a necessary implication of a persistent disorder. Not until 1914, *after* the introduction of the idea of a quantum of energy, *after* the apparent successes of the Bohr atom, and *after* the majority of his theoretical colleagues, did Planck accept that the law of entropy increase is not an absolute law. Still, Planck praised Boltzmann even in his early work and despite his antipathy to atomism for emancipating the concept of entropy from anthropomorphic ideas of machines and the art of human experimentation.

Planck's radiation law

Recall that a blackbody absorbs all incident radiation and emits a spectrum of radiation characteristic of its temperature, while at thermal equilibrium it will emit and absorb radiation at the same rate. In the 1890s, researchers at the PTR had accumulated reliable data on the blackbody spectrum, the distribution $\rho(\nu, T)$ of electromagnetic energy as a function of frequency and temperature, within the range of all temperatures then experimentally accessible. The first task was to find a model of the interaction between radiation and oscillating charges that fit the experimental data. Planck employed a simple fictional construct of matter emitting and absorbing radiation used by Hertz some six years before in his own studies of the interaction of matter and radiation. The model treated the inner surfaces of the blackbody cavity as composed of charged masses that, in response to the electromagnetic radiation incident upon them, oscillate or vibrate at a certain definite frequency, each interacting with one, and only one, color of light. Since an oscillating dipole emits electromagnetic radiation, these simple harmonic oscillators or "resonators", as Planck (following Hertz) called them, were ideal abstract processes that could be justified by Kirchhoff's law, that the radiation distribution at equilibrium is independent of the particular type of system interacting with radiation. So Planck could regard his resonators as merely the simplest conceivable material systems that can be in equilibrium with electromagnetic radiation but as lacking any specific physical meaning and in particular as independent of any controversial molecular or atomic hypotheses. As the equilibrium spectrum did not depend on the specific thermalizing system, the model could be considered sufficient to determine the frequency distribution of blackbody radiation.

Planck attacked the problem of deriving the spectral energy density of electromagnetic radiation (i.e., determining the function $\rho(v, T)$) by relating the energy of the emitted electromagnetic radiation to the (average) energy $\bar{E}_v$ at a given temperature of emitting and absorbing resonators of frequency v. By 1899 Planck had derived from Maxwell's theory and from a random-phase assumption for the resonators the following relationship between the distribution function $\rho(v, T)$ and $\bar{E}_v$,

$$\rho(v, T) = \left(\frac{8\pi v^2}{c^3} \right) \bar{E}_v (T).$$

Planck was almost done: in order to have the explicit form of the distribution law, he needed only to determine $\bar{E}_v(T)$, the average energy of a resonator of frequency v at temperature T. He assumed the resonator energy $\bar{E}_v$ to have an exponential form needed to retrieve Wien's distribution law for the density $\rho(v, T)$. At this point, Planck was confident that he essentially had realized the chief aims of his grand program: a proof of the irreversible evolution of thermal radiation in a cavity, and a determination of the equilibrium spectrum of this radiation.

Alas, in the autumn of 1900 new experiments at longer wavelengths by Heinrich Rubens and Ferdinand Kurlbaum showed that Wien's law matched observations in part but broke down at the low-frequency (infrared) end of the radiation spectrum. Planck then recognized a certain arbitrariness in his previous entropy considerations and sought the simplest generalization of Wien's law. On October 19, 1900 he proposed what turned out to be the empirically correct radiation distribution law for all temperatures then technologically realizable in the laboratory. Using notation introduced several months later, Planck's law of the distribution of blackbody radiation by frequency is

$$\rho(v,\ T) = \left[\frac{8\pi v^2}{c^3}\right] hv \ /\left(\exp\left(hv\ /\ kT\right) - 1\right) \text{ (Planck)}$$

where h is a new extremely small physical constant subsequently named after Planck and k is the constant Planck named after Boltzmann.[18] Planck effectively obtained this expression by introducing h as a parameter in an interpolation formula designed to fit the experimental results of Rubens and Kurlbaum. Between October 19 and December 1900, Planck undertook the important next step of determining the formula's "true physical meaning", deriving it from existing theory. He did this by analogically following Boltzmann's 1877 paper, in which the second law was first given a statistical interpretation. As noted above however, Planck himself did not accept the statistical interpretation of the second law until 1914, so his use (or misuse) of Boltzmann was both partial and selective. The rather twisted path by which Planck derived his radiation law can be presented briefly here.

At thermal equilibrium, it was possible to consider the average energy $\bar{E}_v$ of an individual resonator as constant and characteristic of the resonator, acting only on radiation of frequencies very close to its own resonance frequency and related to the spectral density $\rho(v,\ T)$ at this frequency through the above-mentioned relation,

$$\bar{E}_v = (c^3\ \rho)/8\pi v^2.$$

Planck could have calculated $\bar{E}_v$ from the equipartition theorem of statistical mechanics. Planck, however, was even less inclined than Rayleigh to apply equipartition to this problem. Now, rather than considering the relationship between energy and temperature, Planck focused on the relation between the entropy and the energy of a resonator that had played a central role in his earlier treatment of irreversible radiation processes. At this point, he appealed to the relation between entropy and probability and turned to the formula that Boltzmann never wrote down even though it adorns his grave in Vienna,

$$S = k \log W.$$

Whereas for Boltzmann, the probability W of a thermodynamic state is just the number of complexions (atomic configurations and speeds) corresponding to that macrostate, Planck sought to determine W analogously by considering the number of "complexions" of his set of resonators. Deeming it sufficient to consider a system of N resonators all of frequency ν and energy $\bar{E}_\nu$, the total energy E_N of the N resonators is $E_N = N\bar{E}_\nu$. Additivity of entropy meant that the entropy of the resonator system S_N is just N times the entropy of each resonator S, $S_N = NS$. S_N is then set equal to the probability W via the constant of proportionality k,

$$S_N = k \log W.$$

Crucially, Planck assumed, as Boltzmann's combinatorial method required, that any complexion was as likely to occur as any other. In order to obtain finite values for W, Boltzmann's combinatorial formula required that the total energy to be shared among the N resonators be treated as consisting of an integral number of equal finite parts. So without using the term "quantum", Planck introduced a discrete energy element ε, dividing the total energy E_N by ε to get a whole number P of energy elements (he allowed rounding off to the nearest integer). P then had to be distributed over the finite number N of resonators,

$$E_N = P\varepsilon.$$

Thus Planck's key step was to restrict the energy of each resonator so that it had to be an integral multiple of the energy unit ε, a tiny discrete amount. He insisted that his energy elements were strictly mathematical fictions, yet he determined a constant of proportionality h linking the energy element ε to the frequency ν of radiation,

$$\varepsilon = h\nu$$

that would become a fundamental equation of the quantum theory. Then the resulting expression of W, Boltzmann's relation between S and W and the relation between the energy of a resonator and the spectral density of radiation together implied Planck's law.

In Planck's original view, these energy elements were important in defining the entropy of the resonators. But they did not imply an intrinsic discontinuity of the energy of the resonators, and they did not affect the interaction between radiation and resonators that Planck still treated by means of Maxwell's equations. Ever the reluctant revolutionary, and rather than surrender the classical laws of electrodynamics as the basis for his theory of heat radiation, beginning in 1911 Planck sought to treat emission and absorption of radiation asymmetrically, in what became known as his "second theory". Absorption occurred continuously, in accordance with classical laws. Discontinuity entered with emission of radiation, which occurred in discrete leaps. Planck supposed that his resonators could therefore take on any energy value and so did not intrinsically possess discontinuous energies, but could emit energy only when reaching some threshold in values of $nh\nu$. In this way, Planck was able to re-derive his radiation law of 1900. The explanation of discontinuity in the emission process was left to hypothetical details, differing in the different versions of the second theory, of the interacting microstructure of the matter comprising his resonators. When Bohr's atomic theory of 1913 required discontinuous emission and absorption of energy, Planck's second theory was rejected.[19] However, one of its implications remains today: the existence of zero-point energy. Quantum objects and processes possess a non-zero average rest energy even at absolute zero temperature.

In any case, despite the apparent triumph of deriving the radiation law that bears his name, Planck regarded the constants k and h appearing in the law as of far greater significance. First of all, the values he gave them are remarkably accurate, within a few percent of their current values. But more importantly, Planck showed that they could be combined, together with Newton's gravitational constant G, and c the speed of light in *vacuo*, to form a system of units of length and time that were completely "natural" in that they contain no reference to human measures. By multiplying and dividing different combinations of G, c, k, and h, one obtains what is now called the Planck time (5.39106×10^{-44} seconds) and the Planck length (1.616199×10^{-35} m), the tiny scale that current theory regards as the domain of quantum gravity. The same absolute measures could be found by any careful investigators anywhere in the universe, and in this sense are completely non-anthropomorphic. Planck's striving for the absolute, though it didn't work with the second law, paid off here in spades in giving rise to the ideal of a completely de-anthropomorphized physics of law and of fundamental natural units. This would be a unifying theme of Planck's philosophy of science for the rest of his life.

In 1918, the last year of the Great War, the Nobel Prize in physics was awarded to Planck for "the advancement of Physics by his discovery of energy quanta". In his Nobel address in June 1920, Planck modestly recalled his reluctance to accept the physical reality of the energy quantum. After all, it could be interpreted in two ways, first as just a "fictional quantity" in which case "the whole deduction of the radiation law was in the main illusory and represented nothing more than an empty non-significant play on formulae". Or, secondly, one could regard "the derivation of the radiation law as based on a sound physical conception". Planck continued:

> "Experiment has decided for the second alternative. That the decision could be made so soon and so definitely was due not to the proving of the energy distribution law of heat radiation, still less to the special derivation of that law devised by me, but rather it should be attributed to the restless forward-thrusting work of those research workers who used the quantum of action to help them in their own investigations and experiments. The first impact in this field was made by A. Einstein".[20]

Indeed, Einstein's introduction of the quantum of action into his own theoretical investigations, beginning in 1905, proved instrumental in establishing the quantum. These investigations in early quantum theory are intimately related to the use of a guiding principle that Einstein termed "Boltzmann's principle", taken up in the next chapter.

Notes

1 "On the Theory of the Energy Distribution Law of the Normal Spectrum", as translated from *Die Verhandlungen der Deutsche Physikalische Gesellshaft Berlin* Bd. 2 (1900), in D. ter Haar (ed.), *The Old Quantum Theory*. Oxford, UK and London: Pergamon Press, 1967, pp. 82–90; p. 84.

2 On the *vis viva* controversy, see Hankins, Thomas L., "Eighteenth Century Attempts to Resolve the *Vis viva* Controversy", *Isis* v. 56, no. 3 (Autumn, 1965), pp. 281–97; and Smith, George E., "The *Vis Viva* Dispute: A Controversy at the Dawn of Dynamics", *Physics Today* v. 59, no. 10 (2006), pp. 31–9.

3 Kuhn, Thomas, "Energy Conservation as an Example of Simultaneous Discovery", as reprinted in Thomas Kuhn, *The Essential Tension: Selected Studies in Scientific Tradition and Change*. Chicago: University of Chicago Press, 1977, pp. 66–104.

4 Mach, Ernst, *History and Root of the Principle of the Conservation of Energy*. Translated from 1872 German edition by Philip E.B. Jourdain. Chicago: The Open Court Publishing Co., 1911.

5 In fact, the conservation of energy cannot be derived from the impossibility of perpetual motion only, as Mach acknowledged under criticism from Planck. The impossibility of perpetual motion allows the possibility of *producing* work in a cycle of operations. But in order to get energy conservation or the first law of thermodynamics, one must also assume the impossibility of *annihilating* work.

6 The idea that matter consists of invisibly small indivisible particles originated with the Greek atomists Leucippus and Democritus. During the Middle Ages atomism fell into disrepute on grounds of its close relation to atheism. In the mid-17th century, French priest and scientist Pierre Gassendi (1592–1655) revived ancient atomism while arguing its compatibility with Christian doctrine. Gassendi's influence extended to the founders of mechanics in the 17th century, Galileo and Newton, who both assumed an atomistic theory of matter.

7 Thomson, William, later Lord Kelvin, first proposed an absolute scale in 1847–1848 yet still assumed the conservation of caloric. The later 1854 definition drops this assumption. The guiding idea is that a temperature scale must be independent of any particular substance and it must relate intervals of temperature to specific quantities of heat and the equivalent mechanical effect produced. Its null point is absolute zero (-273.15 K, or -459.67° F), theoretically the temperature at which all classical thermal motion ceases. Absolute temperature is generally denoted with a capital T; unless otherwise indicated, in this book, "temperature" refers to this absolute scale.

8 Often written $dS/dt \geq 0$, although entropy is classically ill-defined in intermediate stages.

9 "Thus molecular science teaches us that our experiments can never give us anything more than statistical information, and that no law deduced from them can pretend to absolute precision". "Molecules", *Nature* v. 8 (25 September 1873), pp. 437–41; reprinted in William D. Niven (ed.), *The Collected Scientific Papers of James Clerk Maxwell*, vol. 2, New York: Dover Publications Inc., 1965, pp. 361–77; p. 374.

10   The constant k already implicitly appears in the kinetic-theoretical interpretation of the "ideal gas" law of Boyle & Gay-Lussac, pressure = (R × temperature)/volume. Where N denotes the number of molecules, the kinetic energy E
of translational motion (in each of the three directions of space) per molecule
is $3kT/2$, where $k = R/N$.

11   We must linger a moment upon a necessary detail that will prove essential to
Einstein's conception of probability. To extend this toy model to a real gas of
molecules, Boltzmann had to consider the case in which the molecular-kinetic
energies are continuous, not discrete. He did this by first representing the entire
gas in a fictional "energy space". The energy space is portioned into cells, so
that the kinetic energy of each molecule is stipulated to lie within a cell ranging
from 0 to ε, ε to 2ε, 2ε to 3ε, and so on. All N molecules are distributed among
these energy cells so that the same combinatorial computation as above can be
used. As Boltzmann further showed, if each complexion is to be equally probable, one must use what today is called "phase space" rather than the energy
space. In phase space, every monoatomic molecule moving according to Newton's laws of motion (in Hamilton's formulation) is completely described by
six generalized coordinates, three spatial components q_i, and three momentum
components p_i. One can then represent the state of a gas of N particles in phase
space in two distinct but related ways: at the macro- or thermodynamic level,
the state of the entire gas is represented at a given instant by a single system
point in a 6N dimensional space Γ; at the micro- or complexion level, the state
of the entire gas at that instant is represented by N points distributed in disjoint
"cells" in a six-dimensional space μ (the distinction is due to the Ehrenfests).
(Ehrenfest, Paul, and Tatiana Afanasyeva, *The Conceptual Foundations of the Statistical
Approach in Mechanics*; original German edition (1912), English translation by
Michael Moravcsik. New York: Dover Publications, Inc., 1990.)

    The connection between the two spaces is established as follows. Exhaus-
tively partition the μ space into $r+1$ discrete cells ω_0, ω_1, ω_2,..., ω_r, each rect-
angular in the position and momentum coordinates and of equal small volume
$dV = dpdq$. These volumes also can be characterized by molecular energy (proportional to velocity). Then to each distribution of particles among the cells
in μ, i.e., specifying the number of particles whose molecular state lies within
each cell ω_1, there corresponds a portion of Γ. Boltzmann now introduced
probability through the *a priori* stipulation that equiprobable distributions (or
macrostates) have equal volumes in μ space. Accordingly, the probability of any
macrostate represented in Γ is proportional to the volume of the corresponding portion of the μ space. The notion of probability thereby acquires a more
abstract character; a transition from probability as relative number to probability as occupied volumes in μ phase space. In his use of Boltzmann's principle,
Einstein in 1904 will exploit an analogous relation between the two spaces
to define the probability of a thermodynamic state in an in-principle *empirical*
way: if one observed a system for a very long time, the probability of each
state specified by a portion of Γ is defined in terms of the fraction of *time spent*

in this portion in the long run. Thus Einstein shunned defining the probability of a state in terms of counting complexions and the involved *a priori* stipulation about equal probabilities, and in this way avoided Boltzmann's implicit assumption of identical but distinguishable particles. So too did Boltzmann in a memoir of 1881 that Einstein presumably did not read.

12 The logarithmic function log W increases with W more slowly than any other exponent, including a fractional power like $W^{1/2}$.

13 Olivier Darrigol has observed that, from a more critical point of view, these conclusions are problematic since in 1877, Boltzmann merely assumed, but did not prove, that his combinatorial probabilities had physical meaning. The proof only came later in 1881. But by then, the method of 1877 had been subordinated to the more powerful ensemble-based approach to statistical mechanics that Boltzmann also pioneered (building on Maxwell). For this reason, the 1877 paper plays little role in Boltzmann's subsequent writings (see below). Ironically Planck, originally one of the strongest critics of the statistical interpretation of the second law, by adopting the 1877 method in 1900 made the 1877 paper so famous.

14 $\rho(v, T)$ can be defined in terms of the universal function above as $J(v, T) = (c/8\pi) \rho(v, T)$, the constant of proportionality containing c the velocity of light and 8π a conversion factor.

15 In 1871 Rayleigh argued that the angle at which sunlight is scattered varies inversely with the fourth power of the wavelength; hence light at short wavelengths (blue end of the visible spectrum) is enhanced much more than light at longer wavelengths.

16 More exactly, the relevant number is the number of quadratic terms of the Hamiltonian of the system, which is equal to the number of degrees of freedom if the energy is entirely kinetic in nature. If the gas molecule of mass m has (only) translational motions, there are three degrees of freedom $\times$ kinetic energy $\frac{1}{2}$ mv; in this case the total energy is $\frac{3}{2}$ kT. In general, for f quadratic terms in the energy of a molecule, the average total energy is f $\times$ $\frac{1}{2}$ kT. For N particles, it is N $\times$ f $\times$ $\frac{1}{2}$ kT.

17 Darrigol, Olivier, "Statistics and Combinatorics in Early Quantum Theory", *Historical Studies in the Physical and Biological Sciences* v. 19, no. 1 (1988), pp. 17–80; p. 41.

18 Planck's constant h has the modern value 6.63 x 10^{-34} joule-seconds (units of energy multiplied by units of time, hence "action"); the so-called "reduced" Planck constant $\hbar$ = h / 2π was introduced later by Dirac.

19 See Kuhn, Thomas, *Black-Body Theory and the Quantum Discontinuity, 1894–1912.* Chicago: University of Chicago Press, 1978; "Afterword: Revisiting Planck", pp. 349–70.

20 Planck, Max, "The Genesis and Present State of Development of the Quantum Theory", *Nobel Lecture*, 2 June 1920, available at www.nobelprize.org/nobel_prizes/physics/laureates/1918/planck-lecture.html

Further reading

Brush, Stephen G. with Ariel Segal, *Making 20th Century Science: How Theories Became Knowledge*. New York: Oxford University Press, 2015.

Klein, Martin, "Thermodynamics and Quanta in Planck's Work", *Physics Today*, November, 1966, pp. 294–302.

Kuhn, Thomas S., *Black-Body Theory and the Quantum Discontinuity, 1894–1912*. Chicago: University of Chicago Press, 1978.

Lindley, David, *Boltzmann's Atom: The Great Debate that Launched a Revolution in Physics*. New York: The Free Press, 2001.

Three

Contributions to the old quantum theory

"It is . . . my opinion that the next stage in the development of theoretical physics will bring us a theory of light that can be understood as a kind of fusion of the wave and emission theories of light".

Einstein, Salzburg, September 1909[1]

Introduction

Since the debut of quantum mechanics in 1925–1926, it is customary to distinguish classical from quantum physics. Just *how* or even *when* the previous body of accepted physical theory – the mechanics of Newton, Lagrange and Hamilton, thermodynamics, the kinetic-molecular theory of heat, Maxwell's electromagnetic theory, both relativity theories of Einstein – became "classical" (today, that means "non-quantum") is less clear. It cannot be quite right to point to the advent of quantum mechanics because a further distinction is required, among theories well-established at the end of the 19th century (all those listed above, with the exception of relativity), the so-called "old quantum theory", and what is now known as quantum mechanics. Retrospectively, quantum theory stems from Max Planck's "discovery" (better, computational use) of discrete energy elements in 1900. It developed as breakdowns of classical theories became increasingly apparent when applied to emission and absorption of radiation but as well in the anomalous low temperature specific heats of different substances. Neither exactly classical nor quantum, nor really a theory at all, the old quantum theory is rather an evolving, sometimes amorphous, collection of computational rules, insights, analogies, and approximations, "a groping with-out correct foundation" as Einstein put it in 1912. The very next year Bohr's model of atomic structure (and its subsequent refinements) provided the platform of an open theory on which to build, some parts of which could be considered well-established and others much more tentative. Yet as an inconsistent mix of classical and non-classical components, the Bohr atom, described below, remains an exemplar of the old quantum theory's neither-this-nor-that character. The old quantum theory might be then deemed a *post-classical* theory, but properly such a designation can be applied only retrospectively, i.e., after the rise of quantum mechanics.

Einstein stands in antipodal relations to the two eras of quantum theory. His response to quantum mechanics (see Chapter 4) is famously critical and skeptical. Yet it is fair to say, as leading historians of physics have argued, that no one contributed more to the development of the old quantum theory. The groundwork for these contributions lies in a largely unremarked trilogy of papers on statistical physics in 1902–1904 where, with the advantage of hindsight, one can recognize that a strategy is fashioned for attacking the problem of the quantum. Einstein cut his teeth as a theoretical physicist in these early works on the statistical foundations of thermodynamics, independently reaching important results in statistical mechanics obtained previously by both Boltzmann and Gibbs, giants in the field. In addition, Einstein developed his own distinctive theory of fluctuation phenomena, of the apparently random departures of thermodynamic quantities from the mean values that correspond to the statistically predicted behavior of large ensembles of atoms or molecules.

His study of fluctuation phenomena provides initial confirmation, as well as further evidence beyond the much more studied case of relativity theory, of an uncanny ability to extract revolutionary insights from simple considerations resting on the affirmed validity of certain general physical principles, postulates deemed reliable in the absence of any trustworthy underlying theory. In this case, the relevant postulate is Boltzmann's definition of the entropy of a thermodynamic state in terms of the state's probability. Termed by Einstein "Boltzmann's principle", it was understood and employed in distinct manner critical of its use by Planck, and indeed implicitly critical of Boltzmann's statistical method on which Planck relied. The novel treatment reaped a rich physical harvest, enabling among much else an identification of the limits of classical thermodynamics and of Maxwell's electromagnetism, the hypothesis of light quanta, and a clarification of the meaning of Planck's theory of blackbody radiation – all ultimately pointing to the need for a new non-classical theory of matter and of radiation. The ensuing contributions to the old quantum theory, nearly all statistical in character, span two decades from 1905–1925 and fall clearly into two periods punctuated by a hiatus (1911–1916) of almost exclusive preoccupation with the general theory of relativity.

First love: Thermodynamics

By 1900, theorists could point to various reasons for accepting the atomic structure of matter: the late 19th-century rise of an experimental microphysics of new radiation phenomena (electrons, X-rays), the success of newer atomistic theories like Lorentz's electrodynamics, and the predictions of the kinetic theory matched, in macroscopic experiments and in the mean, those of thermodynamics. Many attempted to explain the bulk properties of matter, whether in the gaseous, liquid, or solid state, on the basis of the kinetic-molecular theory originated by Maxwell and Boltzmann. Einstein's early papers on the kinetic-molecular theory also presuppose atomism, rather than directly argue for it. They evince an unstated but firm commitment to the assumption that physical systems are composed of tiny particles obeying the laws of mechanics, and that macroscopic thermodynamic properties of systems (temperature, pressure, density, entropy, etc.) result from motions and collisions of vast numbers of such particles whose overall behavior can be statistically predicted with high accuracy. Einstein's conviction may well have been acquired by reading Boltzmann's *Lectures on Gas Theory* (1896–1898), probably beginning in the summer of 1899. In September 1900, Einstein reported about these studies to the woman he would marry:

> The Boltzmann is magnificent. . . . I am firmly convinced of the correctness of the principles of the theory, that is, I am convinced

that in the case of gases it is really a matter of the motion of discrete mass points of definite, finite magnitude that move in accordance with certain conditions. . . . It is a step forward in the dynamical explanation of physical phenomena.[2]

As previously discussed, however, a still-smoldering controversy surrounded kinetic theory's alleged triumph: the statistical explanation of the empirically well-established second law of thermodynamics. The paradox of the irreversibility of the second law indicated a gap in theoretical understanding that gave succor to opponents of atomism and to skeptics of the putative explanatory achievements of the molecular-kinetic theory. Besides Max Planck, these included prominent physicists and chemists, among them Ernst Mach, Wilhelm Ostwald, and Pierre Duhem. The latter two in particular reasoned that if physical observations pertain only to manifestations of energy and if a physical system could be adequately characterized by phenomenological description of transformations of energy permitted by the laws of thermodynamics, atomic hypotheses in the theory of matter would be seen to be extravagant. By adhering strictly to thermodynamics, or better, to an "energetics" purporting to give unified representation to physics and chemistry, misleading pictures of underlying mechanical processes would be altogether unnecessary. In this way the statistical explanation of the second law (reversibility?) could simply be bypassed while avoiding also the epistemological difficulties of atomism in addition to its traditional inconsistencies and paradoxes (indivisibility?).

Despite an admiration for Mach that will influence both the special and general theories of relativity, the young Einstein was not among the critics of atomism. At no point did he evince a trace of hesitation concerning the physical reality of atoms and molecules, nor of the essential correctness of the kinetic theory and statistical mechanics. Well aware of the conflicted theoretical situation, he turned his attention to thermodynamics and its still unclear statistical basis in what he called the "general molecular theory of heat". The idea to begin with thermodynamics was in any case a natural one, for taken by itself it is a highly confirmed theoretical structure relating directly to observable thermal phenomena while remaining uncommitted to specific microphysical assumptions. The generality of thermodynamics rested on empirically attested principles constraining the laws of nature: the impossibility of constructing a perpetual motion machine. Applicable to the broadest range of physical phenomena including radiation, it is not obviously mechanical or particulate in nature. Writing to his friend Michele Besso in January 1903, Einstein stressed the *empirical* validity of the second law; in its "most general form" the law was stated not by any formal expression but by a prohibition, the impossibility of a *perpetuum* mobile (of the second kind).[3] Yet ever a firm proponent of Boltzmann,[4] Einstein accepted that the second law *in its statistical interpretation* does not absolutely prohibit never-observed transformations of energy – for example, heat passing from a colder to a warmer body – but merely accords such processes vanishingly small probability.

Boltzmann's principle

Though dating from 1877, Boltzmann's conceptual connection between entropy and probability remained highly controversial into the first decade of the 20th century. On the other hand, beginning in 1904, Einstein began to refer to this connection as "Boltzmann's principle", adopting it as a postulate. The result was a rich harvest of results that led to a re-fashioning of the statistical basis of thermodynamics, extending it beyond classical thermodynamics where only average or mean values of the thermodynamic quantities come into consideration. By the First Solvay Conference in Brussels in October 1911, Einstein had shaped and employed with considerable success a *generalized thermodynamics*, free from residual mechanical presuppositions remaining in Boltzmann's own statistical treatment. Like no one else at the time, the young Einstein perceived that such an approach could provide a secure probe into domains where classical theory had clearly failed. With Boltzmann's principle as guide, an expanded thermodynamics could provide a reliable epistemic footing upon which to build amidst all other theoretical uncertainty. Thus at Solvay while Einstein could still note "the greatest diversity of opinion as regards the content and domain of validity" of Boltzmann's principle, his own view of the principle's validity was unconditional: "we must hold on to its validity without reservation".[5] "Without reservation" signals without restriction to classical thermodynamics. How is this even possible?

According to atomism, underlying mechanical processes drive the evolution of a non-equilibrium thermodynamic system towards equilibrium. Yet despite Boltzmann's unwavering commitment to atomism, his 1877 statistical method sought to define the entropy of a thermodynamic macrostate in terms of its probability without recourse to kinetic assumptions. It is customary to use Boltzmann's connection between entropy and probability to infer the entropy of a state of a system from the probability of that state, the latter derived by simply computing the number of microstates compatible with the given thermodynamic macrostate. Thus for an isolated system of known energy, the entropy of a given macrostate is simply a relative probability, i.e., the ratio of the number of complexions (microstates) consistent with that macrostate to the total number of all complexions possible at the given energy. By appealing to something like the principle of indifference, each of the vast number of complexions compatible with that macrostate is *a priori* regarded as equiprobable; i.e., the total system at a given instant is as likely to be in one complexion as in any of the others indifferently compatible. And in computing the number of complexions, recall that on the assumption of identical but distinguishable particles, simply permuting particles and velocities gives rise to new complexions to be counted as distinct. The previous chapter shows that the latter assumption is *mechanical* in nature, indeed Boltzmann denominated it "the first assumption of mechanics". Although Boltzmann's statistical method is, as claimed, free of kinetic assumptions (regarding collisions, mean free path, etc.), it retained this fundamental assumption of mechanics.[6] It is precisely here that Einstein departed from Boltzmann, and more specifically from Planck, whose radiation theory followed Boltzmann's statistical method in calculating entropy from probability, employing the above definition of equiprobable complexions. Einstein deemed this to be "a certain logical imperfection" of Planck's theory (Planck had followed Boltzmann's 1877 statistical method), since the statistical method can give only an *a priori* definition of equiprobability.[7] As long as there was no independent definition of probability, it would not be clearly established that the entropy-probability principle has a physical meaning.

Einstein correspondingly inverted the usual way Boltzmann's principle was employed, instead using it to reason from the phenomenological entropy of a macroscopic state *to* that state's probability. By starting from entropy, a macroscopic thermodynamic parameter characterizing the behavior of many different types of physical systems, including radiation for which no molecular model existed, Boltzmann's principle then could be invoked to test – and possibly disconfirm – claims derived from classical theory regarding the probability of the different states of a system. To reason in this direction required that the notion of "probability of a state" be given a clearly physical definition so that claims about the probability of certain states could be compared with observation. Such a definition would allow straightforward hypothetico-deductive testing: any theory yielding a certain value for the probability of a particular state differing from the value ascertained from observation of the system's behavior would be disconfirmed by *modus tollens*. And so from the first mention of Boltzmann's principle in 1904, Einstein employed a phenomenological, empirically ascertainable characterization of the notion of "probability of a state" as occupation frequency, i.e., the fraction of total time ("time spent") by the system in a certain region of phase space (see further below).[8]

The path he then carved out using Boltzmann's principle has four recognizable steps. The first is in 1904–1905 when Boltzmann's principle was used to reveal the possible existence of energy fluctuation phenomena that, should they be shown to be actual, could not be explained by classical thermodynamics. In his first paper on Brownian motion (May 1905), Einstein proposed that such fluctuations are inherent to atomism and gave the first precise theory of these random motions. Although others (Gouy, Poincaré, Boltzmann) had previously suggested that Brownian motion was a kinetic-theoretical consequence of molecular motion, Einstein proposed what was effectively a test of the atomistic hypothesis, that the effects of statistical fluctuations in the thermal motions of a liquid's constitutive molecules might be actually observed under a microscope if suitably large particles, such as pollen grains, are suspended in the liquid. The predictions of this paper, including a determination of Avogadro's number,[9] were confirmed by experiments of Jean Perrin in Paris in 1908. Meanwhile in April 1905, Einstein advanced the hypothesis of light quanta on the basis of his inverted use of Boltzmann's principle. Doing so enabled setting up a formal analogy between results

stemming from the statistical techniques of physics and those pertaining to blackbody radiation, the apparent domain of Maxwell's
theory, and to the problem of the interaction of matter and radiation
where the equipartition theorem of classical statistical mechanics
failed. Exactly one year later, he demonstrated that Planck, deriving
the radiation law in 1900 that now bore his name, had himself made
implicit use of this hypothesis. Third, pursuing the consequences of
the revolutionary quantum hypothesis from the theory of radiation
into the theory of matter, Einstein in 1907 affirmed the need for a
new mechanics of particles to account for known difficulties in the
theory of specific heats of solids. Finally in 1909, once again on
the basis of Boltzmann's principle, Einstein's analysis of energy and
momentum fluctuations in Planck's theory of blackbody radiation
introduced the wave-particle duality into physics. Small wonder that
in summarizing the path he had followed since 1904 to the world's
leading physicists gathered at the First Solvay Conference in Brussels in
October 1911, Einstein observed "the quantum theory of today is . . .
a helpful tool but . . . not a theory in the usual sense of the word",
noting however that it was to Boltzmann's principle that "we owe
the first glimmer of light".[10]

While there is little disputing the importance to of Einstein's
contributions to the early quantum theory, the philosophical significance of the novel methodology should be underscored. In this
chapter we shall see that nearly all of Einstein's contributions to early
quantum theory are the result of a decade of insights provided by
wielding Boltzmann's principle as a probe into the obscure realm
of the quantum. By elevating it to the status of a postulate, Einstein
pursed fundamentally the same methodological strategy that produced the two theories of relativity. With application of the methods
of statistical mechanics to electromagnetism, the early work of Einstein comes full circle, since Maxwell's theory is the chief concern
of the special theory of relativity, presented, essentially fully formed,
in 1905. To be sure, Einstein's own unqualified expression of faith
in Boltzmann's principle rested upon supplying a physical definition
of probability coupled with the novel use of turning it around. A yet
further benefit will not appear for another two decades. By rejecting
Boltzmann's definition of the probability of a macrostate (by counting "complexions"), Einstein avoided assuming the "first axiom of

mechanics" which that definition presupposed. Having developed a statistical thermodynamics free of that axiom, in 1924 Einstein is poised to recognize the distinguishing aspect of quantum statistics where that axiom is expressly denied.

The statistical trilogy

In a letter to physicist Johannes Stark in 1907, Einstein dismissed his two initial published papers as "worthless". The first in 1901 was a failed attempt to explain the phenomena of capillarity, the observable changes in surface tension of a liquid enclosed in a narrow enclosure such as tube, by a simple force law of attraction between the molecules of liquid and solid. The second, in April 1902, used an extension of the second law of thermodynamics to investigate the nature of molecular forces in electrochemistry, in particular, systems consisting of metal electrodes immersed in dilute saline solutions. Though it is true that each was an offshoot of a subsequently abandoned project toward a doctoral thesis, Einstein's later assessment is an overstatement. Whatever their flaws, these first papers reveal an unstated but firm commitment to the kinetic theory, and so to the real existence of atoms and molecules.

The next three publications are of altogether different caliber. Like the first two, they appeared in the prestigious *Annalen der Physik* (then edited by Max Planck and Wilhelm Wien). Published between June 1902 and March 1904, the broad project was to find a *general* molecular foundation for thermodynamics. While the kinetic theory provided a molecular foundation for the behavior of gases at thermal equilibrium, Einstein sought a more general basis for systems (not necessarily gases) depending minimally on mechanical assumptions of the kinetic theory. The trilogy laid the foundations of what became by 1910 essentially a statistical thermodynamics extending beyond "classical thermodynamics", i.e., a generalized thermodynamics applicable to any system, not necessarily mechanical, capable of storing energy or releasing it as radiant heat. These papers reveal that, at the very onset of his career Einstein was already an adept of the kinetic-molecular theory of heat, i.e., statistical physics. Remarkably, in undertaking this project and in obtaining the results he did, Einstein had only limited knowledge of what Boltzmann and Gibbs had previously achieved in this regard. The three papers contain but two references to works other than Einstein's; both from volume 2 of Boltzmann's *Lectures on Gas Theory*.

Historians disagree regarding just which other works of Boltzmann were familiar to Einstein. In September 1901 Einstein reported (in a letter to Marcel Grossmann) that he was presently "thoroughly occupied" in study of Boltzmann's kinetic theory of gases. Without use of a university library, and after February 1902 working full-time in the patent office at Bern, the available evidence shows that he studied only the second volume of Boltzmann's lectures. Coming after nearly three decades of publications on kinetic gas theory and then statistical mechanics, Boltzmann's lectures might be thought to present a unified overview of statistical physics. They do not. Instead they reflect Boltzmann's response to critics in the 1890s of his earlier claim that the H-theorem is a result of pure mechanics. As a result, the lectures are almost entirely concerned with the kinetic method of the early 1870s. But in the 1890s Boltzmann had to admit that derivation of the H-theorem requires a statistical hypothesis of molecular chaos or disorder, essentially that pairs of gas molecules prior to collision should be regarded as uncorrelated (statistically independent). Of course this is a time-asymmetric assumption. Significantly, the conceptual innovations of his 1877 statistical method are given scant mention, prefaced by a remark, "I must content myself to indicate (this method) only in passing", and followed by citation of the 1877 paper. So it is not clear what Einstein in 1902–1904 knew of Boltzmann's statistical method. Perhaps he had not studied its details but was aware of it only from the reference to Boltzmann's 1877 paper in Planck's works on blackbody radiation.

At the time, many European physicists remained unaware of Yale mathematical physicist Josiah Willard Gibbs' monumental treatise *Elementary Principles of Statistical Mechanics* (1902), an abstruse work that even Poincaré conceded as being "a little difficult to read". Einstein was no exception. With a touch of misplaced modesty, Einstein later commented that had he known of Gibbs' book, he would "not have published these papers at all". Yet the very generality of Einstein's aspiration distinguishes these three papers from ostensibly similar and far more widely known programs to base thermodynamics on statistical mechanics. The essential thing was to show just how limited are the mechanical assumptions involved in deriving the laws of equilibrium, the second law, and other thermodynamic relations. Generality is key, for it could license application of molecular statistical reasoning also to phenomena that lay outside the purview of mechanics.

The first paper (June 1902, "The Kinetic Theory of Thermal Equilibrium and the Second Law of Thermodynamics"[11]) proclaimed the programmatic need to find a satisfactory molecular basis for the "general theory of heat", i.e., thermodynamics. Asserting the existence of a "gap" in the basis provided by the kinetic theory of heat, in particular, the theories of Maxwell and Boltzmann, Einstein declared his intent "to close this gap". In this paper, however, it is not clear what "gap" is meant. In fact, the paper considered only systems of constant energy and temperature, i.e., *reversible* systems at thermal equilibrium. Unaware of Gibbs' equivalent notion of a microcanonical ensemble, Einstein supposed a fictional ensemble of very many ("infinitely many") systems, all with identical, constant energies.[12] In the sense of statistical mechanics, energy E is defined as the only function of the state variables (the molecular positions and momenta of the particles) yielding a conserved quantity constant in time. This is equivalent to the assumption that the value of E determines the distribution of states in a stationary ensemble. Under these special conditions, Einstein independently derived the Maxwell-Boltzmann probability distribution (also, Gibb's canonical law for a relatively small subsystem) with the exponential factor, exp $(-H/kT)$, where H is the energy function of the subsystem. Finally, he proved the existence for a canonically distributed system subjected to slow deformations (such as those involved in the motion of a piston).

Einstein concluded that the second law, here restricted to reversible, quasi-static processes, "appears as a necessary consequence of the mechanical world picture". And he pointedly observed that the restricted derivation presented was "not at all limited to gases". In particular, the expression for energy H did not have the form of mechanical energy, i.e., a sum of potential and kinetic energy. Taking some satisfaction at this result, he remarked that "this fact suggests that our results are more general than the mechanical model used" in the derivation. He ends the paper by asking how much of the argument can be "disengaged" altogether from its mechanical origins, stating his aim of going beyond the mechanical underpinnings of the theory of heat. While critics of atomism would emphatically agree with this goal, Einstein was not among them. Few if any of his contemporaries were thinking along similar lines.[13]

The second paper of the trilogy (submitted January 1903, "A Theory of the Foundations of Thermodynamics"[14]) opens with the statement that it consists of "annotations" to the first, with the declared aim of investigating the sense in which the "kinetic" part of the theory of heat is in fact essential. Considering now *irreversible* systems, Einstein turned to the vexing problem of the approach to equilibrium and the irreversibility of the second law. Recall the supposed "gap" in its derivation. What it is or how it is to be filled it is never quite clearly specified. Some scholars have understood Einstein's assertion of a "gap" in the derivation of the second law to be a declaration that since Maxwell and Boltzmann were unable to show that the entropy of a closed system *never* decreases, Einstein would do so. Thus filling the gap would require demonstrating the second law to be an absolute law, not merely a probabilistic one. If that is the alleged difficulty, it is addressed in §7 of the second paper with a *petitio*: "we will have to assume that more probable distributions always follow less probable ones, that W (the probability of the distribution) always increases until the distribution has become stationary and W has reached a maximum". A mere assertion of what is to be demonstrated, this certainly fills no "gap" in the derivation of entropic irreversibility. The third paper, treating energy fluctuations in thermodynamic systems, reveals that Einstein intended altogether another "gap", that the second law in Boltzmann's statistical interpretation applies to fluctuation phenomena not describable by classical thermodynamic quantities and that these phenomena are of non-negligible significance, both observationally and theoretically.

Paper two contains a development that proved fundamental in the subsequent use of Boltzmann's principle relating entropy and probability. Rather than use Boltzmann's 1877 combinatorial – statistical definition of probability of thermodynamic macrostates, Einstein introduced a temporal definition of probability that he believed to be original (though in fact, it was the one most favored by Boltzmann). Whereas statistical mechanics *à la* Boltzmann (1877) and à la Gibbs (who coined the term "statistical mechanics") reasoned from probability of macrostates to system entropy, Einstein would use Boltzmann's principle to reason from the system's entropy (thermodynamically specified in paper one in differential form as dQ/dT) to the probability of that macrostate. This inverse use of Boltzmann's principle requires a different, in principle empirical definition of a state's probability in place of an a priori characterization in terms of sheer numbers of microphysical complexions. Recall that Boltzmann's statistical definition of probability in terms of numbers of complexions rests upon a principle of indifference in treating all complexions as a priori equally probable. In terms of phase space, this works in the following way. The phase space Γ allows representation of the thermodynamic macrostate of the complete system at each moment of time as a single system point in a 6N-dimensional space. The probability of a macrostate is determined by the corresponding volume in μ phase space.

Einstein's rather different idea is to let the natural motion of a system determine the probabilities of its macrostates.[15] He defined the probability of a macrostate as the *relative time* t in which the system point occupies the corresponding region of Γ phase space if the system is observed for a sufficiently long time, i.e., letting t $\rightarrow$ ∞. The existence of this limit was assumed, and also (the so-called "weak ergodic assumption") that it did not depend on the initial microstate of the system, in conformity with the empirical fact that the observable equilibrium property of a system does not depend on a precise choice on its initial microstate so long as this choice is compatible with the macroscopic constraints (e.g., total energy). The probability of a state is then proportional to the occupation frequency, the fraction of the total time that the system spends in that state, a physically well-defined value. No special a priori assumptions about equiprobable states are needed. Logically the two notions of probability ("counting complexions", "time spent") should agree. Under the weak ergodic assumption, they do. Einstein (and Boltzmann) tended to regard this assumption as empirically plausible and did not attempt to derive it from the microdynamics. Attempts to do so in general have failed to this day.

This physical definition of "probability of a state" enabled Einstein to use Boltzmann's principle to reason about energy fluctuations, phenomena lying outside the bounds of classical thermodynamics. In the third and last paper of the trilogy, he focused on the theoretical possibility of phenomena of random fluctuations from the average values of macroscopic parameters that alone are taken into classical thermodynamic consideration. The generality of the fluctuation phenomena was key: they occurred not only in atomistic and molecular systems (as Einstein would soon show, the so-called Brownian motion) but also in systems that were not at all mechanical in nature, e.g., in electromagnetic radiation. His insight was that such fluctuation phenomena are characteristic of any statistical theory and moreover that they need not be treated as negligible, as was then widely supposed, but could be significant conceptually and might even be observed.

Fluctuation phenomena

The third paper of the trilogy (submitted in March 1904 and entitled "On the General Molecular Theory of Heat"[16]) opens with a modest statement that it contains merely "a few addenda" to the previous paper. It is however the capstone of the bunch, and a portent for what is to follow. As summarized to his friend Marcel Grossmann, Einstein deemed the paper a further attempt to develop "the atomic theory of heat without the kinetic hypothesis", i.e., without assuming the laws of classical mechanics. Its stated intent is somewhat more oblique: to give a precise physical interpretation to k, the constant of proportionality in Boltzmann's connection between entropy and probability. In its first part, Einstein derived an expression for the entropy of a system "completely analogous" to that "found by Boltzmann for ideal gases and assumed by Planck in his theory of radiation", Einstein's initial and still only indirect reference to Boltzmann's principle. It is also the first published acknowledgment of his awareness of Planck's work on blackbody radiation. Einstein will show that k sets the scale of energy fluctuation phenomena, that is, it determines the "thermal stability" of a physical system. Why is this important and above all, why be concerned with fluctuation phenomena?

In what Einstein will subsequently term *classical thermodynamics*, the physical variables related by law – pressure, volume, energy, temperature, entropy – correspond to average values in the statistical theory. Fluctuations around the average values are ignored. The two founders of statistical mechanics, Boltzmann and Gibbs, regarded the fluctuations occurring in this theory as insignificantly small, at least for systems of macroscopic size. Gibbs even wrote that neglecting fluctuations was essential for a "rational foundation of thermodynamics". Einstein however chose to emphasize the obvious, that a statistical theory in physics implies that measured values may well depart from average or mean values. It is just in the departure from average values that the conceptual significance of fluctuation phenomena lies. For if such phenomena could be observed, the existence of natural processes not amenable to classical thermodynamic treatment would be demonstrated. This would show that classical thermodynamics is not universally valid.

Einstein considered a system composed of two connected parts, one with variable energy E in thermal equilibrium at absolute temperature T with a second much larger system. The laws of thermodynamics govern the system's average behavior. Letting E designate the equilibrium energy of the first subsystem, E can be considered as consisting of two parts, the mean (or average) energy $\bar{E}$ and fluctuations ε from the mean

$$E = \bar{E} + \varepsilon.$$

The fluctuations ϵ are just the quantities Boltzmann and Gibbs considered negligible. From the canonical distribution of the subsystem, he derived the quadratic average of the fluctuation,

$$\overline{\varepsilon^2} = kT^2 d\bar{E} \, / \, dT.$$

This formula (known already to Gibbs, in slightly different form) enabled Einstein to interpret k as characterizing the "thermal stability" of a physical system. This means the constant k effectively sets the scale of fluctuation phenomena and hence the extent to which the average values of the appropriate quantities of a macroscopic system related by the laws of thermodynamics were a reliable guide to predicting the system's behavior. Einstein summarized the discussion by remarking that nothing in the equation refers to the underlying statistical mechanics from which it had been derived.

He then turned to ask whether such fluctuations might be observed. Observations of fluctuations would provide an exact determination of k and as well determine the limits of validity of classical thermodynamics. Fluctuations might be observed, he suggested, if the scale of the fluctuations and the scale of the mean energy are roughly of the same order of magnitude, i.e.,

$$\overline{\varepsilon^2} \approx (\bar{E})^2$$

Noting that energy fluctuation phenomena should also occur in a cavity filled with blackbody radiation, Einstein immediately pointed out that in this case, the fluctuations became non-negligible in the limiting case of a cavity of roughly the same scale as the radiation wavelengths at the peak of the blackbody spectrum. In this way he could roughly relate Boltzmann's constant to the constants occurring in Planck's blackbody radiation law. Even if actual fluctuations might not be observable, he concluded by stating that the fluctuation formula above could be used to make order-of-magnitude estimates for deviations from classical thermodynamics.

More than two years later (in December 1906 [published in 1907]) Einstein returned to the subject of the limits of classical thermodynamics.[17] Here he effectively showed that most thermodynamic parameters do not actually possess the fixed equilibrium values prescribed by classical thermodynamics. These values are not constant but according to statistical mechanics ("the molecular-kinetic theory of heat") must fluctuate about the thermodynamic value as an average. Einstein argued that the probability of a departure from the average value diminished with the magnitude of this departure. This was shown in a simple example.

Consider the entropy difference ΔS between two states, one with a parameter whose equilibrium value is l_0 and the other whose value is l differing very little from l_0. Using Boltzmann's principle, one gets

$$\Delta S = S(l) - S(l_0) = k \log [W/W_0].$$

Here k is interpreted, as above, as determining the scale of fluctuation phenomena, W_0 is the probability of the state in which the parameter has the equilibrium (average) value l_0, and W the probability of the fluctuation state where the parameter is found to have value l. Then the probability of the fluctuation state is

$$W = \alpha \exp(\Delta S/k)$$

where α is a constant of proportionality. Beyond the broad presupposition that the parameter's fluctuations are due to random thermal molecular motions, no detailed molecular assumptions are required in order to derive the statistical distribution of the fluctuation states. To Einstein, this testified to the general validity of Boltzmann's principle when applied even to non-mechanical systems. The key is the principle's inverse application, i.e., beginning with empirically determinable entropy and then finding the probability of a state. As Einstein had to explain to Poincaré at Solvay in 1911,

> "What is characteristic of this standpoint is that one uses the (temporal) probability of a state defined purely *phenomenologically*. In this way one gains the advantage of not having to base the analysis on any specific elementary theory (e.g., statistical mechanics)".[18]

The above example is a clear illustration of how Einstein employed the inverse application of Boltzmann's principle to test – and possibly disconfirm – claims derived from classical theory regarding the probability of the different states of a system.[19]

March 1905: The light-quantum hypothesis

In 1905 nearly every physicist "knew" that light was a wave; the phenomena of interference and diffraction, demonstrated by Thomas Young (1773–1829) and Augustin Fresnel (1788–1827) a century earlier, seemed otherwise impossible to explain. In consequence, Newton's particle (or so-called "emission") theory was considered obsolete by the mid-19th century. Meanwhile the most fundamental physical theory in 1905, Maxwell's electromagnetism, provided a unified theoretical treatment of light, magnetism, and electricity,

all on the assumption that electromagnetic phenomena were disturbances propagating as waves through an ether. For these reasons,
only the light-quantum paper, of all remarkable papers published
in his *annus mirabilus* 1905, received Einstein's commendation (in a
letter to Conrad Habicht of that year) as "very revolutionary". The
paper is indeed visionary; the jury on the existence of light quanta
would remain out until 1925. Einstein's (only) Nobel Prize in 1921
(awarded in 1922) was awarded largely on its basis, though the
Nobel citation, surely reflecting lack of consensus, remained guarded
regarding the existence of light quanta. Recognizing Einstein's "services to Theoretical Physics", an indirect acknowledgment of the
two theories of relativity, it commended Einstein "especially for his
discovery of the law of the photoelectric effect" with no actual mention of light quanta. As we shall see, explanation of the photoelectric effect is but one of three experimental supports adduced in the
paper for the hypothesis of light quanta.

The argumentative structure of "On a Heuristic Point of View
Concerning the Production and Transformation of Light" reveals a
great deal about the innovative reasoning of a young physicist as
well as about the unsettled state of theory in the early stages of
the quantum revolution. In the "Autobiographical Notes" (p. 49)
Einstein recalled, "The success of the theory of Brownian motion
showed again conclusively that classical mechanics always offered
trustworthy results whenever it was applied to motions in which
the higher time derivatives of velocity are negligibly small". Adopting this point of view paid dividends also in his study of light and
radiation. The introduction to the light-quantum paper points out
that observations of interference and diffraction phenomena involve
not momentary values but time averaged behaviors of an optical system over comparatively extended periods; this fact enabled classical
explanation of light as a wave phenomenon. But what happened in
the case of the emission and absorption of light by matter, processes
that take place almost instantaneously? Could this behavior also be
explained classically? Framing his attention to the theory of light in
this way reveals that Planck's law of blackbody radiation, with all its
attendant complications, was at the forefront of Einstein's concern.
To be sure, there were other known, and simpler, absorption and
emission phenomena, where the limits of the wave theory might be

probed. In fact in the latter sections of the paper Einstein discussed and theoretically accounted for three: 1) in certain substances, light re-emitted after absorption is always of lower frequency (according to the so-called Stoke's rule); 2) the photoelectric effect, discovered by Hertz in 1886; and 3) the ionization of a gas by ultraviolet light. Each involved anomalies that could not be easily explained within the Maxwellian theory of light and radiation. Nonetheless and appropriately, Einstein began the paper by considering a fictional system *analogous* to blackbody radiation in order to reach the conclusion that blackbody radiation appears to consist of energy quanta. The light-quantum hypothesis was subsequently invoked to explain the just-mentioned anomalous phenomena.

The initial system is a perfectly reflecting cavity at a given absolute temperature T containing gas molecules, free electrons, electrons elastically bound to the walls of the cavity, and radiation; the latter is absorbed and emitted by the elastically bound electrons. By introducing gas molecules into the fictional system, the problematic assumption that the methods of statistical mechanics can be applied directly to thermal radiation could be avoided, an insightful maneuver. The contentious nature of this assumption is indicated in the paper's first sentence:

> There exists a profound formal difference between the theoretical conceptions physicists have formed about gases and other ponderable bodies, and Maxwell's theory of electromagnetic processes in so-called empty space.[20]

The gas molecules and free electrons are regarded as moving around in the cavity, colliding, according to the kinetic theory of gases, with each other and with the elastically bound electrons or "resonators". The entire system is assumed to be in thermal equilibrium.

Einstein first considered the gas–resonator subsystem. According to the equipartition theorem[21] of statistical mechanics, the mean kinetic energy of a gas molecule is $3kT/2$; since the gas is in dynamical equilibrium with the resonators, this implies an average energy $U = kT$ for each resonator.[22] Einstein next required equilibrium between the resonators and the radiation field. From Planck, he knew this equilibrium required the relation $\rho(v, T) = [8\pi v^2 / c^3] U(T)$

between the spectral density $\rho(v, T)$ and the average energy $U(T)$ of a resonator at the same frequency. Combining this relation with the previous expression $U = kT$, Einstein obtained the Rayleigh-Jeans law. A major difference with Lord Rayleigh's earlier reasoning in 1900, this law now appeared to be inevitable for any frequency under commonly accepted kinetic-theoretical and electromagnetic assumptions. According to this formula, the total radiation energy in the cavity volume increased without bound, becoming effectively infinite.[23] As noted in the previous chapter, Planck's derivation did not use the equipartition theorem, and Einstein observed that Planck's formula for the spectral density function $\rho(v, T)$ in a blackbody cavity was "sufficient to account for all observations made so far". But he also pointed out that the equipartition theorem, as well as the electromagnetic theory of light, though failing for short wavelengths and low radiation densities, remained valid for long wavelengths and high densities, the limiting regime of the Rayleigh-Jeans law.

This conclusion ends the negative, destructive part of Einstein's argument, according to which the received theory of radiation necessarily led to absurdities when combined with the kinetic-molecular theory. Einstein then went on to investigate how the empirically known form of the blackbody spectrum could be used to investigate the structure of radiation. Despite the glaring theoretical asymmetry between the gas and the radiation systems, and "without establishing any model for the emission and propagation of radiation" (i.e., purely from the standpoint of thermodynamics), he observed a striking formal analogy between the two subsystems. The analogy rests on using Boltzmann's principle associating the entropy of a state with its probability:

> If it makes sense to speak of the probability of a system's state, and if, furthermore, every increase of entropy can be understood as a passage to a more probable state, then the entropy S of a system is a function of the probability W of its instantaneous state.[24]

A well-known thermodynamic result for reversible processes is that the entropy of an ideal gas at constant temperature T varies as the logarithm of the volume of the gas. By Boltzmann's principle, this fact can be used to compute the probability of a fluctuation in which the entire gas is confined within the fraction V/V_0 of the total accessible volume (where V_0 is the initial volume). Where N is the number of individual molecules of the gas, the result is $(V/V_0)^N$, as might be intuitively expected from the fact (assuming the mutual independence of the molecules) that the relative probability for a given molecule to be found in the partial volume V is V/V_0.

What about radiation? Einstein showed that in the Wien limit the entropy of blackbody radiation in a given frequency interval between v and $v + dv$ varies with the available volume just as the entropy of a gas would, if only the number of molecules was taken to be E/hv, E being the energy of the radiation in the given frequency interval. Exploiting this analogy and again invoking Boltzmann's principle, Einstein concluded that light, or more generally,

> monochromatic radiation of low density (within the range of validity of Wien's radiation formula) behaves thermodynamically as if it consisted of mutually independent quanta of magnitude hv.[25]

The extraordinary step taken in reaching this result lies in pursuing the analogy to its end. A standard thermodynamic result, entropy dependence on volume, is extended from the simple case of an ideal gas to the highly complex one of blackbody radiation. Einstein's novel understanding of Boltzmann's principle made it possible to take such a step, consistently using phenomenological information about entropy to then infer what the probability of a state had to be. Referring back to the light-quantum hypothesis at Solvay in 1911, Einstein observed that the use of Boltzmann's principle in setting up the formal analogy had been "essential" in understanding that the hypothesis required the energy of radiation itself to be divided into discrete quanta, rather than, as Planck's calculation suggested, attributing the origin of energy discreteness to material properties of the emitting and absorbing entities (the "resonators").

What is the significance of the "as if"? The paper after all claims to give only "a heuristic point of view" about the nature of light, and Einstein has accordingly hedged his bets. Viewing the paper some two decades later, when the light-quantum hypothesis had finally been accepted, American physicist Karl Darrow noted that the term "heuristic" seems to indicate

> a theory which achieves successes though its author feels at heart that it is really too absurd to be presentable. The implication is, that the experimenters should proceed to verify the predictions based upon the idea quite as if it were acceptable, while remembering always that it is absurd. If the successes continue to mount up, the absurdity may be expected to fade gradually out of the public mind.[26]

The "as if" qualification is an acknowledgment of the utterly heretical nature of the hypothesis: the proposal of light quanta flew into the face of every other contemporary physicist's firm belief about the nature of light. But it is also an admission that the hypothesis would remain a "heuristic view" so long as physicists lacked a proper underlying theoretical basis for blackbody radiation (even for Wien's law, not to mention Planck's).

On the other hand, Einstein offered support for the hypothesis in proposing explanations of the three puzzling physical phenomena cited above; we consider only the best-known, the photoelectric effect. When light falls on a metal, so-called photoelectrons may be ejected from the metal. Readers will know that silicon solar panels harness these electrons into a current that may then be stored in a battery. The effect, the subject of recent celebrated experiments by Hertz's student Philip Lenard in 1902 (who was awarded the Nobel Prize later in 1905 for work on cathode rays) had one very puzzling aspect. For incoming radiation of a frequency below a well-defined threshold frequency (varying with the metal), the effect vanished no matter how high the intensity of this radiation. But the threshold frequency is easily explained under the light-quantum hypothesis. Suppose that a characteristic work P is needed to extract an electron from the metal's surface. Then only light quanta of frequency v greater than P/h will be able to free the electron. According to classical theory, the energy of a light wave can be viewed as continuously distributed along the entire length of the wave. Then the tiny part of the wave first striking the metal's surface carries little energy, mostly insufficient to set electrons in the metal's surface in motion. But if light is conceived with all its energy concentrated in discrete, very small, particles localized in space, then the impact of a light quanta on an electron would communicate a packet of localized energy nearly instantaneously to the electron, setting it in motion with a corresponding speed. By energy conservation the kinetic energy of the extracted electrons should be at most given by

$$E_{max} = h\nu - P,$$

some energy being lost during the extraction process. This is "the law of the photoelectric effect" (in updated notation) for which Einstein "especially" merited the Nobel Prize. Another decade was needed to confirm this formula experimentally.

In a follow up paper the next year, in 1906, Einstein addressed the question of the theoretical basis of Planck's radiation theory.[27] This paper is the basis of Thomas Kuhn's 1978 assessment that Einstein, not Planck, is the true discoverer of the quantum in the precise sense that the quantum hypothesis demanded recognition of the restriction of discontinuity on classically permissible energies and motions. When writing the light-quantum paper, Einstein could not make sense of Planck's derivation of his law for the energy distribution of blackbody radiation. At that time he could only show that Planck could not have derived the empirically adequate formula that he did by employing only the resources of classical theory. On that foundation alone, Planck should have derived the Rayleigh-Jeans law, and so to the absurd result when it was extended to higher radiation frequencies. At that time, he now reported, Planck's theory accordingly appeared as "an opposite (*Gegenstück*) to my work". But recent scrutiny of Planck's derivation has convinced him that

> the theoretical foundation on which Mr. Planck's radiation theory is based differs from the one that would emerge from Maxwell's theory and the theory of electrons, precisely because Planck's theory makes implicit use of the . . . hypothesis of light quanta.[28]

It was seen above that Planck had smuggled in discontinuity as a mathematical dodge to get the known empirically established distribution law. Now Einstein's analysis reveals that "the basis underlying Planck's theory of radiation" lies in a restriction that (in modern notation) holds,

> The energy of an elementary resonator can only assume values that are integral multiples of $h\nu$ by emission and absorption, the energy of a resonator changes by jumps of integral multiples of $h\nu$.

Which is exactly correct. Whereas Planck held (and for several more years continued to hold) that the appearance of discontinuity was a challenge to be addressed in a more conservative fashion by further development of the electron theory of matter, Einstein (together with a paper several weeks later by his friend in Leiden, Paul Ehrenfest) succeeded in changing the physics community's understanding of what Planck had accomplished. The fundamental quantum constraint of discontinuity, as Kuhn urged, should be associated with the name of Einstein, not that of Planck.

Of all Einstein's successes in theoretical physics, the hypothesis of light quanta faced the most sustained resistance. Many of the more reputable physicists, among them those who had become immediate supporters of relativity theory (Planck, von Laue, Wien, Sommerfeld) rejected it for two decades – with apparent good reason. Above all it contravened the universal validity of Maxwell's electromagnetism, the theoretical basis for explaining interference and diffraction phenomena. In 1905 the light-quantum hypothesis might well be deemed a peculiar, perhaps outrageous, conceit of an unknown researcher, possibly a crackpot. Yet certainly after the other papers of 1905 appeared, among the notable physicists of Europe, Einstein was no longer unknown, and he had amply proved he was no crackpot. Even after Einstein had become one of the most respected scientists in Germany, the consensus opinion was hardly more congenial. In a 1914 letter proposing Einstein for membership in the Prussian Academy of Sciences, Planck noted that "(Einstein) sometimes may have missed the target in his speculations, as for example, in his hypothesis of light quanta"; in view of Einstein's other striking successes, such mistakes, Planck continued, "cannot really be held against him".[29] A more extreme case of resistance is that of Robert Millikan (1868–1953). At the University of Chicago, Millikan devoted four years (1913–1916) to experiments on the photoelectric effect in a vain attempt to falsify the light-quantum hypothesis. When his experiments served only to confirm it, Millikan, who would be awarded the Nobel Prize in Physics in 1923 for earlier experiments determining the elementary charge of the electron, nonetheless pronounced the hypothesis unacceptable, as "flying in the face of the thoroughly established facts of interference". Popperians, who dismiss confirmation as an inductivist fallacy, may find solace in Millikan's behavior – although years later Millikan would claim that his experiments indeed showed the correctness of the hypothesis.

In retrospect, period opposition to the light-quantum hypothesis is understandable. First of all, it simply seemed impossible to reconcile the light quantum with the wave theory of light. Even Einstein had second thoughts about light quanta for a few years (1912–1915). Nor perhaps was it really necessary for the explanation of the photoelectric effect; Philip Lenard had proposed an alternative, classical account to do so (so-called "Lenard triggering"). Moreover, only after 1913 was the relation $\Delta E = h\nu$ given a well-defined meaning by Bohr's theory of the atom (see below). On the other side of the ledger, to experimentalists, the high frequency nature of X-rays made the corpuscular character of this kind of radiation more evident. In particular, Johannes Stark (like Philip Lenard, a student of Heinrich Hertz and also a Nobel Prize winner, and with Lenard a co-founder of the nationalist "Aryan physics" [*Deutsche Physik*] that dominated German physics throughout the Nazi period) even invoked the light-quantum hypothesis to explain X-rays. By the 1920s, the paradoxical behavior of X-rays became more and more evident, particularly in Maurice de Broglie's Paris experiments on the photoelectric effect using X-rays. It is amusing that one de Broglie (Maurice) became engaged in showing the corpuscular character of radiation, while to his younger brother (Louis), as we shall see, is owed the notion that matter particles have wave-like properties.

To some, but far from all, major physicists, the eponymous Compton effect, announced in 1923, effectively secured the idea that light also exhibited particle-like behavior.[30] In fact the Compton effect reinforced a growing conviction that light quanta were necessary to interpret some aspects of X-rays. Arthur Holly Compton (1892–1962) at Washington University (St. Louis) demonstrated an increase of wavelength (loss of energy) of X-rays after scattering by free electrons. Compton showed that the energy of the scattered X-ray depended on the scattering angle with its incident trajectory after colliding, billiard-ball fashion, with an electron. That X-rays had definite trajectories, carrying momentum as well as energy, was a clear signal of particle-like behavior. Further evidence came two months after Compton's paper was published when the Scotsman C.T.R. Wilson (1869–1959) reported observing the actual trajectories of the recoil electrons in the newly invented cloud chamber that bears his name. Compton and Wilson shared the Nobel Prize in 1927 for their experiments revealing the particle character of light.

Even so, up to 1925, the light-quantum hypothesis continued to meet resistance from Niels Bohr and others. Accepting his own Nobel Prize in December, 1922, Bohr considered the hypothesis to be "quite irreconcilable with so-called interference phenomena" and as well, "not able to throw light (sic) on the nature of radiation". Whereas Compton explained X-ray–electron scattering by assuming the absolute validity of the laws of conservation of energy and momentum, rather than accept the particulate character of light and radiation, Bohr published a paper in 1924 with H.A. Kramers and J.C. Slater treating these laws as mere statistical averages by denying a direct causal connection between incident X-rays and scattered electrons and X-rays. Repelled by the BKS abandonment of strict causality, Einstein wrote to Max Born (April 29, 1924), in anticipation of remarks he would soon make about quantum mechanics,

> Bohr's opinion about radiation (i.e., BKS) is of great interest. But . . . I find the idea quite intolerable that an electron exposed to radiation should choose of its own free will, not only the moment to jump off, but also its direction. In that case, I would rather be a cobbler, or even an employee in a casino, than a physicist.

In fact, the BKS theory was short-lived, contradicted in 1925 experiments of Walther Bothe and Hans Geiger in Berlin demonstrating simultaneous production of a scattered light quantum and a recoil electron in Compton scattering. This meant that energy and momentum are strictly conserved in individual interactions between radiation and electrons. The Bothe-Geiger experiments ended lingering doubts concerning the validity of Compton's results and of their particle interpretation. A symbolic seal of approval indicating the physics community's acceptance of the light-quantum hypothesis came in 1926 with physical chemist G.N. Lewis's baptism of light quanta as "photons". As we shall see, even after the advent of quantum mechanics, Bohr's acceptance of light quanta went hand-in-hand with his denial of a causal space-time explanation of elementary quantum phenomena.

May 1905: Brownian motion

In a little more than a year following formulation of the theory of fluctuations, Einstein would identify possibly observable fluctuations in the phenomenon known as "Brownian motion". Some eighty years before in 1827, British botanist Robert Brown reported seeing apparently random motions of sufficiently small particles of pollen suspended in water. These mysterious phenomena were studied by a number of scientists throughout the 19th century. By 1900 the molecular thermal origins of the visible erratic motions was widely supposed, though without adequate theoretical understanding. The explanation of these motions was given in May 1905 in Einstein's first paper on Brownian motion, "On the Motion of Small Particles Suspended in a Stationary Liquid Demanded by the Molecular-Kinetic Theory of Heat".[31] The paper predicted that if dispersed throughout a stationary liquid of given temperature and viscosity, insoluble colloidal particles would undergo completely chaotic perpetual motions. The particles are very small yet large enough to be observed under a microscope; their random zigzag motions are the effects of impacts by the molecules of the suspending liquid. Statistically one would expect the thermally moving molecules of the liquid to strike the much more massive suspended particle in approximately equal numbers on all sides, the tiny impacts canceling one other, with the result that the particle should on average remain still. But Einstein argued fluctuations from the thermodynamically derived average kinetic energy (velocities) of the molecules can occur, and from time to time these fluctuations orchestrate a group of molecules to bunch together, moving in the same direction. These group motions can create an atypically large push on one side of the suspended particle, unbalancing the tiny impacts on all sides, with the result that the particle moves in a particular direction.

Einstein calculated the mean square displacement $(\overline{\Delta x})^2$ that a suspended particle (of a given radius) will carry out in any given direction during a given time lapse and found it to be proportional to this time lapse. The displacements in successive intervals are treated as completely independent, corresponding to the predicted chaotic continual motions of the particles. In effect, the velocity of the Brownian particle changes so rapidly in direction and magnitude that it is useless to attempt to measure its average velocity and then to try to explain the motions in those thermodynamic terms. Einstein suggested that if these displacements were observed, they could be regarded as a key confirmation of the "molecular-kinetic theory of heat" (i.e., statistical mechanics, and its assumption of atomism) applied in this case to particles of a size sufficient to be visible.

It is customary to say that Einstein thereby "explained" the mysterious phenomenon of Brownian motion, and in the process definitively established the reality of atoms. This is a bit misleading on both counts. While Einstein remarked that the phenomena considered were possibly identical to those in "the so-called Brownian motion", he also expressly noted that his information regarding the latter, as might be expected from someone who did not have access to a university library, was so scanty that he could "form no judgment in the matter". It is more accurate, as physicist and historian Martin Klein urged, to say that Einstein "invented" Brownian motion, i.e., he derived its physical possibility as a consequence of the molecular-kinetic theory of heat. His predictions of the types of motion that should be observed and his calculations showing how details of the motions depend on the number of atoms present in a given amount of the liquid were confirmed by Jean Perrin's 1908 experiments in Paris. Naturally, the observation of this predicted behavior by Perrin and others served to confirm atomism. Former skeptic and Nobel chemist Wilhelm Ostwald was converted, writing after the announcement of Perrin's results,

I am now convinced that we have recently become possessed of experimental evidence of the discrete or grained nature of matter, which the atomic hypothesis sought in vain for hundreds and thousands of years.[32]

Of course, the existence of atoms, as already noted, had been widely assumed in "the tendency of Boltzmann" to which Einstein belonged.

Developing a theory of fluctuation phenomena as a consequence of statistical mechanics yielded two significant consequences. First, using Boltzmann's constant and the constant R of the ideal gas law ($PV = RT$), Einstein could give a new determination of Avogadro's number N_A and, on account of the relation $k = R/N_A$, a proof of the existence of molecules of a definite size now that he could assess the value of k through fluctuation phenomena. This had been the goal of his doctoral thesis of the same year, where completely different methods (of hydrodynamics) were employed to estimate the size of large molecules in solution. Second, by applying statistical mechanical reasoning to particles large enough to be visible under a microscope, the limits of validity of classical thermodynamics was empirically demonstrated. The observed fluctuation behavior was tangible evidence of thermodynamic quantities deviating from average values, the mean kinetic energies of molecules predicted by the kinetic theory. While Einstein's account of Brownian motion served to confirm the basis of thermodynamics in statistical mechanics, at the same time it demonstrated that basis to be broader, so that classical thermodynamics was no longer "applicable with precision to bodies even of dimensions distinguishable in a microscope".[33]

Specific heats of solids

Success using the theory of fluctuations to probe the bounds of validity of classical thermodynamics applied to the phenomena of radiation prompted Einstein, starting in 1907, to investigate anomalies in the specific heats of certain substances predicted by the kinetic-molecular theory. By the mid-19th century, kinetic theory could demonstrate that all matter has an associated temperature, a measure of the thermal motions of the molecules comprising the matter. Since motion requires energy, heat is a measure of the thermal energy needed to warm a body. The specific heat of a substance is the amount of heat per unit mass (or per mole) required to raise its temperature 1 °Celsius. In Maxwell's time, a problem had already appeared in the kinetic theory's treatment of the specific heat of gases: the measured specific heats were usually much smaller than expected from the equipartition of energy over the various degrees of freedom of the molecules.[34]

On the other hand, through much of the 19th century the specific heats of solids appeared to be on more sound empirical footing. The specific heats of many elements conformed to a uniform empirical rule proposed by Pierre-Louis Dulong and Alexis Petit in 1819: every solid's specific heat is approximately 6.4 calories per unit mass (mole) per degree. Only a half-century later was there an adequate theoretical interpretation of the Dulong-Petit result. In the 1870s Boltzmann derived the Dulong-Petit value, purportedly for "all simple solids", by applying the equipartition theorem of classical statistical mechanics to atoms arranged in a three-dimensional lattice. Boltzmann assumed the thermal motions of the atoms in a solid to be simple oscillations (vibrations due to a harmonic force) about a position of equilibrium and that there are three independent motions per atom. According to the equipartition theorem, each mode of vibration (oscillation in one direction) at temperature T would have an average kinetic energy of kT and the total thermal energy of one mole of a solid amounts to $3N_A kT$ where N_A is Avogadro's number, the number of atoms in one mole. However, there were known exceptions to the Dulong-Petit value. Boltzmann himself noted carbon, boron, and silicon, and he speculated that the anomalies were due to a loss of degrees of freedom as atoms at neighboring lattice points "stuck together" at low temperatures. Lord Kelvin (William Thomson), in an oft-quoted address of 1900, pointed to these anomalies in the Dulong-Petit rule and other failures in the application of equipartition of energy as one of the "clouds over the dynamical theory of heat and light". By 1907 when Einstein, still working as a patent assistant in Bern, turned his attention to the problem it had become clearer than in Boltzmann's time that atoms had internal structure (involving electrons). But understanding of the anomalies in the specific heats of solids had not advanced further.

The very title of Einstein's paper suggests the overriding desire for unity in the foundations of physical theory: "Planck's Theory of

Radiation and the Theory of Specific Heats".[35] What could Planck's
radiation theory possibly have to say about the problem of specific
heats in solid matter? After rehearsing how he understood the quantum character of Planck's distribution law of blackbody radiation,
Einstein came directly to the point:

> I believe we must not content ourselves with this result. For
> the question arises: If the elementary structures assumed in the
> theory of energy exchange between radiation and matter cannot
> be seen in terms of the current molecular-kinetic theory, are we
> then not obliged also to modify the theory for the other period-
> ically oscillating structures considered in the molecular theory
> of heat? In my opinion the answer is not in doubt. If Planck's
> radiation theory goes to the root of the matter, then contradic-
> tions between the current molecular-kinetic theory and expe-
> rience must be expected in other areas of the theory of heat as
> well, which can be resolved along the lines indicated.

Just as Einstein argued in 1906 that Planck's radiation law could
be understood on the assumption that oscillation energies of frequency v can only occur in quanta of magnitude hv, so oscillations on the atomic scale in matter also could be assumed to have
quantized energies. He accordingly proposed a new temperature-
dependent law for the specific heats of solids, one accommodating
the anomalies known to Boltzmann as well as another even more
striking one, that of diamond, a substance also with light atoms
that could be considered to have correspondingly high vibrational
frequencies. Einstein's law yielded the Dulong-Petit rule in the limit
of high temperatures but predicted a well-defined, quantum-based
departure from this rule at lower temperatures. Whereas classically
atoms in a solid can be represented as oscillators of average energy
$\bar{E} = kT$, if the available energies are a discrete set of values, the average energy is given by a modified value[36] that reduces to the classical value when (hv/kT) is very small. At temperatures significantly
above room temperature, and so, much higher than the characteristic oscillation frequency of atoms in a solid, the specific heat will
approach the classical value of Dulong-Petit. Einstein cited diamond
as a dramatic case in point, only approaching the Dulong-Petit value
when heated to more than 1000 °C. More generally, the assumption

that light atoms have higher vibrational frequencies explained why lighter elements have anomalously low specific heats at room temperature. In fact, according to Einstein's theory, a substance's specific heat tends to zero as the temperature approaches absolute zero $(-273.15 \,°C)$, in conformity with Walther Nernst's so-called heat theorem, the third law of thermodynamics proposed also in 1906, that the entropy of all substances approaches zero with absolute temperature.

Just as in the theory of blackbody radiation, Einstein emphasized that in the theory of matter one *can no longer assume* that "the motion of molecules obeys the same laws that hold for the motion of bodies in our world of sense perception". This is to effectively recognize the need for a new, and *non-classical*, mechanics to apply beyond the increasing limits of classical theory. Here again, Einstein was a bit too far in advance of his contemporaries. His quantum-based account of the specific heats of solids was largely ignored until new experiments in 1910 confirmed Nernst's third law. Walter Nernst, as noted above, instigated the First Solvay Conference in Brussels in 1911 and Einstein was specifically invited to report there on the new theory of specific heats, signaling its general acceptance.

A Brownian mirror

In Salzburg in September 1909, at the annual meeting of the German Physical Society, Einstein announced wave-particle duality; ever since it has remained a central (perhaps *the* central) puzzle of quantum theory. Forty years later, on the occasion of Einstein's seventieth birthday, Wolfgang Pauli Jr. characterized the Salzburg address as "one of the landmarks in the development of theoretical physics". Pauli, possibly the most brilliant of the founders of quantum mechanics but also theoretical physics' most severe critic, was not one given to idle praise. For unlike nearly all of his Salzburg auditors, Einstein did not (yet) doubt the reality of the granular quantum structure of radiation (he would do so briefly a bit later on). In pointing out the extent of Planck's departure from classical ideas on radiation, he rhetorically asked whether Planck's law might be derived on some other basis than the "monstrous assumption" of energy quanta. Answering

his own question with a resounding "No", he rehearsed the properties of light that could not easily be explained by the wave theory, as
previously considered in his light-quantum paper of 1905.

He then pointed out that an intuitively clear reason for the origin
of these difficulties lay in the asymmetry in the way the wave theory
had to treat emission and absorption. According to the wave theory,
the emission of light consisted of the production of an expanding
spherical wave by an oscillating electric charge ("ion"). This might
be considered an elementary process in the familiar sense in which
a collision between gas molecules in the kinetic theory of gases usually involved two molecules only. Now the reverse process, possible
according to Maxwell's theory, involves a contracting spherical wave
and so cannot be regarded as an elementary process in this sense
because the production of such a wave would require "an enormous amount of . . . elementary structures" (because the original
field would be spread over a large domain of space). The corpuscular theory of light, on the other hand, could easily treat emission
and absorption as elementary processes and both on a completely
symmetrical footing, a clear advantage. As in the opening paragraphs of his special relativity paper, the wave theory's violation of
an intuitively plausible principle of symmetry (the time reverse of
an elementary process should itself be an elementary process) was
regarded as a count against it. On the other hand, optical phenomena of interference and diffraction apparently could be explained
only by the wave theory.

To show why both wave and particle properties of light had to
be considered, Einstein turned again to fluctuations as treated in
his theory of Brownian motion, and to the problem of blackbody
radiation. The suitability of fluctuation theory is that it required no
specific knowledge regarding the structure of a system to determine
the magnitude of the deviations of its thermodynamic properties
about their average values. Thus one could take thermodynamic
reasoning as far as it would go, until it led to a result in conflict
with observation. In this case, Einstein used it to describe a hypothetical system comprised of a cavity containing blackbody radiation at equilibrium and an ideal gas of low density (relatively few
molecules) at a temperature T fixed by the walls of the cavity. The
cavity also contains a flat solid plate, think of a two-sided perfectly

reflecting small mirror, free to move in a direction perpendicular to its own plane. On the one hand, this mirror is subjected to the collisions of the gas molecules and should thus experience an irregular Brownian motion. On the other hand, Maxwell's theory predicts that radiation, in the form of waves, exerts a mechanical pressure upon both sides of the mirror. If, for the moment fluctuations in this pressure are neglected, the resultant force on the mirror should vanish by symmetry and the mirror remains at rest. But due to Brownian motion the pressure on the two sides of the mirror no longer balance each other and there is a resulting frictional damping on the motion of the mirror. In this way, all the energy of the gas would be transformed into energy of radiation. Consequently, there could not be any equilibrium between the gas, the mirror, and the radiation. What has gone wrong?

As Einstein explained, the fluctuations of the radiation pressure on the mirror must be taken into account. In order to maintain the average kinetic energy kT of the mirror, the quadratic average of these fluctuations must have a well-defined value depending on the distribution of radiation. It is sufficient to consider only that part of the fluctuations in radiation pressure for frequencies in a small interval (between v and $v + dv$) in a volume V. Where Δ is the momentum transferred to the mirror during a given time, the mean-square momentum fluctuations in the radiation pressure $\overline{\Delta^2}$ is given by the sum of two terms,

$$\overline{\Delta^2} = (1/c)\ (h\rho v + \rho^2 c^3 / 8\pi v^2)\ f\ \tau dv,$$

where ρ is the radiation density spectrum taken from Planck's theory and f the surface of the mirror. For present purposes, all that matters in this rather complicated expression is to notice that the right-hand side is a sum of two terms, each signifying a different and independent contribution to the total fluctuations in radiation pressure. By inspection one sees that the first one corresponds to energy quanta, representing the particulate structure of radiation. It dominates in the high frequency – Wien – part of the blackbody spectrum. The second one, as indicated by the term $(c^3/8\pi v^2)$, corresponds to the random interference of waves, and is the only term accounted for by the wave theory of light. It dominates in the low-frequency Rayleigh-Jeans part.

Both are required to describe fluctuations over the entire blackbody spectrum. Einstein concluded that while fluctuation phenomena are useful in showing the limitations of the existing theory of radiation, they are too narrow a basis upon which to erect a satisfactory theory of light that would have to combine aspects of both wave and particle theory.

The Brownian mirror became one of Einstein's favorite tropes; he returned to it in several later publications, describing it one last time in his "Autobiographical Notes" as the principal consideration convincing him that Maxwell's theory could not be completely correct. In a letter to Max von Laue, 17 January 1952, Einstein dates this conviction back to 1905:

> (I)n 1905 I knew already with certainty that (Maxwell's theory) leads to the wrong fluctuations in radiation pressure, and consequently to an incorrect Brownian motion of a mirror in a Planckian radiation cavity.[37]

This might be interpreted to mean that in 1905 Einstein knew only that Maxwell's theory failed for the Wienian part of the blackbody radiation spectrum (since it implied corpuscular fluctuation through entropy-volume dependence), and that subsequently (ca. 1909), he used the more concrete Brownian mirror argument to express this failure of Maxwell's theory in a more vivid form.

Return to Planck's law

Beginning in 1906, Einstein repeatedly remarked that Planck's own derivation of his radiation law contained mutually inconsistent assumptions. The relation between the average energy of a resonator and the spectral distribution function,

$$\bar{E}_\nu = (c^3 \, \rho)/8\pi\nu^2,$$

clearly rested upon classical electromagnetism, On the other hand, Planck's calculated value of $\bar{E}_\nu$, while it agreed with observation, was non-classical: it relied on a quantum prescription. The completion of general relativity by 1916 allowed Einstein to turn his attention once more to Planck's law. However, since he had last considered the

quantum theory of radiation, Bohr's 1913 theory of atomic structure had to be taken into account. His model of the atom became
the new, and fundamentally quantum, starting point for Einstein's
quantum derivation of Planck's law.

The Bohr atom

Bohr had taken over from his teacher Ernst Rutherford the iconic
but misleading image of the atom as a miniature solar system: a
large positively charged nucleus surrounded by planet-like electrons,
smaller and negatively charged, so that the entire atom was electrically neutral. A fundamental problem with this model is that it
is unstable, both mechanically and radiatively. In particular, according to Maxwell's theory, charged moving bodies changing direction
(as in an orbit) will radiate away their energy, eventually spiraling
down to coalesce with the attracting positive nucleus. Bohr sought to
explain the paradoxical stability of the Rutherford atom. He did so by
turning to Planck's theory, simply postulating that electron orbits are
stabilized by a quantum condition involving the quantum of action
h as well as an integer labeling the orbits. Thus an atom can exist in
a series of "stationary states" selected among the classically possible states. As long as an atom remains in a stationary state it neither
emits nor absorbs radiation but in transiting from the ith orbit to the
jth orbit, radiation of angular frequency $v = (E_i - E_j)/h$ is absorbed
if $E_i < E_j$, emitted if $E_i > E_j$. Bohr did not employ Einstein's light-
quantum hypothesis and therefore assumed radiation itself could
only be described classically, by Maxwell's theory. Then transitions
between the stationary states could be related to the frequencies of
emitted radiation, with both emission of radiation and absorption
corresponding to a non-classical discontinuous "jump" between stationary states. In the simple case of the hydrogen atom (with only
one electron), the model was able to predict the distinct line spectrum of hydrogen empirically summarized in the Balmer (1885)
and Rydberg (1908) series, according to which the frequencies of
the visible lines in the hydrogen spectrum stand in the relation

$$v = cR\left(\frac{1}{m^2} - \frac{1}{n^2}\right)$$

wherein m and n are integers and R is the empirical Rydberg's constant. The Bohr model, however, left completely unexplained the reasons for the implied quantum rule as well as any account of why the electron jumps to a particular energy level on emission or absorption of light. Also left open was the question of how a gas of atoms at equilibrium with a radiation field (such as blackbody radiation) maintains the populations of its stationary states.

1916: Quantum derivation of Planck's law

Seeking consistency between the Bohr theory and Planck's radiation theory, in the summer of 1916 Einstein produced the first quantum-theoretical derivation of Planck's law in two papers published that year, with the result republished in 1917 in a journal of wider circulation.[38] As the blackbody law obtains for radiation in contact with matter, Einstein assumed dynamical equilibrium for a gas of molecules immersed in a bath or field of blackbody radiation of yet to be specified spectral density $\rho(v, T)$. At thermal equilibrium the entropy of the system has maximal value; this is the system's most probable state. Now, following Bohr, Einstein assumed that each gas molecule or atom, modeled as an oscillator, has a discrete set of allowed internal energies, the stationary states. This is the only quantum assumption of Einstein's derivation, which is otherwise independent of Bohr's other assumptions.

At equilibrium, the average energies of radiation and matter do not change in time but individual elementary processes of emission and absorption of radiation continually occur. An excited atom, with internal energy above a certain minimum (the atom's "ground state"), can emit radiation, dropping to a lower energy state. Alternately, an atom can absorb radiation of a specific frequency from the blackbody field, jumping up to a higher energy state. At thermal equilibrium, these elementary processes effectively balance: in a given interval dt, on average as many atoms will transit from a state (Zustand) Z_a of lower energy to one of higher energy Z_b under absorption of radiation as transit from higher state Z_b to the lower state Z_a by emission of radiation.

However, Einstein specified two quite distinct processes of emission corresponding to the transition $Z_b \rightarrow Z_a$. First is "spontaneous emission" occurring, as in radioactive decay, without any apparently acting cause. The probability dW that the spontaneous emission process $Z_b \rightarrow Z_a$ occurs in unit time dt is

$$dW = A^b_a \, dt$$

where A^b_a is a coefficient giving the intrinsic ("chance") of a jump from Z_b to Z_a. Spontaneous emission is a quintessentially irreversible "chance" process; Einstein regarded as a "weakness of the theory" that it left "time of occurrence and direction of the elementary processes a matter of 'chance'". This is Einstein's first explicit identification of the indeterministic character of quantum theory.

However, there is also another emission process, "stimulated emission", that Einstein argued unlike spontaneous emission can be viewed as caused. At a rate proportional to the blackbody radiation density ρ, the radiation field can cause an atom in the higher state Z_b to emit radiation at a frequency already present in the field. Accordingly the radiation field can extract energy from the atom by "stimulated emission", a transition $Z_b \rightarrow Z_a$ like the above but in this case dependent on the amount of radiation present. The probability of the transition $Z_b \rightarrow Z_a$ due to stimulated emission in unit time dt is

$$dW = B^b_a \, \rho(\nu,T) \, dt,$$

an expression clearly manifesting dependence on the blackbody radiation field. Much later in the 1950s, the process of stimulated emission formed the theoretical basis of the laser (and the maser).

Finally the radiation field also can induce processes in the other direction, from a lower energy state Z_a to a higher state Z_b. This is absorption, and it is also proportional to the spectral density of the field. The probability that an atom in a lower state Z_a will make a transition $Z_a \rightarrow Z_b$ in time dt is

$$dW = B^a_b \, \rho(\nu,T) \, dt$$

Einstein related the three processes by bookkeeping: At equilibrium, the number of transitions from transition $Z_b \rightarrow Z_a$ must equal the number $Z_a \rightarrow Z_b$,

$$N_a \left(A^b_a + B^b_a\, \rho\right) = N_b \left(B^a_b\, \rho\right),$$

where N_a and N_b are the numbers of atoms in the stationary states a and b respectively. Significantly, Einstein now considered all three processes to be directed, i.e., to be associated with momentum transfer. In particular, in absorption of an elementary quantum of energy $h\nu$, a momentum $h\nu/c$ is given to the atom in the direction of propagation of the incident beam of radiation. Correspondingly in emitting an elementary quantum $h\nu$, the atom recoils in the amount $h\nu/c$ in a direction the theory left open. There is of course an acausal character to Bohr's theory, since the atom, as Bohr emphasized, is given a "choice" to jump to different stationary states, a state of affairs Bohr considered "irrational" yet irreducible. Einstein, to the contrary, ended this paper by stating that while he had full trust in the methods he employed, he nonetheless considered it a "weakness" of the theory presented that it left the time and direction of these elementary processes to "chance". Of course, physicists had been trying to come to terms with the spontaneity of emission processes since the discovery of radioactivity by Henri Becquerel in 1896; already in 1900 Rutherford had proposed a law of radioactive decay (for thorium) containing a spontaneous emission coefficient. But it is not really until after this paper of 1916/17 that Einstein began to express his disquiet with the apparent failure of classical causality in the realm of the quantum.[39] The growing discontent is evident in a letter to Max Born of January 27, 1920:

> That business about causality also afflicts me greatly. Can the quantum character of the absorption and emission of light ever be understood in the sense of the complete requirement of causality, or does there remain a statistical residue? I must confess that there the courage of my conviction fails me. But *complete* causality I renounce only very very (*sehr sehr*) reluctantly. ... (The question whether rigorous causality obtains or not has a clear meaning, although a certain answer at no time.).[40]

Einstein would never reach accommodation with acausal quantum processes, regarding them as acceptable at a phenomenological level only.

On the other hand, whereas at Salzburg in 1909 Einstein had only hinted that emission was a directed process, not that of an expanding spherical wave, he now had shown that a transfer of momentum accompanies the transfer of energy between radiation and matter, that is, that such transfers are directed processes. It had taken Einstein twelve years to associate quanta of momenta $p = h\nu/c$ with quanta of radiation energy $E = h\nu$. Having done so, he wrote to Besso (6 September 1916), "light quanta are as good as established".[41]

1924–25: Quantum statistics

In June 1924 Einstein received a letter from an obscure Bengali physicist, Satyendra Nath Bose with an enclosed paper containing a new derivation of Planck's radiation law. Bose had sent his paper for publication in the *Philosophical Magazine*, a respected British physics journal. But it had been rejected, probably because Bose adopted Einstein's hypothesis of light quanta, treating radiation quanta like the particles of a gas, i.e., as is said today, a "photon gas" of particles. Remarkably Einstein not only read the paper, he translated it into German and arranged for its publication in the *Zeitschrift für Physik*, a journal begun only in 1920 but already the world's leading venue for quantum physics. Ostensibly without making use of classical mechanics or electrodynamics, Bose had employed Boltzmann's relation between entropy and probability to calculate the entropy of the radiation and then obtained Planck's distribution law for blackbody radiation thanks to a special estimate of the relevant probability.[42]

But there was a novelty Bose had not called attention to, and of which, as it turned out, he was not even aware. Rather than follow Boltzmann's statistical method of distributing the individual light quanta over all the microstates (phase space cells) compatible with the total energy of the system, Bose had simply regarded the number of light quanta in each cell as defining the same microstate. For someone blindly following Boltzmann's 1877 characterization of microstates, this was an obvious error. But for someone like Einstein who believed that complexions and combinatorial probabilities should ultimately derive from more fundamental temporal probabilities, Bose's strange but simple combinatorics was worth a try. The departure from Boltzmann's combinatorics is obvious in hindsight: Bose's procedure does not require the assumption that individual particles, though identical, are distinguishable. No doubt because of his long reliance on Boltzmann's principle as a probe of the domain of the quantum, Einstein clearly recognized the originality of Bose's method. As emphasized many times, Einstein's characteristic inverse use of Boltzmann's principle required a physical definition of the probability of a state distinct from Boltzmann's definition in terms of the sheer number of distinct microscopic complexions, and the corresponding assumption that each such complexion was *a priori* equally probable. Since Boltzmann's method of counting complexions depends upon an implicit assumption of identical but distinguishable particles, Einstein was certainly attuned to recognize a statistical method in which this assumption did not hold. Bose (tacitly) and Einstein (explicitly) thereby jointly discovered "Bose-Einstein quantum statistics". This was the initial indication that the assumption of distinct but identical particles cannot be made in counting quantum particles and correspondingly that quantum particles obey different statistics than classical ones. In fact, the anomalous counting of complexions and its relation to indistinguishability was not explicitly addressed by Einstein, but pointed out afterward by Ehrenfest and others. Moreover, the distinction between bosons and fermions had yet to be made.[43] Another type of quantum statistics, "Fermi-Dirac", was discovered the next year (1926).

Perhaps since he believed that complexion counting in any case was not fundamentally well-defined, the new type of statistics was of peripheral interest to Einstein who was mainly impressed that Bose had given, for the first time, a completely corpuscular derivation of Planck's law.[44] Seeing immediately that Bose's method might be applied also to matter, Einstein extended it in three papers in 1924–1925 to the system of an ideal gas of monatomic material particles, i.e., particles with a given mass at a specified volume and temperature. His calculations, based on Bose's statistical method, showed that Bose's quantum treatment predicted a surprising result, a so-called "gas degeneracy" in which at very low temperature, an enormous (macroscopic) number of particles have the same quantum state as opposed to a random distribution over a variety of states. The system as a whole separates into two parts, one a condensate, analogous to the condensation of a saturated vapor, the other a saturated ideal gas. As the temperature approached 0 K, more and more particles would go into the zero energy state, in agreement with Nernst's third law It is the first statistical derivation of a phase transition, effectively to a new state of matter. The result remained largely theoretical until 1995, when low-temperature (less than a millionth of a degree above absolute zero) technologies enabled such a "Bose-Einstein" condensate of a gas of weakly interacting particles to be produced in the laboratory in Boulder, Colorado. In fact, Einstein was only half correct; only particles from a certain family, named for obvious reasons *bosons*, can be placed in the same quantum state.

De Broglie's breakthrough

Einstein's friend and fellow physicist Paul Langevin did not quite know what to make of his student Louis de Broglie's November 1924 Paris doctoral thesis. De Broglie was an unusual student; his older brother was the outstanding experimental physicist Maurice de Broglie, a previous student of Langevin, and also a noble, the 6th duc de Broglie. In his thesis, Louis had explored connections between Einstein's light-quantum hypothesis and the special theory of relativity beginning with Einstein's 1917 paper in which the relation between the energy and momentum of a light quantum is given as $E = hv = pc$ while that between energy and wavelength λ is $E = hc/\lambda$. Seeking to generalize this result, de Broglie argued that Einstein's wave-particle duality for light should be extended to material particles, in particular to electrons. De Broglie therefore hypothesized that the relation $E = hv$ should hold not only for light quanta but also for electrons and that the momentum p of the electron was related to its wavelength λ by the hypothesis, $\lambda = h/p$.

The only way an orbiting electron could be associated with a wave (de Broglie employed the concepts of both wave and particle) bound as it is by an attractive force to an oppositely charged nucleus, required the wave to be a standing wave. Such a wave fits around the nucleus in some whole number of wavelengths; otherwise the wave would self-destructively interfere with itself. Supposing an electron with momentum p to move in a circular orbit of radius r about a nucleus, de Broglie knew the condition on standing waves required $2\pi r = n\lambda$ where $n = 1, 2, 3 \ldots$ is some integer number of wavelengths. Using the classical angular momentum $L = pr$ and the above relation $p = h/\lambda$ led to the relation $L = n\hbar$ ($\hbar = h/2\pi$), i.e., the quantization of angular momentum for circular orbits. With this de Broglie explained what Bohr had simply to assume, the rule selecting the possible electron orbits.

As Einstein was an old friend, Langevin sent the thesis to Berlin, asking for his opinion. It was both immediate and highly favorable. Soon after, in January 1925 when he presented a second note to the Berlin Academy of Sciences on the application of Bose's method to what he called the "quantum theory of an ideal gas", Einstein called attention to de Broglie's 1924 thesis, emphasizing how de Broglie had shown how a scalar wave field can be associated with a material particle. That being the case, he suggested how the two wave trains associated with two particles might interfere if their wave velocities and frequencies were in near agreement.[45] These remarks were prescient for later developments, one theoretical, one experimental. In his third paper on wave mechanics in March 1926, Schrödinger cited this very passage noting,

> My theory was inspired by L. de Broglie . . . (*Theses*, Paris, 1924) and by brief, yet infinitely far-seeing remarks of A. Einstein.[46]

Although de Broglie had mentioned the possibility of matter diffraction, the idea of actually experimentally detecting wave properties of matter particles was novel. In experiments showing the diffraction of beams of electrons by crystals independently conducted in 1927 by G.P. Thomson in the UK and C.J. Davisson and L.H. Germer in the USA, de Broglie's hypothesis of matter waves was confirmed. Whereas we noted above that Louis de Broglie's brother Maurice performed experiments revealing the corpuscular character of X-rays, it is equally ironic that G.P. Thomson was the son of Cambridge physicist J.J. Thomson, who showed that the electron was a particle.

Summary

Using Boltzmann's principle, Einstein developed a statistical theory of fluctuations to identify and then probe beyond the limits of classical theory in the absence of any fundamental quantum theory that in any case only appeared subsequently in 1925/26. He was able to do this through an imaginative transformation of Boltzmann's reasoning. Whereas Boltzmann in 1877, and statistical mechanics more generally, argued from the probability of a state to its entropy (and so from sheer numbers of microstates to the associated thermodynamic macrostate), Einstein employed the principle inversely, arguing from the thermodynamically determinable entropy of a state (macrostate) to the probability of that state.

In effect, Einstein's use of Boltzmann's principle, with the requisite physical conception of probability, freed statistical mechanics of its mechanical presuppositions, even in the supposedly purely statistical form of Boltzmann 1877. In particular, it allowed Einstein to jettison the residual mechanical assumption of Boltzmann's statistical approach, the supposition of identical but distinguishable particles in counting microstates (complexions). And it enabled Einstein to develop fluctuation theory as a conceptually significant further ramification of statistical theory. While classical or "phenomenological" thermodynamics dealt only with average or mean values of macroscopic properties such as energy, no specific knowledge about molecular structure implied fluctuations from the thermodynamic values. Einstein showed that in some cases they are not negligible but might actually be observed. Observation of fluctuation behavior would demonstrate not only the existence of new phenomena beyond the limits of applicability of classical thermodynamics but also could serve as a heuristic probe of the new realm of atomic physics and the quantum.

The fruits of this approach are truly astonishing. From the 1904 inverse use of Boltzmann's principle and fluctuation theory that results, follow in 1905 the hypothesis of the light quantum and suggestion for a quantum theory of radiation, as well as the "invention" of Brownian motion; in 1907 the explanation of the anomaly in the theory of specific heats of solids and demonstration of the need for a quantum theory of matter as well as of radiation; the 1909 proposal for a fusion of wave and particle theories of light and the Brownian mirror showing the inadequacy of basing the theory of radiation on Maxwell's theory; the directed character of the emission of radiation deduced from a molecular simplification of the Brownian mirror example in the latter parts of the 1916/17 paper on spontaneous and stimulated emission of radiation; and finally to the recognition of Bose's non-classical statistics, that together with de Broglie's idea of matter waves, effectively set the stage for Schrödinger's development of wave mechanics.

Notes

1 "On the Development of our Views Concerning the Nature and Constitution of Radiation", *CPAE* 2 (1990), Doc. 60; as translated in English supplement, pp. 379–94; p. 379.

2 Einstein to Milieva Marić, September 13, 1900. *CPAE* 1, Doc. 75.

3 Speziali, Pierre (ed.), *Albert Einstein Michele Besso Correspondance 1903–1955*. Paris: Hermann, 1979, Doc. 1, p. 3.

4 "Today he (Einstein) is one of the most distinguished and recognized of the direction (*Richtung*) of Boltzmann". Friedrich Adler, writing from Zurich to Viktor Adler on July 19, 1908, in Michaela Maier and Wolfgang Maderthaner (eds.), *Physik und Revolution: Friedrich Adler-Albert Einstein, Briefe, Documente, Stellungnahmen*. Wien: Löcker, 2006, p. 66. Einstein is usually and rightly regarded as too much of an iconoclast to fit into anyone's "school", but on June 19, 1908 when Friedrich Adler (1879–1960) wrote this appraisal of Einstein, it was completely on target. The comment occurs in the postscript of a letter to Adler's father, Viktor Adler (1852–1918), then leader of the Austrian Socialist (Social Democratic) Party. The younger Adler studied physics in Zurich at the same time as Einstein, becoming a friend and even a neighbor in the same building.

5 "Discussion" following lecture "On the Present State of the Problem of Specific Heats", November 3, 1911; *CPAE* 3 (1994), Doc. 27, p. 550; English supplement, p. 426.

6 Boltzmann's gas theory assumes classical Hamiltonian mechanics and its implications, in particular, Liouville's Theorem, according to which a system's phase space volume is preserved under temporal evolution. This justifies the assumption of uniform cells in phase space.

7 "On the Present Status of the Radiation Problem", *Physikalische Zeitschrift* v. 10 (1909), pp. 185–93; *CPAE* 3 (1994), Doc. 26, p. 361, English supplement, p. 426.

8 Einstein seems to have been unaware that this way of defining probability of states was also employed by Boltzmann, and in fact, was the latter's preferred definition.

9 Avogadro's number is the number of atoms of carbon in 12 g of carbon-12. Its value is $6.02214129 \times 10^{23}$. A mole of a homogenous substance is a quantity of this substance possessing Avogadro's number of molecules.

10 "'Discussion' following the Lecture Version of 'The Present State of the Problem of Specific Heats", *CPAE* 3 (1994), Doc. 27, English Supplement, pp. 426–37; p. 426.

11 *CPAE* 2 (1990), Doc. 3, pp. 56–75.

12 The notion of such a fictional ensemble appeared as the *Ergode* in the second volume of Boltzmann's *Lectures on Gas Theory* but was not original there. It was discussed previously by Maxwell and Boltzmann.

13 Olivier Darrigol points out that Gibbs shared the aim of making statistical mechanics as independent as possible from any specific mechanical model, and Boltzmann also strove for generality. But both Gibbs and Boltzmann remained within the Hamiltonian framework whereas Einstein was willing to abandon it.

14 *CPAE* 2 (1990), Doc. 4, pp. 76–97.

15 As noted previously Boltzmann also employed this definition of probability, beginning in 1871, and then used it in some of his most important works. Einstein apparently independently invented it.

16 *CPAE* 2 (1990), Doc. 5, pp. 98–108.

17 "On the Limit of Validity of the Law of Thermodynamic Equilibrium and on the Possibility of a New Determination of the Elementary Quanta" (1907), *CPAE* 2 (1990), Doc. 39, English Supplement, pp. 225–8.

18 "'Discussion' following the Lecture Version of 'The Present State of the Problem of Specific Heats", *CPAE* 3 (1994), Doc. 27; English Supplement, pp. 426–37; p. 430.

19 Einstein's method here assumes that the entropy of the non-equilibrium state could be computed by classical thermodynamic means by treating such states as *labile* equilibrium states (quickly becoming equilibrium states); in general there is no way to compute the entropy of a non-equilibrium state.

20 "On a Heuristic Point of View Concerning the Production and Transformation of Light", *Annalen der Physik* v. 17 (1905), pp. 132–48; *CPAE* 2 (1990), Doc. 14, English Supplement, pp. 86–103.

21 The equipartition theorem implies that the mean kinetic energy $\bar{E}$ of a rigid particle is equally distributed among its translational and rotational degrees of freedom, hence there are no preferred directions of motion.

22 There being two quadratic terms in the Hamiltonian of a resonator; see note 16 to Chapter 2.

23 Paul Ehrenfest in 1906 first termed this the "ultraviolet catastrophe".

24 "On a Heuristic Point of View Concerning the Production and Transformation of Light", *CPAE* 2 (1990), Doc. 14, English Supplement, pp. 86–103; p. 94.

25 Ibid., p. 97.

26 Darrow, Karl K., *Introduction to Contemporary Physics*. New York: D. Van Nostrand Co., 1926, pp. 116–17. Cited by Brush, Steven G., "How Ideas Became Knowledge: The Light-Quantum Hypothesis 1905–1935", *Historical Studies in the Physical and Biological Sciences* v. 37 (2007), pp. 205–46; p. 216.

27 "On the Theory of Light Production and Light Absorption", *Annalen der Physik* v. 20 (1906), pp. 199–206; *CPAE* 2, Doc. 34 (English translation).

28 Ibid., p. 192.

29 *CPAE* 5 (1993), Doc. 445.

30 See Stuewer, Roger, *The Compton Effect: Turning Point in Physics*. New York: Science History Publications, 1974. Bruce Wheaton's *The Tiger and the Shark: Empirical Roots of*

Wave-Particle Dualism. (New York: Cambridge University Press, 1983) gives a fuller history of the related physics.

31 "On the Movement of Small Particles Suspended in Stationary Liquids Required by the Molecular-Kinetic Theory of Heat", *Annalen der Physik* v. 17 (1905), pp. 549–60; *CPAE* 2 (1990), Doc. 16, English Supplement, pp. 123–34.

32 Ostwald, Wilhelm, "Preface" to *Outlines of General Chemistry*, fourth edition. Translated from fourth German edition 1908 by W.W. Taylor. London, New York: Macmillan, 1912, p. vi.

33 Einstein, "On the Movement of Small Particles Suspended in Stationary Liquids Required by the Molecular-Kinetic Theory of Heat", *Annalen der Physik* v. 17 (1905), pp. 549–60; *CPAE* 2 (1990), Doc. 16, English Supplement, p. 124.

34 According to the kinetic theory, the ratio of specific heat at constant pressure to that at constant volume is given by

$$\gamma = (n + 2)/n$$

where n is a molecule's number of degrees of freedom (more exactly, the number of quadratic terms in the Hamiltonian of the system). Diatomic molecules, being regarded as rigid bodies, should have three degrees of freedom of translation and three degrees of freedom of rotation. By the 1860s, physicists deemed most molecules in the gaseous state to be diatomic in structure, two mass points bound together by some force. Since each mass point has three degrees of freedom of motion, a diatomic molecule should have six. Hence theoretically γ should be 1.33 whereas experiment showed γ to be approximately 1.4 for air. Maxwell regarded such anomalies as "the greatest difficulty yet encountered by the molecular theory". "On the Dynamical Evidence of the Molecular Constitution of Bodies", *Nature* v. 11 (1875), reprinted in William D. Niven (ed.), *The Scientific Papers of James Clerk Maxwell*. vol. 2, pp. 418–38; p. 433.

35 *Annalen der Physik* v. 22 (1907), pp. 180–90; *CPAE* 2 (1990), Doc. 38 (English Supplement).

36 $\bar{E} = h\nu/\exp(h\nu/kT) - 1$. Recall that Planck's distribution law (of energy densities) for the spectrum of blackbody radiation is $[8\pi\nu^2/c^3]\, h\nu/\exp(h\nu/kT) - 1$, where the first term comes from Maxwell's theory of electromagnetism.

37 Einstein Archive 16-167; as translated and quoted by Robert Rynasiewicz. "The Construction of the Special Theory: Some Queries and Considerations", in Don Howard and John Stachel (eds.), *Einstein:The FormativeYears 1879–1909*. Boston-Basel-Berlin: Birkhäuser, 2000, pp. 159–201; p. 177.

38 The 1916 derivation appears in "On the Quantum Theory of Radiation", *CPAE* 6 (1996), Doc. 38 (English Supplement).

39 Sections 4–6 of this paper apply to molecules immersed in a radiation field (like Bohr atoms, capable of emitting and absorbing radiation) the fluctuation argument previously used in 1909 (in the case of the Brownian mirror) to show the quantum-theoretic character of molecular motions. This fluctuation argument is the strongest one Einstein offered since unlike previous arguments, it could not be criticized for involving complex entities (e.g., a mirror) with possibly complex behavior. Henceforth most competent physicists realized there were only two alternatives: either energy-momentum conservation was violated in

the elementary processes of emission and absorption (Bohr's view, to surface again in the BKS paper of 1924, see below) or the emission of radiation had to be a directed process involving light quanta. Thanks to Olivier Darrigol for his comments here.

40 *Albert Einstein-Max Born Briefwechsel 1916–1955*. München: Nymphenburger Verlagshandlung, 1969, p. 44; *The Born-Einstein Letters*, translated by Irene Born. New York: Walker and Co., 1971, p. 23.

41 *Albert Einstein Michele Besso Correspondance 1903–1955*. P. Speziali (ed.), Paris: Hermann, 1979, Doc. 25, pp. 49–50; p. 50.

42 Bose's paper contained an elementary confusion in applying a combinatorial formula (conflating the roles of particles and boxes).

43 According to the Standard Model, fermions (constituting matter) and bosons (particles carrying forces of interaction) are the fundamental constituents of matter and fundamental force. There are three families of fermions; interactions within each family is mediated by a bosonic particle. On account of the Pauli exclusion principle, two or more fermions cannot occupy the same quantum state.

44 Bose counted cells in configuration space in the manner of the new quantum theory of gases, not by appeal to wave modes in a cavity, thus giving justification to the "Maxwell factor" $8\pi v^2/c^3$ by quantum-corpuscular means.

45 "Quantentheorie des einatomigen idealen Gases. Zweite Abhandlung", *Sitzungsberichte der Preussischen Akademie der Wissenschaften, Sitzung der physicalisch-mathematischen Klasse vom 8 Januar 1925*, pp. 1–14; p. 9. Einstein explained that the fluctuations of the gas in a small sub-volume contained a term that could be explained by random interference of de Broglie waves.

46 Schrödinger, Erwin, "Über das Verhältnis der Heisenberg-Born-Jordanschen Quantenmechanik zu der meinen", *Annalen der Phyik* v. 79, no. 734 (1926); received March 18, 1926; pp. 734–56; translated as "On the Relation between the Quantum Mechanics of Heisenberg, Born, and Jordan, and that of Schrödinger" in Erwin Schrödinger, *Collected Papers on Wave Mechanics*. New York: Chelsea Publishing Co., 1982, pp. 45–61; p. 46, note.

Further reading

Darrigol, Olivier, "The Quantum Enigma", in M. Janssen and C. Lehner (eds.), *The Cambridge Companion to Einstein*. New York: Cambridge University Press, 2014, pp. 117–42.

Klein, Martin J., "Einstein and the Wave-Particle Duality", *The Natural Philosopher* v. 3 (1964), pp. 3–49.

———, "Thermodynamics in Einstein's Thought", *Science* v. 157 no. 3788 (4 August 1967), pp. 509–16.

———, "Fluctuations and Statistical Physics in Einstein's Early Work", in G. Holton and Y. Elkana (eds.), *Albert Einstein: Historical and Cultural Perspectives* (The Centennial Symposium in Jerusalem). Princeton, NJ: Princeton University Press, 1982, pp. 39–58.

Four
Quantum mechanics

"I have thought a hundred times as much about the quantum problems as I have about the theory of general relativity".[1]

"My contemporaries see in me both a heretic and reactionary who has, so to speak, survived himself".[2]

Origins

Quantum mechanics originated with a paper of 24-year-old Werner Heisenberg in early July 1925 seeking "to establish a quantum-theoretical mechanics based entirely on relations between quantities that are observable in principle".[3] Heisenberg sought a mathematical formalism to characterize the phenomenon of spontaneous radiation of an excited atom. He considered only the hydrogen atom in a highly idealized model, as a very simple periodic system ("virtual oscillator"). The guiding idea was to retain Newton's second law of motion for the atom's electron but in a "kinematical reinterpretation", replacing the position coordinate x with a "quantum theoretic quantity" to be found by pursuing the classical-quantum analogy through the corresponding terms of Fourier series describing the orbiting electron. For x Heisenberg substituted a set of quantities, frequencies, and intensities of radiation emitted by the atom, corresponding to Bohr's stationary states and the electron's transitions ("jumps") between them. In this way, information from the observable spectrum of hydrogen would replace kinematical variables of position and period for the unobservable electron orbit. Recognizing the approach required further mathematical development, Heisenberg stated a methodological intention of restricting consideration to observable quantities, letting these dictate the structure, still unknown, of a new quantum theoretical mechanics. He apparently believed he was following Einstein's example (in special relativity) by ridding physics of unobservable quantities (e.g., *absolute simultaneity*). Learning of Heisenberg's intent later on, Einstein is reported to have said, "a good joke shouldn't be repeated too often".[4]

Famously Heisenberg did not realize the multiplication rule required by the new kinematical quantities is equivalent to that for multiplying matrices, a rule in general noncommutative, i.e., $AB \neq BA$. Max Born, professor of theoretical physics in Göttingen and Heisenberg's postdoctoral supervisor, quickly pointed this out and in late September, together with his assistant Pascual Jordan, had cast Heisenberg's theory into the form of a "matrix mechanics". Like Heisenberg, Born and Jordan sought to maintain a close analogy to classical physics; the matrix quantities they constructed for the canonical quantities p and q of Hamiltonian mechanics do not represent momentum and position directly but satisfy equations of motion identical in form to those of classical mechanics. The analogy to classical mechanics also guided P.A.M. Dirac at Cambridge. In early November, Dirac elegantly reformulated the theory of Heisenberg's initial paper,[5] independently deriving the commutation relation $pq - qp = -i\hbar$ ($i = \sqrt{-1}$) previously found by Born in July and the hallmark of Dirac's "symbolic algebra" of quantum observables. Later that month appeared the influential "*Dreimännerarbeit*" of Born, Heisenberg, and Jordan, a comprehensive presentation of matrix mechanics and a first attempt to extend the methods of quantum mechanics to systems with many degrees of freedom, i.e., to fields. Taken together, these papers created a new noncommutative theory of atomic physics set in the frame of Hamiltonian mechanics.

In both matrix and Dirac algebra form, everything appears discontinuous; just as in Bohr's atom theory, there are discrete stationary states with quantum "jumps" between them. Born, Heisenberg, and Jordan conceded the abstract matrix representation of relations between observable quantities to be not at all amenable to a "geometrically visualizable (*anschauliche*) interpretation" of an atomic system; indeed, they rejected any description of electron motions in terms of the concepts of space and time.[6] In contrast, Erwin Schrödinger in

Zurich built on de Broglie's idea of associating matter particles with waves and from January–June 1926 completed six papers developing a theory of atomic systems in terms of a "wave mechanics". Schrödinger employed a mathematical tool much more familiar to physicists, a wave equation based upon a continuous "psi" function $\Psi = \Psi(x,t)$.[7] Assuming stationary states of the electron correspond to the stationary forms of an associated wave, Schrödinger's wave mechanics gave exactly the same results as matrix mechanics for values of the quantized energy levels of the hydrogen atom. But even in the paper demonstrating that "the two theories . . . are completely equivalent from the mathematical point of view", Schrödinger argued that wave mechanics, unlike matrix mechanics, furnished a "guiding physical point of view" since

> to me it seems extraordinarily difficult to tackle problems . . . as long as we feel obliged on epistemological grounds to repress intuition (*Anschauung*) in atomic dynamics, and to operate only with such abstract ideas as transition probabilities, energy levels, etc.[8]

In contrast, Schrödinger claimed wave mechanics could provide an "intuitive" (*anschaulich*) understanding of the emitted frequencies of atomic radiation as "beats", analogous to the fundamental frequency and harmonics of periodic waveforms. By March 1926 Schrödinger demonstrated that the new wave mechanics and matrix mechanics are "completely equivalent from the mathematical point of view" (more exact derivations are due to Dirac, Pauli, and somewhat later, von Neumann). In addition he was able to derive accurate predictions for phenomena that had remained beyond the reach of matrix mechanics, including the behavior of the electron in a uniform magnetic or electric field (the so-called Zeeman and Stark effects). On the other hand, the wave amplitude given by Ψ is a *complex*-valued function of space and time and of course observed quantities are real- (indeed, rational-) valued. How then was the wave function to be physically understood? Schrödinger (see further below) initially supposed a purely wave interpretation of the Ψ-function, to the exclusion of particles. This attempt came at considerable cost: it required him to deny the existence of discrete energy states and of "jumps" between them, and even to speculate that the concept of "energy" is merely a statistical generalization of the more fundamental wave concept of frequency. Ultimately, however, Schrödinger had to retreat from claims of the visualizable character of wave mechanics. Although it might be plausibly argued that a single particle wave function might represent the particle propagating in ordinary three-dimensional space, this view faced insurmountable difficulties when extended to "the poly-electron problem", i.e., to multi-particle systems. In such cases, Ψ is a function defined in a 3N dimensional configuration space, where N is the number of particles. Visualizable (intuitive) interpretation seemed possible only for the simplest of atomic systems.

Particle or wave? Complementarity

The issue of physical interpretation of the wave function was resolved to the satisfaction of most quantum theoreticians by Max Born in a paper of late June 1926. Born applied Schrödinger's new formalism to collision processes, a study that however convinced him of the corpuscular, not wave, nature of electrons. Born argued what is measured is not Ψ but $|\Psi|^2$, i.e., $\Psi^* \Psi$, where Ψ^* denotes the complex conjugate to Ψ. According to what would become known as the "Born rule", $|\Psi(x,t)|^2$ is a "probability density" for a particle to be located within some small region surrounding point x at time t, while the function $\Psi(x,t)$ itself does not represent something physically real but is a mathematical tool representing a "probability current" propagating in time according to the Schrödinger equation's dynamical evolution. With Born's statistical interpretation of Ψ, wave mechanics could answer the question of its physical interpretation, and it quickly proved a more pliable instrument than matrix mechanics. It soon supplanted matrix mechanics, whose methods (as well as those of Dirac's noncommutative symbolic algebra) many physicists found obscure as well as difficult to apply to actual physical problems, in particular, to the helium atom, the next simplest atomic system.

Is the electron *really* a wave packet or is it a point particle? The 1927 experiments on electron diffraction by crystalline solids by Clifton Davisson and Lester Germer in the USA and independently, George P. Thomson in the UK, confirmed de Broglie's hypothesis of the wave character of matter. Heisenberg, for one, did not concede the superiority of wave mechanics despite the existence of matter waves. Instead, the program of quantum mechanics (by which was meant "matrix mechanics") was viewed as requiring liberation from all "intuitive pictures" (*anschaulichen Bildern*). This called for substitution of simple relations between empirically given quantities in place of the kinematic and mechanical descriptions familiar from classical physics. As seen above, that is just what the matrix formulation was developed to do. An ensuing controversy arose concerning in what respect, if at all, quantum mechanics is or could be a "visualizable" (*anschauliche*) theory. Heisenberg correspondingly sought to redefine "visualizability" (*Anschaulichkeit*), arguing that all visualization reasonably can require of a theory is that in all simple cases the theory allow qualitative consideration of its experimental consequences. To illustrate how matrix mechanics remained "intuitive" in the revised sense of *Anschaulichkeit*, he provided, still in 1927, a simple thought-experiment example of light (γ-rays, i.e., light of short wavelength and correspondingly high energy) scattered by an electron (the Compton effect) then observed with a fictional "γ-ray microscope". Underscoring an immediate discontinuous change in the electron's state on impact by a light quantum (and, as Bohr pointed out, despite an erroneous account of the microscope's optics) Heisenberg formulated limitations or uncertainties (Δ) on the accuracy of simultaneously measured values of the electron's position and momentum ($\Delta p \Delta q \sim h$), the uncertainty relations that bear his name. In these relations, Heisenberg claimed a "direct physical interpretation of the equation $pq - qp = -i\hbar$".[9]

Heisenberg's attempt to free quantum mechanics from classical imagery found resonance in Bohr's "complementarity", proclaimed at a conference at Lake Como in northern Italy several months later in September 1927. The philosophy of complementarity was expressly tailored to underwrite the Heisenberg uncertainties. The essence of quantum mechanics lay in what Bohr termed the "quantum postulate", i.e., in Planck's quantum of action h that, to Bohr, symbolized "an essential discontinuity . . . completely foreign to classical theories". On account of the "indivisibility of the quantum of action", Bohr argued for the impossibility of any sharp distinction between systems exchanging energy in an interaction; in particular, between an apparatus of measurement and the quantum system of interest. Both object system and measuring device are "appreciably disturbed" by observation with the result that an intrinsic ambiguity surrounds attribution of properties (distinct states) to individual systems. Overcoming this ambiguity required "renunciation" of a defining characteristic of classical physical theory, the description of physical phenomena through simultaneous use of kinematical (*spatial position, time*) and dynamical (*momentum, energy*) concepts. The complete description of *quantum* phenomena required both to be used, but not at the same time; employing a concept from the first group precludes simultaneous application of one from the other group, and vice versa.

> The very nature of the quantum theory thus forces us to regard space-time coordination and the claim of causality, the union of which characterizes the classical theories, as complementary but exclusive features of the description, symbolizing the idealization of observation and definition respectively.[10]

Whereas Heisenberg argued that experimental conditions limit simultaneous (and unambiguous) use of both kinematic and dynamic concepts in the description of an atomic object, complementarity more broadly enjoined an essential limitation in application of classical concepts to the quantum domain. One cannot say, as in classical physics, that simultaneously precise values of "complementary" kinematic and dynamic concepts are instantiated in the object. Firmly established after 1927, complementarity also brought about something of a reconciliation between "particle" and "wave" approaches; by viewing them as complementary, both concepts, though requiring distinct experimental setups for legitimate application, were required to accommodate the full range of description of quantum phenomena.

The problem of quantum measurement

In use since 1911, the Wilson cloud chamber exhibited tracks of α-particles (helium nuclei) emitted by a decaying radioactive sample as nearly perfectly straight lines.[11] In §3 of the March 1927 uncertainty paper entitled "The Transition From Micromechanics to Macromechanics", Heisenberg considered an analogous situation, the "coming into being" of the classical electron orbit. Quantum mechanically, the electron's position is characterized through the Born rule by a "probability packet" of width λ that freely spreads. But if the electron is repeatedly observed with light of wavelength λ, each act of observation "reduces the wave packet back to its original size". Heisenberg's "reduction of the probability packet" is one of the first attempts to reconcile the probabilistic quantum theory with the fact that outcomes of measurement on quantum systems are definite. Fueled by the continuing successes of wave mechanics, a consensus emerged by the late 1920s that measurement outcomes had to be discussed in the context of the notion of superposition of states, the central concept involved in wave mechanics and the fundamental difference with classical physics.[12]

As the Schrödinger equation is a linear equation, if Ψ_1 and Ψ_2 are two solutions for a system S, then the superposition $\Psi_{12} = \Psi_1 + \Psi_2$ is also a solution, with an equal claim to representing a possible state of S. Then if the total energy of S is known (and so long as no observations are made), the linear Schrödinger equation characterizes continuous propagation of the "probability density". This describes the relative likelihood, if a measurement were made, for S to be found in one of the distinct possible states of a given quantum observable. Broadly speaking, continuous evolution of probability ("unitary evolution") for observable events has the essential features of a causal law; since the Schrödinger equation is a first-order differential equation in time (and again, as long as the system is not disturbed) the state (whatever it is) of S at one time t_1 uniquely determines S's state (whatever it is) at any later time t_2. In his influential 1932 treatment of quantum measurement, John von Neumann termed unitary Schrödinger evolution a "process of type 2" encompassing "automatic changes which occur with the passage of time".[13] On account of its linear nature, however and in sharp contrast, the Schrödinger equation cannot in general produce definite outcomes of measurements on S, the fact that measurement yields a definite state and not a superposition of states: the so-called "reduction of the wave packet". To von Neumann, who based his treatment on Bohr's disturbance account of measurement, definite outcomes were to be explained by the intervention of an external agency, an instrument of measurement M, and an interaction between S and M. For the interaction itself, von Neumann postulated a distinct dynamics, a "process of type 1" breaking the continuous Schrödinger evolution of Ψ and somehow occasioning transition of S into the definite state revealed on measurement.[14] May it be presumed that S was in the definite state revealed by measurement prior to interaction with M? Not according to complementarity. Whatever its nature, the second type of dynamics is "discontinuous, non-causal and instantaneously acting", in accord with the Heisenberg-Bohr emphasis on "discontinuous transitions" in measurement. Von Neumann's two kinds of dynamics and the relation between them gave shape to the "quantum measurement problem" still persisting today.

Into the early 1930s, several dissenting physicists, including Schrödinger, continued to write philosophical papers related to the interpretation of quantum mechanics. Yet by then the interests of nearly all other leading quantum theorists had shifted from philosophical issues, where clear progress no longer seemed likely, to productive generalizations and extensions of quantum mechanics. This took place in several stages, first to relativistic particles and then to the electromagnetic field, and subsequently to the newly discovered complexities of the atomic nucleus following James Chadwick's announcement in May 1932 of the discovery of the neutron. After 1930, Einstein remained virtually alone among theoretical physicists in publicly critiquing quantum mechanics. From that time forward, these criticisms focused not on failures of causality or determinism, but on the incomplete nature of what Einstein termed the wave function's description of "real states" of individual systems.

What is a "complete theory"?

Before considering Einstein's arguments that quantum mechanics is an incomplete theory, we may first ask what a "complete theory" is supposed to be. Logicians know that the notion of a "complete theory" is originally a meta-mathematical one, arising within Göttingen mathematician David Hilbert's program of axiomatization of mathematical theories at the beginning of the 20th century. In 1900 Hilbert explicitly posited an "Axiom of Completeness" pertaining to a set of axioms for arithmetic, to the effect that "the numbers form a system of objects which cannot be enlarged with the preceding axioms continuing to hold".[15] Completeness is subsequently incorporated as an overarching requirement on the system of geometrical axioms in the second (1903) and all later editions of Hilbert's influential *Grundlage der Geometrie*. For both arithmetic and geometrical systems of axioms, completeness is a meta-axiom, a statement about the relevant axiom set, stipulating its sufficiency for derivation of all theorems or truths in the language of the theory. A complete theory is then a maximal consistent set of sentences in the sense that none of its proper extensions are consistent. Logical usage follows Hilbert's meta-level requirement; in logic a complete theory is a theory with a model: all of the provable expressions of the theory are true. Working within the context of Hilbert's program for metamathematics, the logician Kurt Gödel demonstrated a powerful incompleteness result in 1931, just a few years prior to the celebrated EPR paper (see below). Gödel showed that any formal system at least as powerful as elementary number theory will contain legitimate expressions neither provable nor refutable, hence the system is necessarily incomplete. It may well be wondered what any of this has to do with physical theories which are rarely, if ever, sufficiently formalized such that a proof of completeness could be even attempted.

Yet something analogous to the Hilbertian demand for completeness emerged in the discourse of the founders of matrix mechanics, the initial form of quantum mechanics. As noted above, matrix mechanics arose in Göttingen with Werner Heisenberg and then together with Max Born and Pascual Jordan in 1925. All were mathematically inclined physicists among whom Hilbert's influence was considerable and knowledge of his ongoing meta-mathematical program was widespread. Heisenberg, in fact, attended Hilbert's seminars over a two-year period, from 1922–1924. Prompted perhaps by recent demonstrations (of Schrödinger among others) that the very different formalisms of matrix mechanics and wave mechanics are but distinct versions of the same theory, the quantum physicists were understandably eager to adopt an ideology of "finality" regarding quantum mechanics. It is not surprising they were guided by analogy to the meta-mathematical "criterion of completeness" for formally axiomatized mathematical theories.

Early evidence of this rhetoric appears as early as a paper of Heisenberg in 1926 where he notes that matrix mechanics has become a *closed theory* (*geschlossene Theorie*) as the mathematical relations among its observable magnitudes are those of the corresponding relations of classical mechanics, a theme prompted by Dirac's demonstration that the form of the Hamiltonian equations of classical mechanics can be carried over into the new theory.[16] The theme of finality re-emerges at the conclusion of the Born and Heisenberg jointly authored report on "quantum mechanics", given at the Solvay Conference in October 1927:

> By way of summary, we wish to emphasize that we consider *quantum mechanics* to be a closed theory (*geschlossene Theorie*), whose fundamental physical and mathematical assumptions are no longer susceptible of any modification.[17]

Heisenberg's considerable subsequent writing on the notion of a "closed theory" identifies three principal characteristics. A closed theory covers a limited or bounded domain of phenomena; it is a mathematically well-defined, consistent and a perfectly accurate description of those phenomena; and it is final in the sense that its truth content pertaining to the phenomena in its domain holds for all time: all future experience will continue to confirm its laws.[18] Bohr's complementarity is premised on the assumption that in this sense quantum mechanics is a "closed theory" and hence a complete theory of atomic phenomena. Quantum orthodoxy elaborates this with the claim that a unique Ψ-function contains all possible physical information about the state of a quantum system at a given moment of time. This is the sense of completeness of quantum mechanics that Einstein will attack.

Two incompleteness arguments

"Everybody knows" the storied pronouncement "God does not throw dice",[19] the supposed hallmark of Einstein's opposition to quantum mechanics. And *almost* everybody knows that quantum mechanics is correct (more exactly, has never been contradicted by experiment), and therefore the objections are mistaken. The metaphor suggests that Einstein's criticisms target the indeterminism of the theory, the fact that in general, the quantum formalism allows only probabilistic predictions regarding the outcomes of measurements. Thus arises the image of a modern dinosaur, colossus, and master of all surveyed, whose stubborn adherence to strict causality led to *de facto* scientific extinction, exemplified by unwillingness to accept the clear testimony of empirical success. However, while Einstein indeed did not at all care for the irreducibly probabilistic character of quantum mechanics, this remained for him a prejudice, not an argument, against the theory. He labored unsuccessfully for years to find an alternative theory that would be a "model of reality . . . (in) represent(ing) the events themselves and not merely the probability of their occurrence"[20] but that is not an argument against quantum mechanics. We shall see Einstein's *argument* purports to show that quantum mechanical description of individual systems is an incomplete description, and that it takes two quite different forms.

But to first grasp the supposed significance of Einstein's remark, we shall need a bit more precision about quantum indeterminism. The characteristic feature of the quantum mechanical formalism is that the algebra of symbols (technically, self-adjoint linear operators) representing observable quantities is in general noncommutative. Where p and q represent the "canonically conjugate" quantities linear momentum p_x and position q $(= x)$ of a particle, the fundamental equation $pq - qp = - i\hbar$ follows from the quantum formalism.[21] As noted above, Heisenberg proposed a direct physical interpretation of this relation with his celebrated uncertainty (or, indeterminacy) principle, stating that the more precisely one can measure at a given instant the position q of an electron (say, along the x axis), the less precisely one can know the simultaneous value of the conjugate linear momentum p_x (its x-component of motion). In particular, *perfect* knowledge of one of these conjugate variables implies totally indeterminate knowledge of the other. Quantum indeterminacy, according to Heisenberg, thus arises from the fact that "even in principle, we cannot know the present in all detail". The necessarily incomplete knowledge obtainable about a given system at a given time leads to a failure of predictability, and so, Heisenberg concluded, to a breakdown of the law of causality.[22]

In both learned and lay arenas, the familiar story of Einstein's opposition to quantum mechanics recounts a succession of unsuccessful attempts to defeat the Heisenberg uncertainty relations, understood as a prohibition on what can be exactly determined by simultaneous measurements. The narrative owes its origin to Niels Bohr, to whom is due the principal record of epic interchanges with Einstein, public and private, over quantum mechanics.[23] In virtue of the stature of its two protagonists, the Einstein-Bohr debate, as it is known, is often linked to that in the early 18th century between Leibniz and Newton (through the latter's intermediary Samuel Clarke) as one of the truly grand confrontations of genius in the history of modern science. The epic saga contains a dimly perceived element of fact, a seedbed from which subsequent legend sprouted. In the Einstein-Bohr case the fact is that after learning of the uncertainty relations in April 1927, Einstein sought to challenge their scope, arguing in certain heuristically illustrated situations that the stated restrictions on knowledge of simultaneous values of conjugate variables might be overcome. In his 1949 account, widely deemed reliable, Bohr recounts how he was able to show at the Sixth Solvay Conference in Brussels in October 1930 that Einstein's chief argument against the uncertainty relations, though ingenious, nonetheless misfired as it failed to take into account general relativistic effects sufficient to yield the requisite limitations on co-measurements. Not surprisingly, this incident is the acme of Bohr's narrative, acquiring near-epic status for decades afterward as a fitting testament to Bohr's unfailing genius.

The "freedom of choice" argument

Absent from Bohr's account is that perhaps even at Solvay in 1930 but in any case shortly thereafter, Einstein redirected the emphasis of his thought experiment to bring out its salient point without directly challenging the Heisenberg uncertainty relations. Einstein had come to accept that while the relations correctly state in principle restrictions on what can be known by simultaneous *measurements* on two non-commuting quantities, they do not necessarily extend to the stronger claim made by the emerging quantum orthodoxy (encapsulated in Bohr's complementarity), that *independently of any measurement on an object* exact values of both quantities are not simultaneously instantiated in the object. The argument for incompleteness is then reformulated in terms of an observer's *freedom of choice* to make measurements at a given time on one of two (not both) of the quantities. In certain well-defined situations, by measuring *one*, the experimenter could *infer* the precise value of the *other* without in any way interfering with, or disturbing, the latter quantity. Since complementarity does not allow simultaneous assignment of precise values to conjugate variables, the conclusion follows that quantum mechanics provides only an incomplete description of the "real state" of an individual system. This *freedom of choice* argument for incompleteness has in its sights, in particular, Bohr's account of quantum measurement as invariably involving disturbance, an act inevitably introducing "a new uncontrollable element". The best-known version of the *freedom of choice* argument is formulated in the Einstein-Podolsky-Rosen (EPR) paper of 1935, one of the most cited papers in 20th-century physics.[24] However, the argument of EPR also introduced supplementary complications: a vaguely stated "criterion of reality", as well as two additional implicit assumptions, *separability* (regarding the independence of spatially separated systems) and *locality* (prohibiting faster-than-light causal influences), each of which required re-examination after papers of John S. Bell published in 1964 and 1966.[25] But to many observers, EPR is Einstein's argument for incompleteness, and in the sequel to Bell, at least one of the assumptions on which it relies has been shown to fail.

The quantum-to-classical transition argument

Over several decades, Einstein refined another argument that quantum mechanics can provide only an incomplete description of individual systems but in this case he focused only on a single variable, and so neither needed nor invoked an observer's freedom to measure one of two complementary variables. An incipient form of this argument can be tracked from an example introduced by Einstein at the Fifth Solvay Conference in October 1927. According to de Broglie, another heuristic example to the same effect was used informally at the Sixth Solvay Conference in October 1930. In still different form – as a *quantum-to-classical transition* argument – it appears in a graphic illustration in correspondence with Schrödinger in the summer of 1935, in the immediate aftermath of EPR. The most precise form debuts in print only in the early 1950s. This argument disputes that the Ψ-function of quantum mechanics provides a complete description of the "real state" of individual objects, where the emphasis on "real state" underscores Einstein's unwavering belief in the traditional task of physical theory to describe a mind-independent reality without regard to acts of measurement or observation. Already in 1935 Einstein recognized that this argument is most effective when attention is shifted from consideration of states of quantum objects (and the contentious issue of whether "real states" *really* feature exact values of both position and momentum) to states of macro-sized objects about which no one doubts the attribution of "real". On the other hand, macroscopic objects are made of atoms that individually obey quantum mechanics. The *quantum-to-classical transition* argument then targets the purported *universality* of quantum mechanics, the claim that descriptions of "real states" of macroscopic systems can be recovered as the classical limit of the more fundamental underlying quantum mechanical ones. For if quantum mechanics is truly a universal theory, it must allow interpolation from the micro-realm to the macro-realm, i.e., to show that classical mechanical descriptions of the "real states" of macroscopic objects do emerge from the underlying quantum mechanical description in the classical limit of large numbers of particles.

The looming problem in *quantum-to-classical transition* is wave-particle duality and in particular the principle of superposition. Among the quantum theoreticians, Dirac above all accorded recognition to this principle as "the fundamental new idea of quantum mechanics and the basis of the departure from classical theory".[26] It was noted above that the quantum formalism, being linear, requires the superposition principle: two quantum states B, C, each represented by a Ψ-function, can be superposed to form a third state A with its own Ψ-function, and indeed in many cases this can be done in an indefinite number of different ways. On the other hand, Dirac also emphasized that measurement always results in a definite state (*eigenstate*), the Ψ-function of the system corresponding to one of the possible values of the observable measured. This is puzzling even in the case of a single particle: two disjunctive spin states of the particle (e.g., "spin up"

along the z-axis, or "spin down" along that axis) can be combined
to yield a third encompassing both ("spin up" and "spin down") as
long as the particle is not observed yet measurement (say, by passing
the particle through a suitably oriented Stern-Gerlach device) always
registers *either* definite eigenstate "z-spin up" *or* "z-spin down", and
never a superposition of both. The difficulty is that, by itself, the quantum formalism does not say that the superposed state *is not observed*.
The definiteness of measurement outcomes lies outside the mathematical formalism of the theory. Von Neumann's posit of a distinct
dynamics ("Processes of Type 1") simply papered over the fact that
"wave function collapse" is not accounted for by the Schrödinger
equation. Microscopic quantum fuzziness disappears when amplified to macroscopic scale.[27]

A notorious version of the problem is the paradox of Schrödinger's ill-fated cat. A cat is trapped inside a box containing a vial of poison that either has or has not been shattered by a trigger mechanism
coupled to a radioactive substance that either decays or not within
a given time. According to quantum mechanics, until the box is
opened, the cat + box (= trigger + poison + radioactive substance)
system can be regarded as in an indefinite state (half-dead/half-alive
limbo). That is to say, if Ψ_1 (alive) and Ψ_2 (dead) are two states of
the composite cat + box system, the superposition principle permits
a single quantum description Ψ_{12} in which the separate components
become "smeared out in equal measure", losing their disjoint independent character. But the cat is always *observed* to be *either* alive *or*
dead. This example, and an analogous one due to Einstein involving
exploding gunpowder (see below), illustrates an obvious advantage
over the previous *freedom of choice* argument regarding which conjugate
variable to measure on a quantum system. While one may reasonably
dispute the character of states of quantum systems, few doubt that
macroscopic objects at all times do in fact possess "real states", states
characterized by definite values of macroscopic properties, whether
these are exclusive, as "alive" or "dead" (for cats) or jointly possessed, as with precise values of position x and momentum p_x. Moreover, this second argument for incompleteness neither relies on, nor
requires, problematic assumptions of *separability* and *locality* introduced in EPR. Instead it raises the specter of an observer-dependent
reality: Does "looking" bring about macroscopic distinctness? Or,

as Einstein asked of physicist and future biographer Abraham Pais in the early 1950s, "Does the moon exist only when observed?"[28] If this were so, it would be a new type of "disturbance", acting not only at a distance, but, as it were, immediately and always. The quantum-to-classical transition argument has yet to find an adequate response, though the program of decoherence promises to do so.

Emergence of the two arguments: Initial responses 1926–1927

What is known of Einstein's first reactions to quantum mechanics is found in correspondence with several of his closest friends and colleagues. Immediate recognition is accorded to the magnitude of theoretical achievement. To Max Born's wife Heidi, also a personal friend, Einstein wrote (March 7, 1926):

> The Heisenberg-Born ideas hold everyone breathless and have made a deep impression on all theoretically oriented people. In place of dull resignation, a singular tension appears in us thick-blooded ones.[29]

In the meantime Einstein was reading Schrödinger's papers on wave mechanics as they appeared. On May 1, in a letter to H.A. Lorentz, he stated that Schrödinger's conception of quantum rules had "made a great impression" and "seemed to me a part of the truth".[30] By the summer of 1926, enthusiasm for Schrödinger had somewhat waned; writing to Munich physicist Arnold Sommerfeld on August 21, Einstein conceded Schrödinger's approach to quantum laws to be "the best", but he nonetheless expressed a wish that the wave fields propagating according to the Ψ-function could be transplanted from N-dimensional configuration space to physical spaces of three or four dimensions.[31] Just a week later he confided to Paul Ehrenfest that the Ψ-function's setting in configuration space was "indigestible" (EA 10–144), and by January 1927, he had altogether soured on Schrödinger's wave mechanics, writing to Ehrenfest that "My heart does not warm toward the Schrödingerei (Schrödinger stuff) – it is non-causal and altogether too primitive". (EA 10–152)

Yet apparently he still viewed the wave mechanical program as a fruitful arena for further theoretical development. In the spring of 1927, Einstein thought he had definitively answered the question, purposively left open by Born in advancing the statistical interpretation in July 1926, of whether wave mechanics was *necessarily* a theory yielding only probabilistic predictions of a given system's evolution. At the May 5 session of the Prussian Academy of Sciences, he presented a paper "Bestimmt Schrödingers Wellenmechanik die Bewegung eines Systems vollständig oder nur im Sinne der Statistik?" ("Does Schrödinger's Wave Mechanics Determine the Motion of a System Completely, or Only in the Sense of Statistics?"). Its premise is that Born's statistical interpretation would be *unnecessary* if the Schrödinger equation somehow could be considered to *completely determine* the group motion of an individual (although many-particle) quantum system. Not surprisingly, Einstein was quick to alert Born of his purported achievement, writing to Born a week later that he had shown how "one can coordinate *entirely determinate* motions to Schrödinger wave mechanics, without any statistical interpretation".[32] Louis de Broglie in Paris was thinking along similar lines at the same time, and in fact what Einstein presented on May 5 is a hidden variable trajectory theory (i.e., a theory in which at all times particles have positions, so-called *hidden variables*, since they do not appear in the wave function) broadly analogous to de Broglie's "pilot wave" theory introduced at the Fifth Solvay Conference later in October. The core idea was to *complete* wave mechanics by finding solutions of the Schrödinger equation for an N-point particle system by *supplementing* that equation with an imposed velocity field assigning a definite direction at each point of the system's 3N-dimensional configuration space, directions at all points corresponding to the given instantaneous value of the Ψ-function. By summing over the components of the velocities of all N-particles, a law of motion results for the *system point* in the 3N-dimensional space. Informally in conversation, Einstein dubbed this added velocity field a "ghost field" (*Gespensterfeld*); it essentially plays the same role as the much simpler "pilot wave" of de Broglie's 1927 theory and "quantum potential" of Bohm's theory in the 1950s.

It was customary at the Prussian Academy that after presentation in a session, the paper would be sent for publication in the Academy's *Proceedings*. But sometime shortly after May 5, Walther Bothe (of

Bothe-Geiger fame) pointed out to Einstein a flaw in the scheme
if applied to compound systems, i.e., systems whose overall state
may be represented by a single product of the wave functions of
each of its subsystems. This is possible just in case the subsystems
are independent, i.e., *non-interacting*. Appended to the original manuscript in the Einstein archives, there is an appended note intended
for the printed version ("*Nachtrag zur Korrektur*") citing Bothe's observation, together with a remark that therefore the scheme does not
meet what Einstein regarded as a general condition on any law of
motion, namely, a compound system's motion should be comprised
entirely by the *independent* motions of its non-interacting component
subsystems. Fully accepting that many-particle wave mechanics can
express the notion of independence in terms of product states, Einstein showed, with the simple illustration of a two-body system, that
the *Bestimmt* scheme required the velocity of one particle to immediately depend on the distant (in configuration space) coordinates of
the other. But on account of this dependence, the general solution
of the Schrödinger equation for a composite system could not be
written as a product state

$$\psi = \psi_1 \psi_2$$

of the wave functions of the individual particles, as would be expected
if the particles are physically independent. Nonetheless, Einstein
added optimistically that his assistant at that time, Jakob Grommer,
had come up with a fix that should work, though no details were
provided. In any case Einstein withdrew the paper on May 21, apparently telephoning the publisher just before the paper was going to
the press.[33] Afterwards, Einstein never referred to the paper in public,
and it was quickly forgotten. Knowledge of the paper's existence and
content remained anecdotal until recently.[34]

Why did Einstein reject his scheme to complete wave mechanics?
According to quantum mechanics, the wave function for a compound system composed of independent subsystems should be a
product state of the component wave functions, but the mutual
dependence of hidden parameters in his theory failed this condition of independence. So it must have appeared to him that the
attempt to complete wave mechanics by giving particles trajectories

was inconsistent with wave mechanics itself. As has recently been pointed out, there are other flaws in Einstein's construction (notably, it cannot reproduce the empirical content of quantum mechanics), but this is not one of them.[35] The *Bestimmt* scheme uncovered a holist and highly non-classical aspect of trajectory theories related to a much better known, also non-classical, holist feature of ordinary quantum mechanics: that in certain cases, subsystems of a composite system that have once *interacted* then separated *cannot* be represented as independent even when they are subsequently *spatially* distant from one another. This aspect, to be given the name *entanglement* by Schrödinger in the immediate aftermath of the famous EPR paper, will be discussed below. But in trajectory theories, of which modern de Broglie–Bohm theory is the prominent example, the *context* (environment) of a particle is always relevant to its motion. Particle trajectories of a *non-interacting* many-particle quantum system are correlated since all depend upon the entire environment in a way ultimately choreographed by a mathematically represented (in configuration space) *wave function of the universe*. In a hidden variables trajectory theory, each individual particle of a compound system is then only *statistically* (not absolutely) independent of its environment. This permits compound systems to be written as product states: the product of the state of a single particle with a state that is generally the configuration of hidden variables in the rest of the universe. The de Broglie–Bohm theory is a theory of this kind, a "contextual hidden variable theory" that is, remarkably, able to reproduce the empirical content of ordinary quantum mechanics.

Little of this was apparent in 1927. But Einstein surely recognized that the apparent instantaneous dependence "at a distance" in the *Bestimmt* scheme was a violation of relativistic causality as well as the principle of local action or action by contact (*Nahewirkung*) lying at the basis of field theory. Whether he attributed that flaw to an unknown error in his construction or regarded it as inherent to quantum mechanics (Schrödinger's wave mechanics) is not known. Nonetheless, the result must have been disagreeable to him. It may well have been the reason why, less than a month later, he backed out of a promise (made to the venerated H. A. Lorentz) to contribute a report on quantum statistics to the upcoming Fifth Solvay Conference in October. In a letter written on June 17, 1927, Einstein gave

Lorentz some idea of the pathos he felt at failure to produce a viable alternative to "the purely statistical way of thinking on which the new theories are founded".

> I kept hoping to be able to contribute something of value in Brussels; I have now given up that hope. I beg you not to be angry with me because of that; I did not take this (task) lightly but tried with all my strength.[36]

1927　October (Fifth Solvay Conference, Brussels)

Complementarity, born a month earlier at Lake Como, was in the air.[37] Born, Heisenberg, Jordan, Pauli, and most of the next generation of quantum physicists would all subscribe to it in one or another version. Yet complementarity failed to resolve the lingering controversy concerning whether quantum mechanics was a fundamentally non-causal and indeterministic theory, a principal topic of debate at the Fifth Solvay Conference. Here Bohr, Born, Pauli, and Heisenberg categorically supported the purely probabilistic non-causal interpretation they had developed. Dirac agreed, though with the proviso that the fundamentally statistical aspect of the theory arose from experiments.

An old guard comprised of H.A. Lorentz, Einstein, Schrödinger, and Paul Langevin resisted. A causal interpretation of the Ψ-function had been presented at Solvay. In de Broglie's "pilot wave theory" a free particle's motion is determined by the complex phase of its wave equation. But under sharp criticism from Pauli, de Broglie's causal theory failed to win public support. Although Einstein privately encouraged his efforts, de Broglie quickly abandoned this approach; it was utterly forgotten until David Bohm (1952) revived pilot wave theory a quarter of a century later.

Until quite recently, much of what was reported of Einstein's first critical dialogues about quantum mechanics with Bohr at Solvay 1927 and Solvay 1930 was derived from Bohr's retrospective account, published in Schilpp (1949) where Bohr reports on private discussions with Einstein at these conferences.[38] In Bohr's narrative, Einstein's arguments exclusively attempted to circumvent the restrictions on simultaneous measurements stated by the uncertainty (or, indeterminacy) relations. Many scholars regard Bohr's account as reliable. For example, in the otherwise admirable *Philosophy of Quantum Mechanics* (1974) by historian and physicist Max Jammer we find this characterization of Einstein's objections:

> (Einstein's) aim obviously was to refute the Bohr-Heisenberg interpretation by designing thought-experiments which show that the indeterminacy relations can be circumvented and that, in particular, the transfer of energy and momentum in individual processes can be given a fully detailed description in space and time. . . . Bohr's masterly report (reference to the above) of his discussions with Einstein on this issue, though written more than 20 years after they had taken place, is undoubtedly a reliable source for the history of this episode.[39]

A more recent 2006 account by physicist Andrew Whitaker follows suit:

> I don't think (Bohr's account) of what I shall call the first two rounds (Solvay 1927; Solvay 1930) . . . is really open to much dispute. . . . In this first round of the Bohr-Einstein debate, Einstein was determined to show that both position and momentum could be measured simultaneously for a particular system.[40]

In fact, the published proceedings of the 1927 conference record only some remarks by Einstein in the general discussion following Bohr's report (in the published proceedings of the conference, the text of Bohr's Como lecture was substituted for his report). The intervention concerned a topic seemingly unrelated to the Heisenberg relations yet still bearing on the question of whether quantum mechanics is a statistical theory. Einstein presented a thought experiment, based on the recent findings of Davisson and Germer of diffraction of matter waves by nickel crystals that confirmed de Broglie's hypothesis of the wave nature of matter. A beam of electrons strikes a screen with a small hole; passing through, the electron wave is diffracted, spreading out uniformly in all directions into a hemispheric enclosure lined with photographic film. A detection event is registered at some particular point X on the sphere. There are then "two points of view" regarding how this outcome should be interpreted (Figure 4.1).

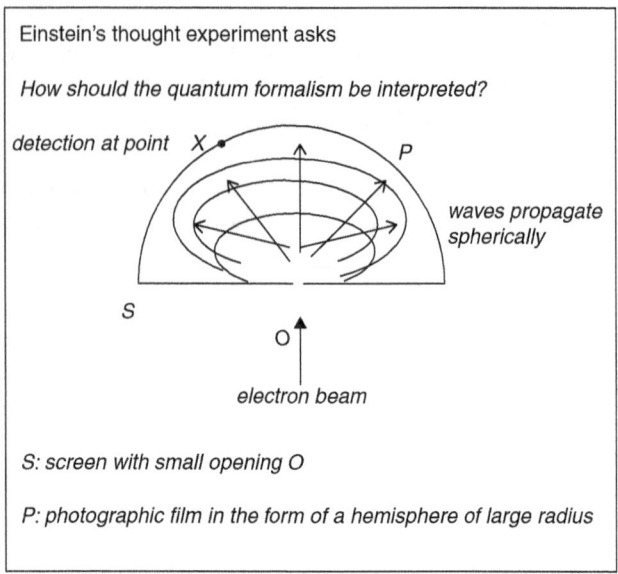

Einstein's thought experiment asks

How should the quantum formalism be interpreted?

detection at point X

P

waves propagate spherically

S

O

electron beam

S: screen with small opening O

P: photographic film in the form of a hemisphere of large radius

Figure 4.1

Viewpoint 1. The Ψ-function ("De Broglie-Schrödinger wave") describes not a single electron but "a cloud of electrons extended in space". Then Ψ represents a kind of ensemble average over the motions of a large number of particles, and $|\Psi|^2$ the relative probability (i.e., statistical probability) that any particular particle of the ensemble hits a given point X on the screen. Implicit here is the idea that each electron follows some definite trajectory to P. The Born probability $|\Psi|^2$ represents ignorance of the exact trajectories of all the particles. This interpretation is *incomplete* because it fails to describe the actual path taken by each electron.

Viewpoint 2. "The theory claims to be a complete theory of individual processes", i.e., a statement of quantum mechanical orthodoxy that the Ψ-function suffices to represent all physical aspects of a particle. According to Einstein, the Ψ-function denies electrons have definite trajectories but rather must assume every electron at each instant to be spread out in a spherical wave without any privileged direction towards P. Then the Born probability $|\Psi|^2$ expresses "the probability that at a certain point a given particle is found at a given time" all along the propagating wave front.

The two conceptions are not entirely physically equivalent. Einstein noted that only in the second are the conservation laws valid for energy and momentum of individual ("elementary") processes (as demonstrated by Geiger and Bothe). The latter is also a fair characterization of contemporary views of Jordan and (later) Heisenberg that $|\Psi|^2$ represents a spatially dispersed *potentiality* for particle localization; the potentiality becomes localized reality by a detection event at X.

In the second interpretation, up to the instant when localization occurs, the electron is regarded as potentially present all along the propagating wave front and so with almost constant probability to be detected anywhere on P. However, once localization happens (the detection is always at a definite point X on P – so-called *collapse of the wave packet*), "an entirely peculiar mechanism of action at a distance" must be assumed "which prevents the wave continuously distributed in space from producing an action in two places on the screen".[41] In brief, localization at X manifests a causal dependence upon other separated potential localization events X', X", . . . all along P where the particle *might have been registered*. Einstein then objected that interpretation 2 "contradicts the postulate of relativity since, as P can be *arbitrarily large*, the purported causal dependence can occur between events at *space-like* separation, i.e., events that cannot all be included within the light cone structure (see Chapter 5) of the event of a single particle's passage through slit O. The only way to avoid this violation, Einstein hinted, would be to supplement the Schrödinger equation with something that "localizes the particle during propagation" (precisely what he attempted earlier in the year but now) a path he encouraged de Broglie to follow. Otherwise the thought experiment poses a dilemma, also arising in EPR, between relativistic locality (no action at a distance) and completeness. It is possible to have either one or the other, but not *both*. Einstein accordingly states his preference for the ensemble interpretation 1, which is, however, an incomplete description of individual processes.

From Louis de Broglie[42] it is known that Einstein around the same time employed a different thought experiment to pose essentially the same dilemma. This example (though interestingly with a macroscopic ball instead of a microscopic particle; see further below) also appears in a letter to Schrödinger of June 19, 1935. It concerns a particle placed inside a box in one location; subsequently the box is somehow divided into two equal parts, and the two parts separated. Colloquially, this thought experiment is known as "Einstein's Boxes"[43], and it only implicitly involves the problematic attribution of trajectories to individual particles. The original locations of the different boxes are unknown, placed here in London and Tokyo for vividness (Figure 4.2).

In Princeton a particle is placed within a box **B** with impermeable walls; its associated Ψ-function is confined entirely within **B**. According to viewpoint 2 above, the particle is "potentially" present in all of **B** with essentially equal probability $|\Psi(x,t)|^2$ at each point x within **B**'s volume. Now suppose a partition in the box is somehow inserted, dividing **B** into two equal boxes $\mathbf{B}_1$ and $\mathbf{B}_2$ subsequently separated. No one has looked in either box so the particle has not yet appeared. According to the superposition principle of quantum mechanics, the Ψ-function can have the form

$$\Psi = c_1\,\Psi_1 + c_2\,\Psi_2$$

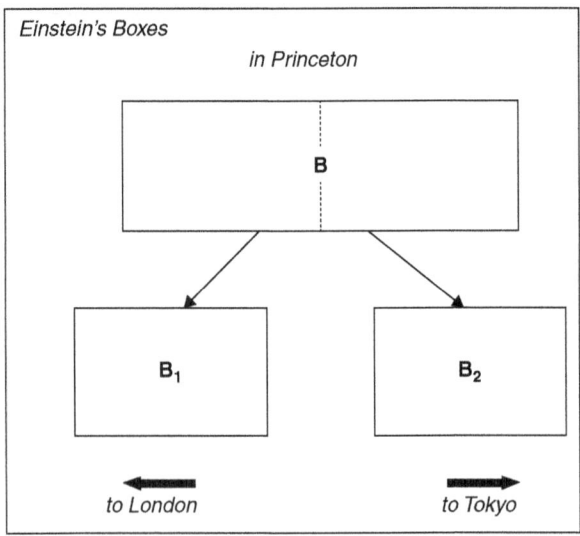

Figure 4.2

where Ψ_1 has the particle located within $\mathbf{B}_1$, Ψ_2 within $\mathbf{B}_2$ and $|c_1|^2 = |c_2|^2 = \frac{1}{2}$. This says that the particle has probability $\frac{1}{2}$ of being found in $\mathbf{B}_1$ and probability $\frac{1}{2}$ of being found in $\mathbf{B}_2$ and zero probability of being anywhere else. The particle remains potentially present in both boxes. Suppose $\mathbf{B}_1$ is opened in London; before opening, the probability of finding the particle is $|c_1|^2$ while the probability of not finding it is $1 - |c_1|^2 = |c_2|^2$. In the usual interpretation (i.e., viewpoint 2), this means that because the particle is potentially present jointly in both boxes, a positive result in London immediately localizes it in $\mathbf{B}_1$. Conversely, a negative result in London immediately localizes the particle in $\mathbf{B}_2$ in Tokyo. Again, this would seem to involve "an entirely peculiar mechanism of action at a distance". In the alternative interpretation (corresponding to viewpoint 1, hence implicitly assuming particle trajectories), prior to its observable localization in $\mathbf{B}_1$ the particle was already in *either* $\mathbf{B}_1$ *or* $\mathbf{B}_2$, prior to looking one is simply ignorant as to which. Finding it in $\mathbf{B}_1$ implies that it was there prior to opening $\mathbf{B}_1$ and similarly for finding it in $\mathbf{B}_2$.

In sum, until a box is opened, quantum mechanics allows that there is a non-zero probability that the particle is in *both* boxes. Since the Ψ-function does not tell us definitely *which* box the particle is in, though it is always definitely found in one or the other, it is *either* an incomplete description of the state of each box *or* it involves some weird relativity-violating action at a distance. In this primordial argument for incompleteness, it is relevant for the latter disjunct that at the moment when one box is opened for inspection it is spatially distant from the other box. Spatial separation underscores that what is found on opening one has immediate implication for the contents of the other box, no matter how far away it may be. This suggests that thought experiment of separated boxes already has in mind Bohr's account of measurement as invariably involving an "uncontrollable disturbance". Hence the boxes illustrate a kind of "indirect measurement": observation on one system (box), allows inferring the definite state of another distant system that is correlated with it without observing or performing a measurement on the latter. The concept of "indirect measurement" introduced in this example will become the heart of the critique of quantum mechanical completeness posed by EPR. EPR will indeed highlight spatial separation of the systems at the time of measurement to emphasize that measurement on one cannot reasonably be thought to influence the distant other system.

As in the previous example of electrons traveling towards a detection screen, "Einstein's boxes" explicitly consider only one observable property (*position*). However it may be pointed out that in both thought experiments, the viewpoint that the Ψ-function provides a complete description of the *position* observable for an individual particle is unfavorably contrasted with an alternative statistical interpretation (Viewpoint 1) that is an incomplete description of individual particles. In the latter, as in classical statistical mechanics, the details of individual particles are replaced by averages over an ensemble of particles. But doing so tacitly assumes particles possess at all times determinate trajectories, hence also linear momentum. And this can be viewed as an implicit attack on the Heisenberg uncertainty relations.

1930 October (Sixth Solvay Conference, Brussels)

As noted above, Bohr's retrospective account of Einstein's criticisms at the Sixth Solvay Conference was, until recently, taken to be authoritative. As related by Bohr, Einstein presented a thought experiment targeting the so-called time-energy uncertainty relations

$$\Delta E \, \Delta t \geq \frac{\hbar}{2},$$

roughly, the statement that the more precisely the energy of a quantum object is known (the smaller is ΔE), the less precisely is the time known (the larger is Δt) during which the object actually has this energy. It may seem curious that Einstein targeted these relations since *time* is not an operator in quantum mechanics and so not a quantum observable but merely a parameter. But Einstein's choice of target perhaps is explained by the fact that in Bohr's basic quantum postulates, an atom's *energy* is privileged over all other properties of its motion (e.g., *position, momentum*) since a numerical value for energy can always be given.[44] And in the Como lecture introducing complementarity, Bohr laid particular stress upon the time-energy uncertainty relations, deriving the relation

$$\Delta p \, \Delta q = \Delta E \, \Delta t \approx h$$

through an analogy between the Fourier analysis of a spatially extended wave and that of a time signal, and then using the de Broglie relations to convert between the two distinct sets of non-commuting quantities.[45] For this reason, both sets of non-commuting quantities can be referred to as pairs of "complementary variables".

The 1930 thought experiment concerned a beam source of photons produced inside a box and a shutter that opens and closes a hole in the box controlled by a timer, or clock, also inside the box. Bohr reproduced a crude diagram to which the photon source γ has been added (Figure 4.3).

The shutter can be opened and closed quickly enough to allow a single beam photon to escape the box as determined by the clock mechanism. The challenge to the uncertainties between measurements of energy and time is then that 1) the time when the photon leaves the box can be precisely registered by the clock, and 2) the box can be weighed *immediately before* and *after* the photon passes through the aperture, giving the difference Δm in the mass of the box. Notice that the measurements are not simultaneous but are required to follow one another immediately. Using the famous relation $E = \Delta mc^2$ in this gravitational context, the energy of the escaped photon can be precisely determined. And in this way the time-energy uncertainty relation can be defeated.

Bohr, perhaps even with Einstein's help ("At the outcome of the discussion, to which Einstein himself contributed"), pointed out a flaw in the above argument. To weigh the box precisely, say on a spring balance, an additional uncertainty must be considered. For to precisely determine the position of the pointer on a spring balance scale requires a somewhat extended balancing interval T in which the momentum of the box is carefully controlled (no jiggling of the

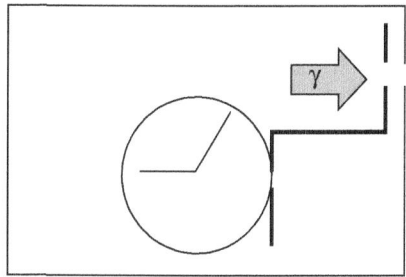

Figure 4.3

box). The exactness of the control of momentum is proportional to T and to the gravitational potential at the location of measurement. But according to general relativity, clocks are sensitive to gravitational fields (see Chapter 6): the stronger the field, the slower the rate of the clock. And the longer the balancing interval T, the longer the clock is displaced in a gravitational field, and so the more its rate will alter depending on $\Delta T/T$ is also proportional to the gravitational potential and to the uncertainty in position of the pointer. From ΔT Bohr could again recover the time-energy uncertainty relations.

There is only Bohr's 1949 report, written some twenty years subsequently, that this is actually the argument Einstein presented at Solvay 1930. If that is so, Einstein was defeated in the way Bohr describes. However, Einstein's close friend Paul Ehrenfest wrote to Bohr on July 9, 1931:

> (Einstein) said to me that, for a very long time already, he absolutely no longer doubted the uncertainty relations, and that he thus, e.g., had BY NO MEANS invented the "weighable light-flash box" . . . "contra" uncertainty relation (sic), but rather for an entirely different purpose.[46]

Indeed, the photon box reappears together with an "amendment" (*Abänderung*) in an Einstein letter to Ehrenfest in 1932.[47] The incompleteness of quantum mechanics can appear without requiring *measurements* of both complementary variables. Instead incompleteness follows by now making *explicit* how one might violate the prohibition that these variables *can possess* simultaneously precise values. For if such values can be shown to *simultaneously exist*, then there are aspects of an individual physical system not described by the Ψ-function. As quantum mechanics holds that *all possible* physical information about the state of the individual system is represented by Ψ, the theory must be incomplete.

The photon box + scale example thus modified introduces the idea of *free choice*. An observer can choose to precisely measure the value of one complementary variable but not both at the same time. She can measure *either* the energy of the photon (by weighing the box) or she can precisely determine the time at which the photon left the box. Then, employing the idea of indirect measurement, measuring just one of the pair enables prediction of an exact value of the other. If the time at which the photon leaves the box is precisely determined, then since photons travel at constant velocity (c, the speed of light), the photon will lie on a light cone sphere propagating away from the box at distance ct. If by weighing the box, the energy of the photon is precisely determined, then its momentum (direction) can be precisely calculated, enabling prediction of both the exact space-time location X of the photon on the light cone as well as the energy it will have at its arrival at X. Therefore depending on which measurement the observer *chooses* to make, it is possible to unambiguously predict either time of arrival at X or the energy of the photon at X. This suggests that properties of the photon that cannot be measured simultaneously are nonetheless simultaneously present at X.

The *either/or* of free choice is exclusive: an observer chooses to measure one physical variable, and the information obtained is then used to infer the precise value of the other conjugate variable. Freedom to choose which complementary variable to measure (obtaining an exact value) suggests that precise values of both are simultaneously present. But then the quantum mechanical prohibition against simultaneous exact values of complementary variables appears to *depend on what is actually measured*. If description by the Ψ-function really is complete, what normally is thought to be a state of a physical system (here, a photon) is not independent of the whim of the observer. This line of criticism, only tacitly present in the augmented photon box example, is explicitly developed in the aftermath of EPR.

EPR 1935 (May 15)

The most elaborate and most familiar of Einstein's arguments for quantum mechanical incompleteness is presented in the article "Can Quantum-Mechanical Description of Physical Reality Be Considered Complete?" It was published in the issue of *Physical Review* appearing on May 15, 1935 under the name of Einstein together with two young postdoctoral associates at the Institute in Princeton, Boris Podolsky and Nathan Rosen. Now universally known under the acronym EPR, it was largely ignored for three decades. In 1964 and 1966, Northern Irish physicist John S. Bell showed that the assumptions of the paper could be tested experimentally. This was done by the early 1980s.[48] As a result, between 1980–2003, the EPR paper quickly acquired the record (average citation age of 59.6 years in 2005) for the longest-lived article with more than 35 citations per year within all the journals published by the American Physical Society.[49] EPR explicitly pointed out, essentially for the first time in print, a puzzling feature of quantum mechanics that especially troubled Einstein. But it was Schrödinger who gave it a name. Stimulated by the EPR paper, and by a subsequent interchange of letters with Einstein, Schrödinger still in 1935 wrote several papers in which both the German term "*Verschränkung*" and its English equivalent "entanglement" first appeared, and the iconic quantum paradox, "Schrödinger's cat", was introduced to the world. We will see that correspondence with Einstein may well have prompted the "cat" which is very similar to an illustrative example posed in a previous letter of Einstein.

Before turning to EPR, several preliminaries are of interest. First, from Einstein's correspondence following the publication of EPR, Arthur Fine established that Podolsky wrote the text, and then submitted it from Cincinnati, Ohio, after he left Princeton to take up a physics position at the University of Cincinnati.[50] Einstein did not see the final version before submission. As published, the EPR argument is unnecessarily complex, even convoluted. Podolsky, supposedly an "expert in formal logic", may well have been responsible. Of Russian origin, Podolsky's authorship presumably explains a puzzling omission of the customary (in English) definite article in the title.[51] Second, the EPR argument for incompleteness is crafted to avoid the doctrine of measurement disturbance that Bohr had placed at the center of his viewpoint known as "complementarity". The result is the introduction of supplementary considerations of *separability* and *locality* otherwise not a necessary component of Einstein's previous or subsequent arguments for the incompleteness of the theory. Since Bell's work, these considerations have been widely regarded as the basis of the EPR argument, and in one way, they are. Third, in a June 19, 1935 letter to Schrödinger shortly after publication, Einstein complained that the essential point of the paper, the grounds for its claim of incompleteness, was "so to speak, buried by erudition

(*Gelehrsamkeit*)". At least in part, this is likely a reference to the complicated wave function constructed in the paper (by Podolsky and Rosen, according to Fine) that served as an example of an entangled system.[52] Following the example of David Bohm in 1952, and Bohm and his student Yakov Aharonov (1959), much simpler entangled ("EPR") states can be constructed, where the observables are not position and momentum but spin or photon polarization angles, which are readily measurable bivalent observables.

Criterion of completeness

But there is another reason why Einstein may have been dissatisfied with the published text of the EPR paper. For in order to set up a contrast to its claim that quantum mechanics is incomplete, the paper schematically characterizes a "complete theory". It does this by providing a necessary condition for completeness. The EPR abstract begins with the declaration, "In a complete theory there is an element corresponding to each element of reality". In the body of the paper, the condition of completeness that "seems necessary" is given by the statement (original italic emphasis), that *"every element of the physical reality must have a counterpart in the physical theory"*.[53] The characterization is remarkable by mirroring the traditional realist definition of truth as correspondence of representation to reality, a definition that in turn echoes the Scholastic dogma *veritas est adaequatio intellectus ad rem* ("truth is adequation of the intellect to the thing"). The expression "element of the physical reality", while vague, has a strong ontological cast, while the requirement that every such element have a "counterpart in the physical theory" is that of a one-to-one correspondence.

The EPR necessary condition for completeness may well be a garbled attempt to state a much weaker thesis for the Ψ-function that Einstein presumably assumed was generally accepted quantum mechanical orthodoxy. Namely, quantum mechanical completeness requires only that *one* and *only one* Ψ-function can be associated (at a given time) with a "real state of the real system" whereas the statistical character of quantum mechanical measurement outcomes results from measurement interactions (von Neumann's "Process 1").

Writing to Schrödinger on June 19, 1935, he expressed the quantum orthodoxy this way:

> One would now very much like to say the following: Ψ is univocally coordinated to the real state of the real system. The statistical character of measurement outcomes is exclusively due to the measuring apparatus, or the process of measurement.[54]

Similarly, discussing the EPR example in the article "*Physik und Realität*" written shortly after the EPR paper (see Chapter 9), Einstein remarked that the EPR example shows, by coordinating several Ψ-functions to the same physical state of an individual system, that description by the Ψ-function cannot be interpreted as a (complete) description of that physical state (1936, p. 341). And, rehearsing the EPR argument in a letter two years later (March 16, 1937) to philosopher Ernst Cassirer, the requirement of a complete quantum mechanical description is stated as the univocal (unambiguous) coordination of a Ψ-function to the physical state of a specified system.[55]

   The above notion of completeness, univocal coordination of a Ψ-function to each physical state, not to mention the stronger notion of a one-to-one correspondence, is not unambiguous. It may appear to be a mere methodological requirement: any satisfactory theory assigns one, and only one, state description to each physical state. Such a requirement is widely obeyed, allowing that two "state descriptions" can be mathematically distinct yet remain physically equivalent, for example, under a symmetry transformation. But what exactly does the methodological requirement mean as pertains to the Ψ-function, and to quantum states? Does a given Ψ-function purport to *represent* a unique quantum state as something real? In terms of Harrigan and Spekkens (2010), this is the "ontic" view that the Ψ-function describes "real" quantum states.[56] The difficulty with this reading is to understand what quantum orthodoxy (at least at the time of EPR) could possibly mean by a "real quantum state". Certainly any imputation of a one-to-one correspondence between states and Ψ-functions *presupposes* that quantum states have their own identity, and can be individuated, independently of Ψ-functions. A one-to-one correspondence, after all, is a unique correspondence between the *individual elements* of two sets. But Bohr's complementarity does not permit assigning properties to quantum systems independently of the specific experimental context by which one or another property can be measured and the resulting value of the property obtained. The term "real state", independent of context, is then just meaningless jargon. By characterizing quantum orthodoxy as requiring "Ψ is univocally coordinated to real state of the real system", Einstein appears to foist an ontic view on it that is unwarranted, even question-begging. At the same time, however, the purported reference of the term "real state of a real system" is, for Einstein, quite different than the intended reference of a Ψ-function under any ontic interpretation. The former presumably presupposes an underlying classical field-theoretic physical reality, the latter most definitely a fundamentally quantum one.

Other remarks of Einstein suggest quantum orthodoxy holds a completely different sense of quantum mechanical completeness. Recall Born and Heisenberg's account of quantum mechanics as

> a closed theory (*geschlossene Theorie*), whose fundamental physical and mathematical assumptions are no longer susceptible of any modification.[57]

The Born-Heisenberg characterization evolved (with the assistance of complementarity) to become the claim (orthodox today) that the Ψ-function contains all physical information about an individual quantum system at a given time. In the terminology of Harrigan and Spekkens (2010), this is an "epistemic" view of the Ψ-function; it pertains to, and bounds, what is *possible* to know about individual quantum systems. This sense is indirectly identified in Einstein's previous attempts to argue that the Ψ-function is to be understood as the statistical description of an ensemble of systems; as a description of individual processes, it is incomplete. At Solvay in 1928 in the ensemble picture ("Viewpoint 1"), the Born probability "$|\Psi|^2$ then represents ignorance of the exact trajectories of all the particles". Similarly, in the case of "Einstein's boxes", incompleteness and locality were posed as the two horns of a dilemma,

> Since the Ψ-function does not tell us which box the particle is definitely in, though it is always definitely found in one or the other,

it is either an incomplete description of the state of each box or it involves some weird relativity-violating action at a distance.

Einstein never really clearly distinguishes between the two senses of quantum mechanical completeness and in fact, often runs them together. On the other hand, despite the posed "condition of completeness" in the text of EPR, and its ontic echo in other texts of Einstein, it is the quite distinct sense of epistemic completeness that Einstein successfully challenged.[58] Note, however, that to restrict attack to the epistemic sense of completeness, it is crucial, as Einstein surely recognized, that the supposed *information left out* of description by the Ψ-function, hence rendering that description *incomplete*, does not beg questions against quantum mechanics. In the "boxes" example, the information is inferred after observation of a macroscopic system (opening one box). As will be seen, the argument of EPR shows that there is *information about the unmeasured system obtainable by using the quantum formalism* that is not represented by the available Ψ-function; hence it is an incomplete description.

Criterion of reality

Podolsky, who had studied mathematics and physics at Caltech, may well have crafted the above condition of completeness as a suitable physical analogy to the notion of a "complete theory" in the sense of Hilbertian metamathematics. The intent, surely, was to hoist quantum physicists on their own petard, targeting claims bruited about since 1926 that quantum mechanics was a complete theory not admitting of modification. But how could this be done without setting quantum mechanics against a complete theory that EPR does not propose but, in its final sentence, merely affirms the possibility? The answer was a "criterion of reality" whose purpose is to suggest an epistemic route to the metaphysical notion of "element of physical reality" and so to the indicated sense of completeness.

> If, without in any way disturbing a system, we can predict with certainty (i.e., with probability equal to unity) the value of a physical quantity, then there exists an element of physical reality corresponding to that quantity.[59]

Though the ontological term "element of physical reality" recurs here, the criterion need not be thought controversial in the same way as the necessary condition of completeness, for a prediction can be judged certain if the predicted event invariably occurs – in this case, the predicted value of a physical quantity is the one actually (or, potentially) measured. It would seem to follow – though this is a fallible causal inference – that some "element of reality" producing the measured value exists. Then precise prediction of observed values becomes a bellwether for claims regarding the physical reality or "real state of a real system". EPR will show that one can use the quantum mechanical formalism to predict simultaneously precise real values for both position x_2 and momentum p_{x_2} for one particle of a two-particle composite system. Of course, quantum mechanics enjoins that these cannot be simultaneously elements of reality since the position and momentum operators for a single particle do not commute. Yet since both of these quantities can be predicted "with certainty" and "without in any way disturbing" the particle, the corresponding "real state" of the particle exists that quantum mechanics does not, and cannot, describe. Hence it is an incomplete theory. This completes the dialectical stage-setting.

The EPR argument

EPR supposes a system composed of two partial systems (particles 1 and 2) that briefly interact then separate, say along the x-axis. According to quantum mechanics, if before interaction the composite system is described by a single wave function, then Schrödinger evolution will furnish a wave function Ψ_{12} for the composite system after interaction. EPR then constructs a peculiar yet permissible function Ψ_{12} for the state of the composite system that assumes that (relative) distance and total linear momentum are determinate after the interaction. Relative distance Q is determinate because it is just the instantaneous spatial distance $q_1 - q_2$ between the two particles. And since linear momentum is conserved for the composite system, the definite total momentum P is just the sum of components of the two particles $p_1 + p_2 = 0$. Quantum mechanics allows that P and Q commute. But Ψ_{12} is not a simple product of the separate wave functions Ψ_1 for particle 1 and Ψ_2 particle 2; essentially this means that after interacting, the two particles are correlated and no longer independent.

After constructing the wave function Ψ_{12} the EPR argument crucially relies on the notion of an indirect measurement implicit in the "criterion of reality". A measurement on the first particle together with knowledge of the composite system wave function Ψ_{12} allows one to assign a wave function to the distant second particle "without in any way disturbing" it. Then if one knows Ψ_{12} and Ψ_1 (following a measurement on particle 1), it is possible to infer Ψ_2 without measuring or otherwise interfering with particle 2. Suppose a precise measurement of position q_1 is made on particle 1 at time t. Then, with the information of Ψ_{12} it is possible to *predict with certainty* the determinate value q_2 of position of particle 2 at t without in any way interfering with it (no measurement is made on particle 2). Note that prediction with certainty of position value q_2 of particle 2 at time t does not require an actual measurement on particle 2 but rather licenses the inductively testable counterfactual: *if a measurement of position were to be made on particle 2 at time t, the result would invariably be the predicted value q_2*.

Alternately, instead of position, suppose a measurement of linear momentum p_1 (in direction x) of particle 1 is made at time t. Since the wave function Ψ_{12} also contains definite information about the total momentum $p_1 + p_2$, a measurement of p_1 together with knowledge of $p_1 + p_2$ similarly allows the prediction with certainty of a determinate value for p_2 (in direction x) at t, again without in any way interfering with particle 2. Prediction with certainty of value p_2 of momentum of particle 2 at time t again does not require an actual measurement on particle 2 but licenses the counterfactual: *if a measurement of momentum were to be made on particle 2, the result would invariably be the predicted value p_2*.

The notion of indirect measurement (if one knows Ψ_{12} and Ψ_1, then one can predict Ψ_2 without measuring or otherwise disturbing system 2) then sets the stage for application of the "criterion of reality". For as either q_2 or p_2 can be predicted with certainty on particle 2 at t depending only on the *free choice* of the observer regarding measurement on particle 1, both q_2 and p_2 may be inferred to exist at t, i.e., both predicted observables have "simultaneous reality" corresponding to "elements of physical reality" of particle 2 at *that* time. As EPR point out, doing so violates the quantum mechanical prohibition

on assigning simultaneous sharp values to non-commuting complementary quantities:

> The usual conclusion . . . in quantum mechanics is that when
> the momentum of a particle is known, its coordinate (i.e., position) has no
> physical reality.[60]

But just here the "criterion of reality" bites, bringing in the idea of
an objective "real state" of particle 2 independent of the choice of
measurement made on particle 1 and because the two subsystems
are spatially separated at the time of measurement, independent of
any disturbance occasioned by measurement on particle 1. Making either of two possible exact measurements on the first system
(the type of observable measured need not be specified) together
with knowledge of Ψ_{12} allows one to assign a wave function to the
second system. To be sure, which wave function that is depends on
the type of measurement made on the first system, a point that we
return to below. But the main result of EPR is to have shown that "it
is possible to assign two different wave functions . . . to the same reality (the second
system after the interaction with the first)". Now on the assumption
that quantum mechanics is complete, only one Ψ-function, not two,
can be assigned to the second system at a given time. Hence "the
quantum-mechanical description of physical reality given by wave
functions is not complete".[61]

"Spooky action at a distance"?

In fact, EPR identifies one way whereby this conclusion might be
avoided. Recall that the wave function Ψ_2, assigned to the second system, depends on the type of measurement made on the first system.
This, of course, does not violate the quantum mechanical condition
of completeness; one and only one wave function Ψ_2 is assigned to
the second system at the time of measurement on the first. But it
is to imply that the reality of the second system "depend(s) upon
the process of measurement carried out on the first system which
does not disturb the second system in any way". Unsurprisingly EPR
affirms, "No reasonable definition of reality could be expected to
permit this".

Already at Solvay in 1927 Einstein allowed quantum mechanical completeness to be salvaged by invoking a relativity-violating influence immediately propagating between spatially separate regions; in EPR such an alleged influence propagates from the measured first system to the unmeasured second system, determining the latter's state. And ten years after EPR, responding to an article on EPR by Caltech physicist Paul Epstein, Einstein writes that the assumption of action at a distance propagating with superluminal velocity

> is of course logically possible, but it is so very repugnant to my physical instinct that I am not in a position to take it seriously – entirely apart from the fact that we cannot form any clear idea of the structure of such a process.[62]

The logical possibility of transluminal causal influence is thus the price of retaining quantum mechanical completeness. Yet the essential point of EPR lies not in sorting out what is or what is not a reasonable definition of "reality" nor in proscribing action at a distance but rather in posing a dilemma: Either quantum mechanics is incomplete by allowing two, not one, Ψ-functions to be simultaneously attributed to the "real state" of the second system, or it must allow superluminal causal influences.

To unpack what is going on in the second horn of the dilemma, recall the "no disturbance" clause of the criterion of reality. It was suggested in the above discussion of "Einstein's boxes" that the ideas of indirect measurement – and its correlate, spatial separation between two systems correlated by a measurement on one of them, both primitively introduced there – were fashioned to circumvent Bohr's doctrine that measurement invariably involves an "uncontrollable disturbance". This point is made explicitly in the weeks following publication of EPR. Describing (presumably) Bohr as a Talmudic philosopher who identifies the real with what can be observed, and so "doesn't give a hoot for reality" (June 19, 1935 letter to Schrödinger), Einstein rehearsed the boxes example, this time with a macroscopic "ball" not a microscopic particle, stating that

> one cannot get at the Talmudist if one does not make use of a supplementary principle: the "separation principle" (*Trennungsprinzip*).

That is to say, "the second box, along with everything having to do with its contents, is independent of what happens with regard to the first box (separated partial systems)".[63]

Adherence to the separation principle, he adds, is needed to exclude the orthodox viewpoint that (as seen above) before opening any box the Ψ-function for both boxes has the form

$$\Psi = (\tfrac{1}{\sqrt{2}})\Psi_1 + (\tfrac{1}{\sqrt{2}})\Psi_2,$$

so that the ball has probability ½ of being in $\mathbf{B}_1$ and probability ½ of being in $\mathbf{B}_2$ and so zero probability of being anywhere else.

Arthur Fine, Don Howard, and Carsten Held have all argued the "separation principle" entails two distinct but related notions, of the independence of spatially separated systems (*separability*) and related to this, what is usually termed *locality*, explicitly implementing a relativistic prohibition against superluminal physical effects.[64] *Separability*, however, is the more fundamental principle, and it is presupposed by the latter. The fullest characterization of *separability* is given in a letter to Max Born on March 18, 1948; a nearly identical statement is found in Einstein's "*Quantum Mechanik und Realität*", written at nearly the same time for publication in the Swiss journal *Dialectica*. Here Einstein tried to explain what he meant (in a previous letter to Born of March 3, 1947) in stating that quantum theory "cannot be reconciled with the idea that physics should represent a reality in time and space, free from spooky actions at a distance (*eine Wirklichkeit in Zeit und Raum . . . ohne spukhafte Fernwirkungen*)":

> (W)hatever we regard as existing ("actual") [*existierend* (*"wirklich"*)] should somehow be localized in time and space. That is, the real (*das Reale*) in part of space A should (in the theory) somehow "exist" independently of what is thought of as real in the other part of space B. If a physical system extends over the parts of space A *and* B, then that existing at B (in B *Vorhandene*) should somehow have an independent existence (*Existenz*) of that existing at A (in A *Vorhandene*). That actually existing at B therefore (*also*) should not depend on whatever measurement is carried out in part of space

A; it also (*auch*) should be independent in general of whether or not a measurement is made at A at all.

If one adheres to this program, one can hardly consider the quantum-theoretical description as a complete representation of the physically real (*Physicaklisch-Realen*). If one tries to do so in spite of this, one has to suppose that the physically real at B suffers a sudden change as a result of measurement at A. Against this my physical instinct bristles. However, if one renounces (*verzichtet*) the assumption that what exists at different parts of space has independent, real existence, then I cannot see at all what it is that physics should describe. For then what would be thought a "system" is after all only conventional, and I don't see how one might objectively divide up the world so that one could state something about its parts.[65]

Separability is a principle fundamental to Einstein's conception of physical theory, and vital to the "realism as a program" that conception presupposes (see Chapter 9). The very definition of an "individual physical system" stipulates the system's spatial separation from other systems, a condition of individuation requiring the system to occupy a specifically distinct region of space (or space-time), one not overlapping with those occupied by other systems. Location in a particular region of space or space-time then bestows "independent, real existence" (sometimes termed, "being thus") on individual physical systems. Since the only argument marshaled against the requirement is that otherwise "I cannot at all see what it is that physics should describe", Einstein presumably thought discrete localization of systems in space or better, space-time, to be an uncontroversial necessary, and perhaps even sufficient, condition for attribution of physical reality. It might even be thought, referring back to the EPR criterion, to ground the notion "element of physical reality". Except perhaps in cosmological contexts, the requirement may appear equally uncontroversial today. But Einstein is also requiring physical theory to directly *represent* physical systems as localized in space or space-time. Here is an obvious departure from quantum mechanical descriptions since in the latter the state space of wave mechanics is configuration space (or, for quantum mechanics more generally, Hilbert space) and not three-dimensional space or four-dimensional space-time.

The principle of separability enjoins a subordinate requirement on physical theory; composite systems are to be spatially-
temporally represented "without spooky actions at a distance"
(*ohne spukhafte Fernwirkungen*) between their spatially separated parts.
The idea is that causal transmission of influence can propagate in
space and time only from point to point, via next-to-next action,
and so with finite velocity. More specifically, in the *Dialectica* article, Einstein refers to this notion as "the principle of contiguity"
(*"Prinzip der Nahewirkung"*) that "only in field theory is applied consistently". Indeed, the principle is constitutive of any field theory,
classical or quantum. Relativity theory brings a refinement of the
principle by setting a limit velocity to causal influence: Casual
effects propagate only point by point or event by event at a finite
speed, bounded by the speed of light. As seen in Chapter 5, this
is the requirement that all possible causal effects at a point-event
p are transmitted by time-like curves within p's past light cone,
while p's possible causal influences on other point-events are represented by timelike curves lying within the future light cone at
p. This is then the principle of *locality*, a relativistic and a field-
theoretic constraint.

Separability and *locality* are then principles through which Einstein
implements a *programmatic* requirement that "physics should represent a reality in time and space, free from spooky actions at a distance". It is worth pointing out that this requirement on physical
theory is flagrantly violated by complementarity. Einstein alludes
to this in the above quotation ("if one renounces [verzichtet] the
assumption that what exists at different parts of space has independent, real existence"). Bohr in particular stressed that it is only
via measurement outcomes in particular experimental contexts
that quantum mechanical descriptions of atomic phenomena *can*
be given representation in space and time. And such representation must be in accordance with complementarity's restrictions on
simultaneous application of kinematic and dynamic concepts. Using
Bohr's own phrase from the Como lecture, such attenuated representation requires "a renunciation as regards the causal space-time
coordination of atomic processes". To Einstein, this manner of indirect connection to spatio-temporal representation is of course deeply
unsatisfactory and even incomprehensible. For it is a renunciation of

his conception of what it is that a physical theory is required to do (see Chapters 8 and 9).

"On the classical level (quantum mechanics) corresponds to ordinary dynamics"

With fundamental disagreement over whether or how causal and spatio-temporal concepts can be applied in physical descriptions of phenomena, there would seem to be an unbridgeable crevasse between Einstein and Bohr, a clash of paradigms not so much over quantum mechanics *per se*, but at the *meta*-level regarding the very charge or mission of physical theory. However, in the aftermath of EPR and then nearly up to the end of his life, Einstein explored and exploited another line of argument that is able to shift the dispute again to a more neutral ground. This argument for incompleteness targets the claim that quantum mechanical descriptions go over to classical ones in the limit of large object masses. One such a claim, from which this section takes its name, was made by Yale physicist Henry Margenau in the 1949 Schilpp volume, to which Einstein replied

> This is entirely correct – *cum grano salis*, and it is precisely this *granum salis* which is significant for the question of interpretation.[66]

The alluded-to sticking point is a broad *methodological* requirement on all physical theories, classical or quantum, that the "real state" of a system at any one time must be represented univocally, i.e., the theory assigns one and only one description to each state. This says nothing about representation in space and time, or systems existing independently of other systems. It applies both to quantum and to classical theories. And as the theory of special relativity already taught, it is not to say there cannot be distinct but physically *equivalent* descriptions of one and the same state; one can give an entire class of descriptions of an inertial system, one for each inertial frame of reference, but all are equivalent via the Lorentz transformations. Similar remarks pertain to other physical symmetries where there can be multiple mathematically distinct representations of an invariant physical situation.

The more general methodological requirement suggests that the principles of separability and locality are really not necessary for an argument that quantum mechanical description is incomplete. To this end, in the immediate aftermath of EPR, Einstein proposed an "exploding gunpowder" system described in his correspondence with Schrödinger in the summer of 1935, an example shifting the charge of incompleteness from quantum descriptions of systems in the microscopic realm to single macroscopic systems where separability and locality are superfluous. The macroscopic realm is also far more fertile ground for the "realism as a program" type of description that informs Einstein's conception of physical theory; it is standardly taken as obvious that such systems are, at all times, in unique real states. The argument does require an assumption of the universality of quantum mechanics; in particular, that quantum mechanical description of a system goes over to the classical description when the system mass is sufficiently large. However, as the quote from Margenau demonstrates, quantum theorists widely assume the universality assumption to be true. Einstein surely thought he was on secure ground.

For this more general strategy to work, it is incumbent to specify in some way the notion of "the real state of a real system" without any reliance on principles or concepts that beg the question against quantum mechanics. In his contribution to a *Festschrift* for Louis de Broglie in 1953, Einstein states a "thesis of reality" intended to apply uncontroversially to both macroscopic and microscopic systems:

> *There is such a thing as the "real" state* of a physical system existing independently of any measurement or observation that in principle can be described by the means of expression of physics.[67]

He added: "no one doubts this program within the realm of the macroscopic" and slyly suggests that even quantum theoreticians adhere to it, "so long as they are not discussing the foundations of quantum theory". The issue is then whether, in the classical (i.e., macroscopic) limit, quantum mechanical description yields a complete description of this "real" state, i.e., provides a univocal description. The initial heuristic illustration appears in a letter to Schrödinger of

August 8, 1935; it concerns a definite macroscopic event, the explosion, or not, of a charge of gunpowder:

> The system is a substance in chemically unstable equilibrium, perhaps a charge of gunpowder that, by means of intrinsic forces, can spontaneously combust, and where the average life span of the whole setup is a year. In principle this can quite easily be represented quantum-mechanically. In the beginning the Ψ-function characterizes a reasonably well-defined macroscopic state. But, according to your equation, after the course of a year this is no longer the case at all. Rather the Ψ-function then describes a sort of blend (*Gemisch*) of not-yet and of already-exploded systems. Through no art of interpretation can this Ψ-function be turned into an adequate description of a real state of affairs; in reality (*Wahrheit*) there is just no intermediary between exploded and not-exploded.[68]

If quantum mechanics cannot give a complete description of the state of an individual macroscopic system, then the other alternative – it is the statistical theory of an large ensemble of identical copies of gunpowder systems each monitored for a year – can provide an empirically adequate description of the ensemble wherein each individual system definitely has or has not exploded, the quantum mechanical description providing the average number of each. The statistical description is therefore an incomplete description of any individual gunpowder system. Analogy to the famous cat of Schrödinger is unavoidable but apparently coincidental, for in Schrödinger's reply dated August 19, 1935, he says,

> In a lengthy essay I have just written I give an example that is very similar to your exploding powder keg,[69]

proceeding to give the details of the cat + box system described above. The problem, once again, is the quantum mechanical principle of superposition, the fact that the Schrödinger equation allows an evolution of an initial Ψ-function for the gunpowder system to a superposition

$$\Psi_{12} = c_1 \Psi_1 + c_2 \Psi_2$$

where Ψ_{12} is a *Gemisch* of "exploded" (Ψ_1) and "unexploded" (Ψ_2) with the coefficients giving the respective probabilities of each, $|c_1|^2 + |c_2|^2 = 1$. Once again, few will doubt that macroscopic systems at all times do in fact possess "real states", states characterized by definite values of macroscopic properties, in particular when these are exclusive, dramatically posed as "exploded" or "unexploded". And since the state Ψ_{12} is never observed, the question invariably arises whether perceptible macroscopically distinct states arise merely from "looking", the *reductio ad adsurdum* of Berkeleyian subjective idealism. The threat of idealism, of course, is a mere rhetorical aside. Rather the pertinent issue is not mind-created reality but that macroscopic objects do possess "real states of affairs" or "real situations", independently of observation or measurement, or what Leggett and Garg call "macroscopic realism".

> A macroscopic system with two or more macroscopically distinct states available to it will at all times *be* in one or the other of these states.[70]

If quantum mechanics is actually a universal theory allowing extrapolation from the micro-realm to the macro-realm, it must respect the considerable empirical evidence for "macroscopic realism".

In a 1953 paper in a *Festschrift* in honor of Max Born,[71] Einstein returned, for essentially the last time, to his longstanding criticism that quantum mechanics requires a significant departure from the "program" of description that had remained an unquestioned basis in the development of physical thinking up to the quantum theory. According to this program, the freely created concepts of theory pertain to spatial-temporal objects and their lawful relations (see Chapter 9). Such objects (the moon is given as example) are situated at all times in a particular locus (i.e., referred to a particular coordinate system) independently of whether they are observed. Positivism and also quantum mechanics Einstein insinuates – he is not careful here to distinguish positivist from other aspects of Bohr's complementarity – deny this. While the program of classical description is not an *a priori* necessity of thought, it cannot be dismissed merely as a violation of positivist strictures. Rather the sole justification of the program's validity lies in whether the resulting description is confirmed in experience. If so, the phenomena are "explained", i.e., that the projected order of the phenomena is in agreement with observation.

The charge of incompleteness is that the quantum theory is unable to furnish a complete description of an individual "real state". Bohr had retorted that on account of the non-distinguishability between object system and measuring apparatus occasioned by the quantum postulate, it was necessary to "renounce" the usual sense of classical descriptive terms, "real state of an individual system", or "physical reality"; in their place, only mutually exclusive spatial-temporal or causal-dynamical descriptions might be employed. As before, Einstein now shifted the ground of attack. First, quantum mechanics is a universal physical theory: in principle, it should apply to macroscopic objects. Furthermore, no one, he again asserts, doubts that macroscopic objects have "real situations" at all times, i.e., the definite positions and velocities (or momenta) assumed by classical physical theory. The question then becomes whether quantum mechanics implies the description of the real situation of macroscopic objects provided by classical mechanics. Unlike in the previous gunpowder/cat example, Einstein made another strategic move, implicitly invoking Bohr's correspondence principle that quantum mechanical descriptions must go over to classical ones as quantum numbers become larger and larger.

Einstein's example draws upon the one-dimensional quantum mechanical problem of a "particle in a box", a somewhat physically unrealistic but permissible case. A particle of mass m is trapped in a box of width L between impenetrable walls. Collisions with the wall are perfectly elastic, so this is a stationary state, i.e., a system of constant energy. The potential is zero for all the particle's center-of-mass positions $x < L/2$, but approaches infinity at the wall and beyond. The wave function Ψ_Q for these boundary conditions is one describing the interference of two plane waves, each with uniform momentum in opposite directions, one to the right and one to the left. From the Born rule for the probabilistic interpretation of the wave function, one can derive the fact that two and only two equal and opposite values of momentum are possible, and that both have equal probability. However, as a stationary state, the probability that the center-of-mass coordinate x of the particle lies within a given interval Δx does not change with time. This means that the particle does not correspond to the classical picture of a ball moving to and fro within the box.

At this point Einstein makes a demand similar to "macroscopic realism", though the term is a bit misleading. In a note attached to a letter to Born (January 12, 1954) Einstein states the requirement that the quantum mechanical description must yield a "localization theorem" for the macroscopic object, "every system is at any time (quasi-) sharp in relation to its macro-coordinates",[72] i.e., that it is always possible to describe a macroscopic object in macro-coordinates pertaining to its localization in space and time. The question then becomes whether Ψ_Q entails a "localization theorem" for objects when the pertinent object described by Ψ_Q enlarges to macroscopic size.

Accordingly Einstein asked whether Ψ_Q with fixed energy represents a possible description of the physical state of the particle when it is enlarged to a 1 mm sphere (i.e., regarding the particle's one "macro-coordinate" x) in passing to the macroscopic limit. This limit is characterized by the requirement that the de Broglie wavelength $\lambda = h/p$ of the particle is vastly smaller than the width of the box; essentially this means that with increasing mass m of the particle, it is possible to neglect h (the quantum of action) and consider the particle momentum classically, as peaked around two non-overlapping values, corresponding to the case of a classical particle moving with fixed energy either to the right or to the left. The question becomes whether, in the macroscopic limit, the wave function of the particle give rise to at least a "quasi-localization" x of the 1 mm sphere at any arbitrarily selected particular time. Einstein answered that it does not: according to the probabilistic interpretation of the wave function Ψ_S the center of the sphere is just as likely to be in one position as in any other along L.

Decoherence

Decades later EPR produced an enormously fruitful and continuing development in the study of entangled quantum systems. Einstein's last arguments for quantum mechanical incompleteness had a similar fate, also anticipating much later currents in the foundations of quantum mechanics. The problem of quantum-to-classical transition is that of reconciling the essential feature of quantum theory, the superposition principle, with the claim that quantum mechanics is truly a fundamental theory. If so, then quantum mechanics must justify classical physics by attempting to explain how it is that only very special states of objects (i.e., the "classical" ones) are ever observed. From a contemporary point of view, the objects need not be "macroscopic" in size; superconductors are objects that possess classical "macroscopic quantum states". From a historical point of view, it is highly interesting that theoretical proposals for characterizing quantum-to-classical transition began only in the late 1960s and early 1970s; it is as if the founders of quantum mechanics (Schrödinger aside) thought the problem unworthy of investigation.

In the contemporary research program of *decoherence*, macroscopic objects are considered to continually interact with the surrounding (quantum mechanical) environment; for example, such objects are bombarded by photons and other particles, such as molecules of the Earth's atmosphere. For essentially the same reason that the quantum dynamical evolution of a system is abrogated by measurement, the Schrödinger equation does not hold for macroscopic objects. Nonetheless the observed classical "definiteness" these objects possess (e.g., localization in space) are not inherent states, but are produced by irreversible and continuous interaction with the environment. In particular, objects appear localized in space because of the typical dependence of these interactions on position. Why this is so is not clearly understood, and can be studied mostly only in toy models. Nor is there any theoretical derivation of the classical limit that begins with the Schrödinger equation.

Summary

The sense in which quantum mechanics claimed completeness might be challenged by implementing an underlying determinism in two quite distinct ways. As will be seen, Einstein deemed quantum mechanics to be a perfectly acceptable statistical theory of outcomes of measurements on ensembles of particles, all experimentally prepared to be in the same state, but denied it is a theory fully characterizing the behavior of *individual* particles. Such an underlying theory did not as yet exist, but he aspired (for three decades) to find one by generalizing his gravitational theory to include electromagnetism. In the program of a "unified field theory" (see Chapter 10) particles are then to appear as solutions to the highly complex nonlinear and non-quantum field equations of the "total field". Altogether another type of challenge is a "hidden variables" agenda within quantum theory. Presumably it is this kind of program that Born and Heisenberg intended to rule out by calling quantum mechanics a "closed theory", i.e., no additional or "hidden" (not appearing in the orthodox quantum formalism) parameters can turn its intrinsically probabilistic treatment of individual processes into a deterministic description. This sense of completeness originated after the demonstration of the formal equivalence of matrix mechanics and wave mechanics by Schrödinger, John von Neumann, and others. It would become axiomatic in the quantum mechanical literature, following von Neumann's 1932 supposed "proof" of the impossibility of any theory empirically equivalent to quantum mechanics formulated by supplementing the quantum mechanical state vector with hidden variables. Since the work of John Bell, it is known that the von Neumann result is flawed, and the pilot wave theory of de Broglie (1927) and the deterministic theory of Bohm (1952) stand as counter-examples to completeness in this sense. In the language of Bell, both are highly non-local theories, violating the classical field-theoretic constraints of Einstein's approach.[73]

Notes

1 Reported remark of Einstein to Otto Stern in letter of Res Jost to A. Pais, August 17, 1977, quoted in Abraham Pais, *"Subtle Is the Lord ..."The Science and the Life of Albert Einstein.* New York: Oxford University Press, 1982, p. 7.

2 Einstein to Solovine, March 29, 1949 in *Albert Einstein: Letters to Solovine,* with an Introduction by Maurice Solovine. New York: Philosophical Library, 1987, p. 110; p. 111.

3 "Quantum-Theoretical Re-Interpretation of Kinematic and Mechanical Relations", *Zeitschrift für Physik* v. 33 (1925), pp. 879–93; as translated in B.L. van der Waerden (ed.), *Sources of Quantum Mechanics.* Amsterdam: North-Holland Pub., 1967, pp. 261–76.

4 Reportedly Einstein made this remark in reply to Phillip Frank, who was in agreement with Heisenberg's method. Schaffner, Kenneth, "Outlines of a Logic of Comparative Theory Evaluation with Special Attention to Pre- and Post-Relativistic Electrodynamics", in Roger Steuwer (ed.), *Historical and Philosophical*

Perspectives on Science, Minnesota Studies in the Philosophy of Science, vol. 5, Minneapolis: University of Minnesota Press, 1970, pp. 311–64; p. 362.

5 Dirac showed that Heisenberg's results correspond to a noncommutative generalization of the Poisson algebra of canonically commuting variables of classical Hamiltonian mechanics.

6 Born, Max, Werner Heisenberg, and Pascual Jordan, "Zur Quantenmechanik II", *Zeitschrift für Physik* Bd. 35, pp. 557–615; translation in Waerden, *Sources of Quantum Mechanics*, pp. 321–85; p. 322.

7 The (time-dependent) Schrödinger equation for a particle of mass m is
$-\frac{\hbar^2}{2m}\nabla^2\psi + V\psi = i\hbar\frac{\partial\psi}{\partial t}$ where ∇ is the Laplacian, V is the potential energy (generally a function of both space and time) and $\hbar$ is Planck's constant h divided by 2π.

8 Schrödinger, Erwin, "Über das Verhältnis der Heisenberg-Born-Jordanschen Quantenmechanik zu der meinen", *Annalen der Physik* Folge no. 4, Bd. 79 (1926); translated in E. Schrödinger, *Collected Papers on Wave Mechanics*. New York: Chelsea Publishing Co., 1982, pp. 45–61; p. 57; p. 59.

9 Heisenberg, Werner, "Über den anschaulichen Inhalt der quantentheoretischen Kinematik und Mechanik", *Zeitschrift für Physik* Bd. 43 (1927), pp. 172–98; translation in John A. Wheeler and Wojciech Zurek, *Quantum Theory and Measurement*. Princeton, NJ: Princeton University Press, 1981; pp. 60–84; p. 64.

10 Bohr, Niels "The Quantum Postulate and the Recent Development of Atomic Theory", *Nature* v. 121, pp. 580–90; as reprinted in Wheeler and Zurek (1981), pp. 87–126; pp. 89–90.

11 A Wilson cloud chamber is an enclosed cavity containing supersaturated water vapor; an electrically charged particle, such as an α-particle (helium nucleus) ionizes the vapor when passing through it, with condensation appearing around the ionized water molecules. The tiny droplets scatter light, leaving the visible appearance of a track.

12 Dirac, Paul Adrien Maurice, *Principles of Quantum Mechanics*. Oxford: Clarendon Press, 1930, pp. 7–11.

13 Neumann, John von, *Mathematische Grundlagen der Quantenmechanik*. Berlin: J. Springer, 1932; translation by R.T. Beyer, *Mathematical Foundations of Quantum Mechanics*. Princeton, NJ: Princeton University Press, 1955; Chapter V.

14 In the language of Hilbert space used by von Neumann, this is collapse into an *eigenstate* of the operator $\hat{O}$ corresponding to the chosen observable O.

15 Hilbert, David, "Über den Zahlbegriff", *Jahresbericht der Deutschen Mathematiker-Vereinigung* Bd. 8 (1900), pp. 180–3; p. 183.

16 Heisenberg, "Quantenmechanik", *Die Naturwissenschaften* v. 14, no. 45 (November 5, 1926), pp. 989–94; p. 990.

17 Born, Max and Werner Heisenberg, "Quantum Mechanics", as translated in Bacciagaluppi and Valentini, *Quantum Theory at the Crossroads: Reconsidering the 1927 Solvay Conference*. New York: Cambridge University Press, 2009, pp. 372–401; p. 398.

18 Cf. Bokulich, Alisa, "Heisenberg Meets Kuhn: Closed Theories and Paradigms", *Philosophy of Science* v. 73, no. 1 (January 2006), pp. 90–107.

19 Initially in a letter to Max Born, dated December 4, 1926: "Quantum mechanics is certainly commanding of attention. But an inner voice tells me that it is not yet the true Jacob (*der wahre Jakob*). The theory delivers a lot, but hardly brings us closer to the secret of the Old One. I, at any rate, am convinced that *He* is not throwing dice (*dass der nicht würfelt*)". (*Albert Einstein-Max Born Briefwechsel 1916–1955*. München: Nymphenburger Verlagshandlung, p. 127). The phrase *the true Jacob* alludes certainly to biblical story of Jacob who coerced his firstborn brother Esau to sell his birthright (*Genesis* 25:19–34). But it also may be an allusion to *Der wahre Jakob*, the name of a widely read satirical social-democratic newspaper, published in Germany from 1879–1933. Colloquially, the phrase has the meaning "the real McCoy".

20 Einstein, "On the Method of Theoretical Physics", pp. 163–9; pp. 168–9.

21 *Conjugate* variables are pairs of variables that are Fourier transform duals of each other; position q and momentum p, the canonical variables of Hamiltonian mechanics, are conjugate.

22 Heisenberg mistakenly assumed failure of predictability implies the breakdown of causality, a mistake pointed out by Cassirer, Ernst, *Determinismus und Indeterminismus in moderne Physik* (1936); translated as *Determinism and Indeterminism in Modern Physics*. New Haven, CT: Yale University Press, 1956. From a contemporary perspective, chaotic deterministic systems are a clear counterexample to Heisenberg's assumption.

23 Bohr, Niels, "Discussion with Einstein on Epistemological Problems in Atomic Physics", in Paul A. Schilpp (ed.), *Albert Einstein: Philosopher-Scientist*. Evanston, IL: Northwestern University Press, 1949, pp. 199–241.

24 Einstein, Albert, Boris Podolsky, and Nathan Rosen, "Can Quantum-Mechanical Description of Physical Reality Be Considered Complete?", *Physical Review* v. 47 (May 15, 1935), pp. 777–80.

25 Bell, John S., "On the Problem of Hidden Variables in Quantum Theory", *Reviews of Modern Physics* v. 38 (1966), pp. 447–52; "On the Einstein-Podolsky-Rosen Paradox", *Physics* v. 1 (1964), pp. 195–200. Reprinted in J.S. Bell, *Speakable and Unspeakable in Quantum Mechanics: Collected Papers on Quantum Philosophy*. New York: Cambridge University Press, 1987, pp. 1–13; pp. 14–21.

26 Dirac, *Principles of Quantum Mechanics*, p. 2.

27 It has been observed in quantum systems of just a few ions, and can appear at so-called *mesoscopic* scale.

28 "I recall that during one walk Einstein suddenly turned to me and asked whether I really believe the moon exists only when I look at it". Pais, Abraham, "Einstein and the Quantum Theory", *Reviews of Modern Physics* v. 51, no. 4 (October 1979), pp. 863–914; p. 907.

29 *Albert Einstein-Max Born Briefwechsel 1916–1955*. München: Nymphenburger Verlagshandlung, 1969, p. 125; *The Born-Einstein Letters*, translated by Irene Born. New York: Walker and Co., 1971, pp. 88–9.

30 Einstein, letter to H.A. Lorentz, May 1, 1926, in A.J. Kox (ed.), *The Scientific Correspondence of H.A. Lorentz*, vol. 1. New York: Springer, 2008, p. 602.

31 Hermann, Armin (ed.), *Albert Einstein/Arnold Sommerfeld Briefwechsel*. Basel/Stuttgart: Schwabe & Co. Verlag, 1969, p. 108.

32 *Albert Einstein-Max Born Briefwechsel 1916–1955*, pp. 133–4, cf. p. 96 (Eng). Similarly, writing to Ehrenfest on the very day of presentation, May 5, Einstein disclosed that he had shown how it is possible to unambiguously associate to every solution of the Schrödinger equation a unique motion of a multi-particle system in configuration space, hence making a statistical interpretation unnecessary (EA 10–162) as quoted in "'Nicht Sein Kann Was Nicht Sein Darf', or The Prehistory of EPR, 1909–1935: Einstein's Early Worries About the Quantum Mechanics of Composite Systems", in Arthur Miller (ed.), *Sixty-Two Years of Uncertainty: Historical, Philosophical, Physical Inquiries into the Foundations of Quantum Physics*. New York: Plenum, 1990, pp. 61–111; p. 89.

33 Pais, *"Subtle Is the Lord ..."*, p. 444.

34 Born comments on the 1927 paper after the letter cited in note 17, "I cannot remember it now; like so many similar attempts of other authors, it has disappeared without a trace". (Ibid) To my knowledge, Pais (1979, p. 901, note 83) is the first published mention of the paper's existence in the Einstein archives; An English translation is in Belousek, Darrin, "Einstein's 1927 Unpublished Hidden-Variable Theory: Its Background, Context and Significance", in *Studies in History and Philosophy of Modern Physics*, v. 27 (1996), pp. 437–61. The account here is based on the detailed analysis in Holland, Peter, "What's Wrong with Einstein's 1927 Hidden-Variable Interpretation of Quantum Mechanics", *Foundations of Physics* v. 35, no. 2 (February 2005), pp. 177–96.

35 Misled by classical analogies, Einstein imposed restricted energy conditions on his construction that do not permit the scheme to reproduce all the possible predictions of quantum mechanics, and so on that ground alone can be rejected. See Holland, "What's Wrong with Einstein's 1927 Hidden-Variable Interpretation of Quantum Mechanics".

36 Pais, *"Subtle Is the Lord ..."*, p. 432.

37 Bohr at Solvay 1927, as quoted in Bacciagaluppi and Valentini, *Quantum Theory at the Crossroads*, 2009, p. 175–76: "I do not know what quantum mechanics is. I think we are dealing with some mathematical methods which are adequate for description of our experiments.... (We must realize) that we are away from that state where we could hope of describing things on classical theories. (I) understand the same view is held by Born and Heisenberg.... The whole foundation for causal space-time description is taken away by quantum theory, for it is based on assumption of observations without interference.... excluding interference means exclusion of experiment and the whole meaning of space and time observation ... because we (have) interaction (between object and measuring instrument) and thereby ... a quite different standpoint than we thought we could take in classical theories".

38 Bohr, "Discussion with Einstein on Epistemological Problems in Atomic Physics", pp. 199–241.

39 Jammer, Max, *The Philosophy of Quantum Mechanics*. New York: John Wiley, 1974, p. 120.

40 Whitaker, Andrew, *Einstein, Bohr and the Quantum Dilemma*. Second edition. New York: Cambridge University Press, 2006, pp. 205–10; pp. 217–19.

41 Bacciagaluppi and Valentini, *Quantum Theory at the Crossroads*, p. 441.

42 Broglie, Louis de, *The Current Interpretation of Wave Mechanics: A Critical Study*. Amsterdam: Elsevier, 1964.

43 Fine, *The Shaky Game*, p. 37. See also Travis Norsen, "Einstein's Boxes", *American Journal of Physics* v. 73 (2005), p. 164.

44 Heisenberg, "Quantenmechanik", note 9, in J.A. Wheeler and W. Zurek (eds.), *Quantum Theory and Measurement*. Princeton, NJ: Princeton University Press, 1981, pp. 77–8.

45 For discussion of Bohr's "derivation", see Jan Hilgevoord, "The Uncertainty Principle for Energy and Time: II", *American Journal of Physics* v. 66, no. 5 (May 1998), pp. 396–402.

46 Original capital emphasis in the letter (in Bohr Collection at Archives for History of Quantum Physics) cited, with German original text, in Don Howard, "'Nicht Sein Kann Was Nicht Sein Darf', or The Prehistory of EPR, 1909–1935: Einstein's Early Worries About the Quantum Mechanics of Composite Systems", in Arthur Miller (ed.), *Sixty-Two Years of Uncertainty: Historical, Philosophical, Physical Inquiries into the Foundations of Quantum Physics*. New York: Plenum, 1990, pp. 61–111; p. 98, note.

47 Held, Carsten, *Die Bohr-Einstein Debatte: Quantenmechanik und physikalische Wirklichkeit*. Paderborn: Ferdinand Schöningh, 1998, p. 89 ff.

48 Aspect Alain, Philippe Grangier, and Gérard Roger, "Experimental Realization of Einstein-Podolsky-Rosen-Bohm *Gedankenexperiment*: A New Violation of Bell's Inequalities", *Physical Review Letters* v. 49 (1982), pp. 91–4, and other subsequent papers by Aspect's group.

49 Redner Sidney, "Citation Statistics from 110 Years of *Physical Review*", *Physics Today* (June, 2005), pp. 49–54; p. 52.

50 Fine, *The Shaky Game*. According to Israeli physicist Asher Peres (1934–2005), a doctoral student of Nathan Rosen, Rosen claimed, one afternoon during the Institute of Advanced Study's traditional 3 p.m. tea, to have brought to Einstein's attention the issue of interpretation of entangled wave functions. Rosen told Peres that Einstein immediately saw the implications for his longstanding dispute with Bohr. As they were discussing the problem, Podolsky joined in the conversation and then proposed to write an article, to which Einstein agreed. See "The Advent and Fallout of EPR", Kelly Devine Thomas, *The Institute Letter* (Fall 2013), available at www.ias.edu/ideas/2013/epr-fallout

51 Notes of records in the former NKVD archives by former KGB officer Alexander Vassiliev, since 2009 in the Library of Congress, identify Boris Podolsky (1896–1966) as "Quantum" (*Kvant*), an intelligence source that approached the Soviet Embassy in Washington in 1943. Podolsky is alleged to have handed over detailed information about the gaseous diffusion process of separating U-235 from U-238 (then underway at the research facility in Oak Ridge, Tennessee, a crucial part of the "Manhattan Project") in exchange for a payment of $300 ($3,500 in 2008 dollars). See John Earl Haynes, Harvey Klehr, and Alexander Vassiliev, *Spies: The Rise and Fall of the KGB in America*. Boston and New Haven, CT: Yale University Press, p. 73.

52 The constructed wave function is not one that can be arrived at by letting systems interact and then evolve according to the usual Schrödinger dynamics. The entangled EPR state is a momentary state only; its support has measure

zero in configuration space. Only 75 years later was it shown that the EPR wave function can be reproduced in the rigorous C* algebraic formalism as a positive linear functional on the Weyl algebra for two degrees of freedom. By this fact it is legitimately a proper quantum state; see H. Halvorson, "The Einstein-Podolsky-Rosen State Maximally Violates Bell's Inequalities", *Letters in Mathematical Physics* v. 53 (2000), pp. 321–9.

53 Einstein, Albert, Boris Podolsky, and Nathan Rosen, "Can Quantum-Mechanical Description of Physical Reality Be Considered Complete?" *Physical Review* v. 47 (May 15, 1935), pp. 777–80; p. 777.

54 As quoted in Fine, *The Shaky Game*, pp. 71–2. Einstein can be read, as Fine does, as suggesting the quantum orthodox view to be that there is a one-to-one correspondence between Ψ-functions and "real states", a condition Fine calls "bijective completeness".

55 *Den eine vollständige Beschreibung würde eine eindeutige Zuordnung von* Ψ_2 *zum physikalischen Zustand des Punktes 2 erfordern.*

56 Harrigan, Nicholas and Robert W. Spekkens ("Einstein, Incompleteness, and the Epistemic View of Quantum States", *Foundations of Physics* v. 40 [2010], pp. 125–57) use the term "Ψ-ontic" for models in which the quantum state has the status of something real. Harrigan and Spekkens read EPR as showing that the condition of *locality* implies the falsity of any "Ψ-ontic model" whereas they credit Einstein with holding a "Ψ-epistemic" view, that the Ψ-function represents an *incomplete description* of an individual system.

57 Born, Max and Werner Heisenberg, "Quantum Mechanics", as translated in Bacciagaluppi and Valentini, *Quantum Theory at the Crossroads*, 2009, pp. 372–401; p. 398.

58 In fact, the "condition of completeness" expressed in EPR is even stronger than the texts of Einstein suggest, and as well an impossibly strong metaphysical requirement. It appears tantamount to requiring a theory to be *exhaustive* of physical reality, i.e., a final "theory of everything" applicable to and capable of describing "every element of physical reality" known or unknown, including those of objects and processes at the most fundamental level of microphysics, at energies many orders of magnitude beyond those accessible in experiments with current or foreseeable technologies. While currently some argue that string theory holds out a promise of doing this, no theory in 1935 or today meets this criterion. To the physicist, the EPR "condition of completeness" even may seem bizarre, since discovering "elements" (observed matters of fact) in "the physical reality" lacking counterparts in a given theory simply means that the earlier theory be recovered as a limiting case of a more general theory. A familiar example is Newtonian planetary theory: as the observed value for the advance of the perihelion of Mercury exceeds the permissible Newtonian value (i.e., has no counterpart there), Newtonian theory is regarded as incomplete and so a limiting case of Einstein's gravity. But this backwards-inclusion of earlier successful theories is a process that may be viewed as indefinitely continuing and, in any case, it is not the sense of "completeness" affirmed as necessary in the text of EPR.

59 Einstein, Albert, Boris Podolsky, and Nathan Rosen, "Can Quantum-Mechanical Description of Physical Reality Be Considered Complete?" *Physical Review* v. 47 (May 15, 1935), pp. 777–80, original italic emphasis, p. 777.

60 *Ibid.*, original italic emphasis, p. 778.

61 *Ibid.*, pp. 779–80.

62 Einstein, letter to Paul Epstein of November 5, 1945 (EA 10–582, as quoted and translated in Howard *op. cit.*, 1990, p. 102). Similarly, writing to Cassirer in 1937, Einstein stated that his "physical instinct" (*meinem physikalischen Instinkt*) would be violated if measurement on the first system could have a "distant action" on the state of the second system.

63 As quoted in Fine, *The Shaky Game*, pp. 69–70.

64 *Ibid.*, p. 36; see also Howard, Don, "Einstein on Locality and Separability", *Studies in History and Philosophy of Science* v. 16 (1985), pp. 171–201; Held, *Die Bohr-Einstein-Debatte.*

65 *Albert Einstein-Max Born Briefwechsel* 1916–1955, pp. 218–19, cf. pp. 164–5 (English).

66 "Replies to Criticisms" (1949), p. 682.

67 Einstein, Albert, "Einleitende Bemerkungen über Grundbegriffe", in *Louis de Broglie: Physicien et Penseur, collection dirigée par André George* Paris: Éditions Albin Michel, 1953, pp. 4–15; p. 6.

68 As quoted in Fine, *The Shaky Game*, p. 78.

69 *Ibid.*, p. 82.

70 Leggett, Anthony J. and Anupam Garg, "Quantum Mechanics versus Macroscopic Realism: Is the Flux There When Nobody Looks?", *Physical Review Letters* v. 54, no. 9 (March 4, 1985), pp. 857–60; p. 857.

71 Einstein, "Elementare Überlegungen zur Interpretation der Grundlagen der Quanten-Mechanik", in *Scientific Papers Presented to Max Born on His Retirement from the Tait Chair of Natural Philosophy in the University of Edinburgh.* Edinburgh and London: Oliver and Boyd, 1953, pp. 33–49.

72 *Albert Einstein-Max Born Briefwechsel 1916–1955.* München: Nymphenburger Verlagshandlung, 1969, p. 279; The Born-Einstein Letters, translated by Irene Born. New York: Walker and Co., 1971, p. 215.

73 Bell, "On the Problem of Hidden Variables in Quantum Theory" showed that von Neumann's proof relies on an unnaturally strong assumption that corresponds to an unusual situation in quantum mechanics. Until John Bell showed otherwise, quantum physicists generally believed von Neumann had shown quantum mechanics to be "complete" in the sense that hidden variable supplementations were impossible.

Further reading

Bacciagaluppi, Guido, and Antony Valentini. *Quantum Theory at the Crossroads: Reconsidering the 1927 Solvay Conference.* New York: Cambridge University Press, 2009.

Fine, Arthur, *The Shaky Game: Einstein Realism and the Quantum Theory.* Second edition. Chicago: University of Chicago Press, 1996.

Norsen, Travis, "Einstein's Boxes", *American Journal of Physics* v. 73 no. 2 (February 2005), pp. 164–76.

Part II
Relativity

Five
Special relativity

"The introduction of a 'light ether' will prove superfluous, insofar as in accordance with the view to be developed here, no 'space at absolute rest' endowed with special properties will be introduced".[1]

Introduction

Explicit recognition of a relativity principle first came in 1632. In the book that cost him his freedom, Galileo Galilei asserted that the phenomena of nature appear the same to an observer at rest and to one in uniform (non-accelerating) motion relative to the former. This was not a disinterested claim; Galileo used the premise that observation does not reveal the Earth's essentially uniform rotational and orbital motions around the sun to argue against Scholastic physicists who, remarking that such motions would surely give rise to visible effects, denied the Earth's motion as contradicting Holy Scripture. Employing the 17th century equivalent of the familiar example of a train traveling uniformly along straight rails, Galileo countered that by no experiment conducted below decks could an observer tell whether his ship was stationary at calm anchor or sailing with any constant velocity, as long as the motion was steady "and not fluctuating this way and that".[2]

In 1687 Newton set his three laws of motion in the framework of absolute space and time, distinguishing between the "true, absolute motion" of a body with respect to this framework, and the "relative motion" of bodies with respect to one another. Newton's concepts of absolute space and time imply an objective distinction between uniform motion and absolute rest that critics from Leibniz and Berkeley to Mach (see Chapter 6) held to be untenable. Cognizant, however, that bodies can only be observed relative to other bodies, not with respect to absolute space, and of the difficulties (and controversies) in determining the true center of rest for "the system of the world", Newton adopted the relativity principle of Galileo, regarding it as a consequence of his laws of motion. It was stated in Book I of the *Principia* as Corollary V to the laws. In the recent translation it says:

> When bodies are enclosed in a given space, their motions in relation to one another are the same whether the space is at rest or whether it is moving uniformly straight forward without circular motion.[3]

Notice that the "given spaces" for which this statement is true correspond broadly to what we now know as inertial reference frames. Such frames are required for the validity of Newton's laws of motion, and in particular for the definition of the core Newtonian notion of *force*. The above principle of relativity of motion accommodates the fact that in Newtonian mechanics, velocities are relative but accelerations – the rate of change of velocity – are absolute, i.e., are the same regardless of whether an observer is at rest or in uniform motion. Newton himself regarded accelerative motions as motions referred to a background absolute space and time, and only derivatively motions with respect to other bodies. In its modern form, Corollary V is known as the relativity principle of mechanics, or the "Galilean Principle of Relativity". It states,

> [n]o mechanical experiment can distinguish between a "rest frame" S and a frame S' moving relatively to S with a uniform inertial (rectilinear) motion.

The most advanced theory when Einstein initiated his studies of physics was no longer mechanics but an electrodynamics derived from Maxwell's equations of electromagnetism. However Maxwell's equations did not obey the Galilean transformation relations between inertial frames as in Newtonian mechanics. To many physicists this was not an issue worth troubling over. Maxwell, after all, had shown that electromagnetic phenomena propagate through space as waves, a decidedly non-mechanical motion. Already a convinced adherent of the inherent reasonableness of the relativity principle, Einstein in 1905 showed how the Galilean principle above can be generalized to apply to the phenomena of electrodynamics, affirming the principle of relativity (in slightly modernized language):

[t]he laws of physics (mechanics and electrodynamics) are the same in every inertial frame.

Others, such as H.A. Lorentz and Henri Poincaré, sought to implement a relativity principle in electrodynamics through various hypothetical models of the electron. Such models generally retained a privileged inertial frame, a rest frame in which the velocity of light is constant. Einstein correctly judged the state of physical theory premature for attempting detailed accounts of the forces acting on the microconstituents of moving measuring rods and clocks, instruments required for the definition of an inertial frame. Instead he legislated a sharp separation between *kinematics* (pertaining to generic structural features of the notions of space, time, and motion) and *dynamics* (pertaining to specific interactions of physical systems located in space and time). It will be seen that such a clean distinction does not survive in the theory of general relativity, and it remains controversial in special relativity today. Yet it was the key to Einstein's success where others failed.

Overview

The theory of special relativity arose in the context of electromagnetism, but unlike general relativity, it is not the theory of a specific interaction. It is rather a *meta-theory*, now usually regarded as constraining theories of all non-gravitational interactions. Confirmed every day in particle accelerators and other high-energy phenomena such as cosmic rays, it rests on two explicit postulates and one implicit one. Observers regarded at rest or in uniform motion with respect to one another are *inertial observers*. In special relativity, such observers need be equipped only with measuring rods, e.g., meter sticks, to measure lengths, and with clocks to measure time intervals between events. The observer together with his rods and synchronized clocks comprise an *inertial system* or *inertial frame*, i.e., a frame of reference in which units of lengths and times correspond to the readings of rods and clocks at rest in that frame. The two explicit postulates then are:

1. The outcome of any given experiment is independent of the inertial system in which it is performed.
2. The velocity of light in *vacuo* is the same in all inertial systems, and independent of the velocity of the emitting source.

The first statement is a common formulation of the principle of *special relativity*, "special" on account of the reference to inertial systems. It is equivalent to the one given by Einstein in 1905: *the laws of physics are the same in any inertial system*. The equivalence of *all* inertial systems implies that there is no preferred rest system, one truly at rest. As discussed above, such a system was the supposed state of the light ether, a seemingly necessary presupposition in 1905 of the Maxwell-Lorentz theory. The second is also known as the *light principle*; it is a reformulation of the light principle of the ether theories of electromagnetism. The velocity of light c has become a hallmark of relativistic theories. As mentioned above, variation of the electron's mass with its velocity was already known by 1900. And the proportionality between the energy of a physical system and its mass appeared in electrodynamic theory prior to 1905. In September 1905 Einstein made the first attempt at a general proof of the mass-energy relation, presenting a calculation showing that if a body releases an amount of energy E in the form of light, its mass must diminish by an amount E / c^2. The fact that the energy released is in the form of light "evidently makes no difference (ist es offenbar unwesentlich)".[4] Though showing that energy has mass, two years passed before Einstein would claim full equivalence between mass and energy. The equation familiar to everyone today $E = mc^2$ appeared in 1907, together with the observation that it is "a result of enormous theoretical significance".[5] In fact, there are numerous gaps or flaws in various proofs of this relation given by Einstein over the course of four decades; von Laue in 1911 produced the first satisfactory general proof.[6]

A third postulate of special relativity was not overtly stated but was implicitly assumed by Einstein. Explicitly spelled out, it affirms:

> Measuring rods determine the same lengths, and clocks tick at the same (uniform) rate even if – as is permitted by the Lorentz transformations – the inertial frame of reference in which they are regarded as stationary is "boosted" to an arbitrary constant velocity. In other words, rods and clocks, although they are physical bodies or processes, are to be regarded as ideal, or perfect, fiduciary instruments of measurement, suffering no dynamical effects from (instantaneous) changes in velocity.

As might be inferred, this is a rather surprising assumption; its effect, Einstein confessed much later, is to partition physical things into two types: 1) rods and clocks, and 2) everything else.[7] There are far-reaching consequences from its adoption, among them the tacit but erroneous belief that mere coordinate differences have chrono-metrical significance. This belief hindered Einstein for years in his search for a relativistic theory of gravitation (see Chapter 6). But it is this postulate, even more than the first two, distinguishing special relativity as a *theory of principle* not a *constructive theory*. Without it, special relativity cannot provide a new *kinematics* – a new account of the *generic* features of space and time – for all of physics. In this regard, special relativity is completely unlike the *constructive dynamical* alternative theories of Lorentz and Poincaré. For these theories retained the ether as a rest frame in which "true" lengths and durations were defined, whereas the rods and clocks of frames moving with respect to the ether suffer dynamical effects; in particular, the dimensions of their constituent electrons contract in the direction of their motion, the deformation varying directly with their velocity of motion.

The luminiferous ether

By the first decades of the 19th century, the phenomena of light polarization and interference, demonstrated by English polymath Thomas Young (1773–1829) and French physicist Augustin Fresnel (1788–1827), had largely overturned the particle (or, emission) theory of light put forward in *Opticks* (1704) by Sir Isaac Newton

(1643–1727) in favor of the wave (or undulatory) theory. To be sure, the latter's mathematical and physical aspects did not coalesce until around 1830, when wave fronts replaced rays as tools of analysis, while in explanations of optical phenomena, waves in the ether replaced particles of light.

In analogy to sound and water waves, it was naturally assumed that waves require a medium in which to travel. Though unobserved, this space-filling "luminiferous ether" (in Young's phrase) reached out to the most distant visible stars. As a purely theoretical entity, the ether seemingly required the conflicting physical properties of an elastic solid (enormous rigidity per unit mass to transmit waves with the speed of light) and of a completely diaphanous void (for how else could the planets, without slowing down due to resistance, continually orbit the Sun in agreement with Newtonian predictions?). Despite these profound conceptual difficulties, the wave theory of light was empirically too successful to give up, with the result that attention turned to finding observational evidence of the ether; in particular, to experiments that might measure the absolute motion of the Earth with respect to the resting ether. Though Earth's orbital velocity was approximately known, any measurement of its absolute motion had also to take into account the unknown speed and direction with which the Sun, carrying along its planets, was moving with respect to the ether.

One complication to carrying out this program arose already within the framework of Newton's particle theory of light. In 1727 English astronomer James Bradley (1693–1762), attempting to measure stellar parallax as a predicted consequence of the Copernican theory, discovered instead the phenomenon of the displacement or aberration of starlight. Due to the orbital motion of Earth, Bradley found that an observer needed to slightly tilt a telescope in the direction of Earth's motion in order to keep the star in the center of the field of view. Since Earth's orbital direction is continually changing, stars appear to trace out small ellipses over the course of a year. Bradley explained this displacement on the assumption that the emitted particles of light have a large but finite speed, c (in a geocentric system), and, where v is Earth's orbital velocity, he calculated the tilt or "aberration" angle (of order v/c) defining the line of sight at which a telescope's eyepiece can continue to receive particles of light emanating from the star. However, explaining stellar aberration in the wave theory of light presented something of a challenge. If the ether was anything like Earth's atmosphere (as was widely supposed), together with the starlight it supported, it would be carried along inside the tube of the telescope, hence (and contrary to fact) the telescope need not be tilted. Young in 1804 accordingly proposed that unlike the atmosphere, an immobile ether passed freely through the walls of the moving telescope and indeed that "the luminiferous ether pervades the substance of all material bodies with little or no resistance, as freely perhaps as the wind passes through a grove of trees" (1804, p. 12–13). Young thus gave rise to many subsequent attempts to demonstrate Earth's motion through the ether through the measurable detection of an "ether wind".

In 1810 the French physicist François Arago (1786–1853), still assuming the emission theory of Newton, sought to use directional differences in Earth's orbital motion to detect variations in the speed of light from different stars. Arago supposed on Newtonian grounds that the velocity of the source would influence the speed of the emitted light. Noting that the Bradley aberration, the amount of which is found by subtracting the motion of Earth from the incoming ray of light, was too crude to detect differences in the speed of starlight as Earth approached or receded from a star, Arago attempted to quantify these differences by measuring the refraction of starlight passing through a glass prism. He reasoned that the bending of light rays by a prism should be affected by motion of Earth toward or away from (at six-month intervals) the stellar source, the differences providing a natural measure of the differing speeds of light. To his surprise, Arago found that light coming from any star behaves in all cases of reflection and refraction just as if the star were in the place it appears to occupy and Earth were at rest. In other words, Earth's motion had no perceptible effect on the refractive index of glass.

Fresnel in 1818 claimed that Arago's null result could be explained by the wave theory. Fresnel argued that the motion of Earth has no influence on the laws of refraction since a part of the ether – its density in the moving body less its density *in vacuo* – together with the light waves propagating through it, is dragged along with the motion of the optical medium (prism, telescope). Fresnel contended that the refractive properties of transparent media (such as glass, air, or water) accordingly depend on the concentration of ether within them and that the ethereal density in any body is proportional to the square of the body's refractive index. Where n is the body's index of refraction, this meant that the body's relative velocity through the resting ether is not v, but v $(1-1/n^2)$. The term in parentheses, known through the rest of the 19th century as the *Fresnel dragging coefficient*, implied that the relation of the velocity c of light *in vacuo* to the velocity c' of light in transparent materials such as a prism or the tube of a telescope is not $c' = c/n + v$ but $c' = c/n + v(1-1/n^2)$, a difference thought to exactly compensate for the expected optical effects of the motion of Earth through the ether. Fresnel maintained that the ether-drag hypothesis could explain Arago's null result as well as stellar aberration since a medium (such as inside a telescope) moving through a stationary ether drags the light propagating through it with a fraction of the medium's speed, i.e., a velocity slightly less than the velocity of the medium (in the ratio $n^{2-}1/n^2$), permitting the observed stellar displacement.

Fresnel's ether-drag hypothesis converted Arago to the wave theory of light and further appeared to be confirmed by the water tube experiments of the French physicist Hippolyte Fizeau (1819–1896) in 1851, the most precise measurements of the speed of light yet undertaken. Fizeau crafted a technically ingenious method of comparing the speed of solar light in a tube of water at rest with the speed of light when the water flowed with velocity v relative to the tube. If the luminiferous ether adhered to the water, the velocity of light relative to the water should be the same, regardless of the motion of the water. Accordingly, it should be expected that with the water in motion, the velocity of light relative to the tube is greater by v. From an observable displacement in the interference bands, Fizeau found instead that the water's motion increased the velocity of light relative to the tube not by v but only a small fraction of this value, v $(1-1/n^2)$, just the Fresnel dragging coefficient where n is now the refractive index of water. Although Fizeau himself did not do so, his result could be interpreted to mean that water partially drags the ether with it. Moreover, for media with little or no refractive capacity, the motion of the medium has practically no influence on the velocity of light. Nonetheless, the idea that a body moving through a stationary ether partially drags along both the ether and the light propagating through it with a tiny fraction of the body's velocity added yet another mysterious property to the ethereal substance.

Support for the hypothesis that the luminiferous ether does not participate in the motion of matter came from the Scottish physicist James Clark Maxwell (1831–1879) with his revolutionary dynamical theory of electricity and magnetism (1865). Maxwell simply assumed, as established physics, the existence of an all-pervading ether of small but real density and a certain kind of elastic yielding capable of transmitting motions with great, but not infinite, velocity. His theory made the resting ether the bearer of electric and magnetic fields and so of energy, a credible attribute of physical reality. The equations relating changing electric and magnetic fields showed that the ratio of the physical dimensions of an electric field to a magnetic field is a constant with the dimensions of a velocity, in magnitude very close to Fizeau's experimentally determined velocity of light. Identifying the two, Maxwell unified physical optics with electromagnetism, declaring light waves to be electromagnetic disturbances in the ether propagating with velocity c. In accord with the assumed homogeneity and isotropy of the ether, Maxwell's theory supposed that the velocity of propagation of light has always the same value c in any direction. This so-called light principle was widely accepted on experimental grounds. But the light principle in turn implies an apparent violation of the principle of relativity, since the constancy of the speed of light would seem to hold only in the resting ether. Maxwell's equations accordingly discriminated between resting and uniformly moving systems, having their simplest form in a system considered at rest in the ether. Since in its orbit of the sun the Earth was viewed as moving at a constant velocity v of some 30 km per second through the ether (the unknown motion of the solar system through the galaxy was neglected), Maxwell's equations could not have the same form in terrestrial experiments as in the rest system of the ether. Tied to a preferred inertial frame, Maxwell's theory violated the principle of Galilean relativity satisfied by Newtonian mechanics.

Maxwell's initial attempts to construct heuristic mechanical models of the ether failed, and in its mature version (1873) the theory was presented without them. It was in this form that it encouraged the experimental search to detect the motion of Earth through the resting ether. Numerous "first-order" experiments had been carried out to measure v/c, the expected effect of the motion of the Earth depending on the ratio of Earth's velocity v to the speed of light in the resting ether c. By the late 1870s, however, it was widely acknowledged that measurements of first order in v/c are too crude, and that any observable effect would lie in the more subtle range of second-order effects, v^2/c^2. In a letter to an American astronomer written just before he died in 1879, Maxwell proposed a variant of the discovery in 1676 by the Danish astronomer Ole Rømer (1644–1710) of the finite velocity of light based on observations of the eclipses of Jupiter's moons to argue that the velocity of Earth through the ether could theoretically be determined by timing the eclipses of Jupiter's moons when Earth's orbit is approaching or receding from Jupiter. By observing variations in the pattern of eclipses one could, in theory, infer the velocity of the solar system, and so Earth, through the ether. However, the only possible Earth-based experiment to do this was to measure variations in the velocity of light on a round-trip journey between two mirrors, second-order effects of Earth's motion corresponding to a tiny time interval (about a thousandth of a millionth of a millionth of a second) that Maxwell considered "quite too small to be observed".

The American physicist Albert A. Michelson (1852–1931) read Maxwell's letter, posthumously published in the journal Nature in 1880, and declared the second-order effects "easily measurable" by the method of light interferometry. Michelson's experiments in 1881 in Potsdam and later in Cleveland with the American physicist Edward W. Morley (1838–1923) in 1887 are the most important attempts to detect the predicted second-order effects of the motion of Earth through the resting ether by measuring the speed of light traveling "upwind" (in the direction of Earth's motion, $c + v$) and comparing it with the speed of light traveling "downwind", $(c - v)$. In the 1887 experiment a beam of light produced by a source S was split into two parts by a half-silvered plate of glass M (see Figure 5.1). The two beams were sent along arms MA and MB of equal length, reflected at the mirrors A and B and the beam from B again reflected at M. The now parallel beams travel to C where interference fringes were produced. The interferometer was set up first so that MB was in the direction of Earth's orbital motion ("into

the wind"), then rotated slowly so that M*A* was in this direction
(Figure 5.1).

Michelson and Morley expected to find a shift proportional to v^2/c^2
in the fringe pattern of light recombined at C after traveling round
trip paths of approximately 11 m in the direction of Earth's motion
through the ether and perpendicular to this direction. Famously,
their result was null: no significant shift in the fringe pattern was
observed. The experiment was repeated six months later when Earth,
with considerably less orbital velocity, was on the far side of its orbit;
the same null result ensued. The aberration of light had seemed
to show that ether must flow freely through matter. The failure of
first-order experiments to detect this wind might be explained by

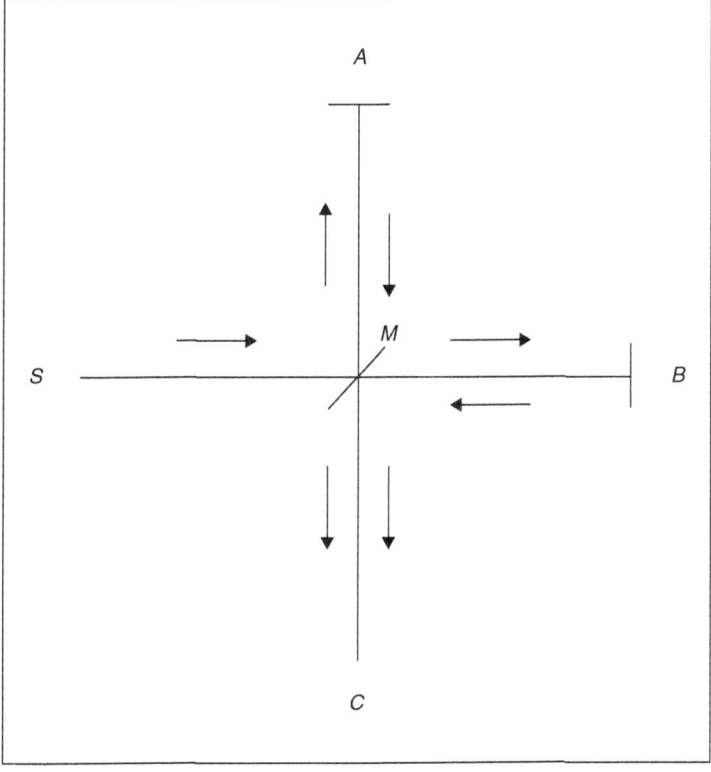

Figure 5.1

Fresnel's ether-drag hypothesis, but second-order effects were far more sensitive. If there was any relative motion between Earth and the luminiferous ether, it was very small – small enough to rule out Fresnel's explanation of aberration via a measurable drag of the ether by Earth. A variety of theoretical proposals were put forward to explain what had gone wrong. The most influential of these sought to explain the failure to detect the relative motion of Earth by a compensating dynamical hypothesis, formulated first in 1889 by the Irish physicist G.F. FitzGerald (1851–1901) and independently by the Dutch physicist Hendrik A. Lorentz (1853–1928) in 1892.

Lorentz's electron theory

At the beginning of the 20th century H.A. Lorentz, at Leiden, was regarded by Einstein as the world's foremost theoretical physicist.[8] Lorentz tentatively supported the "electromagnetic world view", a minority opinion among leading physicists at the turn of the century, e.g., Wilhelm Wien, co-editor with Max Planck of the *Annalen der Physik*, that all forces and mechanics itself might be reduced to electromagnetism.[9] Over the previous decade or so, he had patiently elaborated and refined a theory of electrodynamics, seeking to bring together the mechanics of Newton governing the motions of matter, and Maxwell's theory of electromagnetism describing electric and magnetic fields. Maxwell had employed the term "electrodynamics" for his own theory of electric and magnetic fields augmented to include interactions between field quantities (electric, magnetic, current) and their *ponderable* material sources, objects possessing mass. Lorentz's electrodynamics was rooted in his theory of the "ion" or electron, a classical (pre-quantum) particle largely extrapolated from Maxwell's theory. The goal was to set moving charges – the particulate character of electricity – within the frame of electromagnetic forces acting contiguously in the ether, a hugely ambitious undertaking since the connection of particle and field was even then an outstanding problem of theoretical physics.

The difficulties were formidable. First, Maxwell's theory did not clearly distinguish between charged matter and electromagnetic fields. Moreover, it appeared even to not allow the existence of primitive charged particles. For if electrons were charged point particles, susceptible to Newton's laws, they would, according to Maxwell's equations, possess infinite self energy. But then if they were finite extended bodies with negative charge distributed amidst the parts, then Maxwell's theory implied that such bodies should burst apart from forces of electrostatic repulsion. Finally, Maxwell's theory had nothing to say about why the electron's charge has the discrete value it has. By the beginning of the 20th century, however, Maxwell's elastic solid ether was no longer in vogue. Though retaining one core function of the earlier conception – the receptacle of electromagnetic energy – Lorentz in the early 1890s revived and elaborated Fresnel's conception of the ether as an immobile medium possessing neither velocity nor acceleration. For Lorentz, this ether was the carrier of contiguously acting electromagnetic forces and the medium through which charged particles influenced one another. It permeated all of space, being co-extensive with charged and even uncharged particulate matter. Entirely non-mechanical in nature, one could neither properly speak of the ether's mass nor of forces applied to it. By 1900 this immobile ether, though "undoubtedly widely different from all ordinary matter",[10] had replaced the empty space of Newtonian mechanics as the medium through which forces acted upon matter.

Though not yet by that name, the discovery of the electron was announced by J.J. Thomson at Cambridge in 1897, following upon immediately prior experimental and theoretical results of Pieter Zeeman and Lorentz in Leiden and Emil Wiechert in Königsberg.[11] By around 1900, physicists knew that a moving electrically charged particle carries with it an electrical field that contributes to the particle's inertial mass; more specifically, the particle's mass depends directly on its velocity. This not only implies an absolute difference in mass between a resting charged particle and one in motion, but it meant that mass itself had no longer the Newtonian meaning of "quantity of matter" but only the meaning "resistance to acceleration". Yet while mass retained the meaning of being the ratio between applied force and acceleration, minimally, the velocity-dependence of charged particles necessitated a modification of Newton's force law, $F = ma$. Theorists such as Wien and Lorentz speculated that not only the mass of the electron but mass in general was exclusively electromagnetic, not mechanical, in origin.[12]

Despite its unifying intent, Lorentz's electrodynamics was essentially dualistic in nature, with field and particle concepts both necessary and irreducible to one another. It comprised five equations of two distinct types: Maxwell's four partial differential equations, characterizing electric and magnetic fields seated in the ether, and a fifth equation, an ordinary differential equation, that Lorentz understood as expressing the force of the ether on charged particles. The latter has two force components, an electrical component acting on a charge at rest in the ether, and a velocity-dependent electrodynamic component acting on a charged particle moving with determinate velocity through the ether. This motion-dependent force acted on the spherically charged electrons and the intermolecular electrical forces comprising those bodies. Moving charged particles in turn generated changes in the ether – electric and magnetic fields – propagating outwards with the speed of light. Electrons were thus viewed as fundamental constituents underlying the macroscopic phenomena described by Maxwell's theory, producing all the electrical, magnetic and optical properties of magnets, conductors and indeed matter in general. Lorentz further hoped the theory might also explain chemical and thermal properties of matter. Although subsequent physics jettisoned the electron theory of matter soon after the advent of the quantum theory, Lorentz's contributions to electrodynamics are still celebrated in the *Lorentz force law*, understood as specifying the force a magnetic field exerts on a moving charged particle.

Freely conceding "we know next to nothing about the structure of an electron",[13] beginning in 1892 Lorentz nonetheless sought to explain the failure to detect the motion of the Earth through the ether by a compensating dynamical hypothesis, independently formulated in 1889 by the Irish physicist G.F. FitzGerald. According to the "Lorentz-FitzGerald hypothesis", the dimensions of individual electrons – and so the macroscopic bodies like the arms of Michelson and Morley's interferometer of which they are the microconstituents – are contracted in the direction of their motion through the ether by exactly the second-order v^2/c^2 quantities that Michelson and Morley failed to detect. In particular, Lorentz supposed electrons at rest in the ether to be perfect spheres, while the shape of an electron moving through the ether is deformed to a flatted ellipsoid of revolution, the axis in the direction of motion contracted in an amount varying directly with its velocity v of motion. Should it attain the velocity of light, the electron's limiting form would be a circular disk perpendicular to its line of motion.[14] Since electrons (in Lorentz's sense, all ordinary matter contains small spherical particles called electrons, both "positive" and "negative") are among the microconstituents of macroscopic bodies such as measuring rods, the latter are deformed in the direction of their motion through the ether in exactly the ratio of second-order quantities v^2/c^2 required to satisfy the light principle so that the measured velocity of light is indeed independent of the motion of the observer. Hence using optical means, motion through the ether cannot be detected. It is as if nature conspired to hide the effects of motion through the ether.

In the 1904 version of his electron theory, Lorentz incorporated a new relativity principle, presenting transformation equations between Maxwell's equations in all inertial frames (promptly designated, by Poincaré, the "Lorentz transformations" although they first appeared in a 1900 paper by Cambridge physicist Joseph Larmor [1857–1942]) that showed that electrical and magnetic experiments occur in the same way regardless of the inertial state of motion of an observer through the ether. The Lorentz transformations, as seen below, are also the mathematical core of special relativity. However, Lorentz's 1904 theory and Poincaré's further development of it in 1905 retained the ether, and so gave a different interpretation to the Lorentz transformations than in special relativity. These theories are, in Einstein's 1919 terminology, "constructive" rivals to special relativity. By this is meant that they sought to build up a satisfactory account of the phenomena of electrodynamics (increase of mass with velocity, length contraction, and failure to detect the effects of motion through the ether) from a hypothetical dynamics of the electron itself, regarded as the discrete elementary constituent of matter.

"On the electrodynamics of moving bodies"

Einstein's paper introducing special relativity was received on June 30, 1905 by the *Annalen der Physik*, appearing in September. At that time the *Annalen*, published in Berlin, was the leading journal of theoretical physics. Like its sister 1905 papers, Einstein's choice of publication venue was ambitious and proved to be highly strategic; one of the two chief editors of the *Annalen* was Max Planck, then Germany's leading theoretical physicist. Planck's support was immediate, and proved crucial in persuading other physicists to give Einstein's theory serious consideration. A mythology surrounds the paper: its largely informal style, its lack of textual citations, its acknowledgment only of one person, Swiss-Italian engineer Michele Besso (1873–1955), a friend of its unknown author, working obscurely as an assistant patent clerk in Bern. It is still regarded as a model scientific paper on account of its clarity of exposition, at least in its first, or kinematical, part.

The paper's title reflects not its content so much as Einstein's designation for issues pertaining to the open problem of ether drag, that is, whether when a body moves through the ether (e.g., the Earth in its orbit of the sun), it drags the ether along in its interior regions and those immediately surrounding its surface, or whether it simply traverses its path without any displacement of a stationary ether. This is just the question that the experiment of Michelson and Morley (1887) sought, and failed, to answer. Still there is no mention of Michelson and Morley's work in Einstein's paper (nor references to any published work by others) and, when questioned much later on, Einstein affirmed that in 1905 he had no knowledge of their experiment.[15] This seems highly unlikely since papers of Lorentz that Einstein surely had read and studied refer to Michelson and Morley. Ultimately, however, the issue is of little interest since Einstein's skepticism regarding the very existence of the ether was epistemological and methodological, rather than empirically motivated. It was voiced already in an 1899 letter to fellow student of physics Mileva Marić at the *Eidgenoessische Polytechnische Schule* in Zurich (later the ETH),

> I am more and more convinced that the electrodynamics of moving bodies as currently presented is not correct, and that it should be possible to present it in a simpler way. The introduction of the term "ether" into theories of electricity and magnetism leads to the notion of a medium of whose motion one can speak without, I believe, being able to associate any physical meaning with such a statement.[16]

The dismantling of the ether is precisely what the 1905 paper achieved, and here again, it is easy to draw parallels with both Hume and Mach: Hume for his empiricist deflation of the concept of cause, and Mach for his resolute skepticism of the Newtonian concepts of absolute space and absolute time, neither of which satisfied Mach's positivist scruples on physical meaningfulness. The alleged influences of Hume and Mach are considered below.

The paper is divided into a "kinematical" and an "electrodynamical" part. Attention here is restricted to the first "kinematical" part that is by far the most influential; the "dynamical" part is largely an exercise in algebra, setting both mechanics and the Maxwell-Lorentz theory of electrodynamics within the new kinematics of the principle of relativity. The relativity principle of mechanics is thus generalized to encompass optics and electromagnetism by showing how inertial frames transform according to what are now known as the Lorentz transformations (see below).

Einstein began by drawing attention to an asymmetry in theoretical explanation using a simple example of the relative motions of two bodies, a bar magnet and an electrical conductor, a wire loop (see Figure 5.2). The motions of these bodies with respect to one another produce the familiar phenomenon of electrodynamic induction: a detectable current flows in the wire. But theory in 1905 provided two distinct accounts of why the current was produced, depending on which body was considered at rest in the ether and which in motion. With the magnet moving uniformly through the ether and through the center of a stationary wire loop (case 1), a changing magnetic flux is generated at each point of space surrounding the magnet, giving rise (by Faraday's law of induction, one of Maxwell's

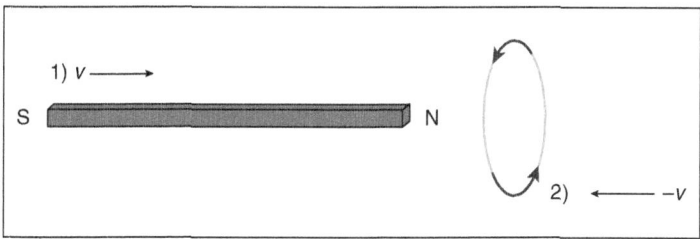

Figure 5.2

equations) to an electrical field, which in turn induces a current in the wire (by Gauss's law, another of Maxwell's equations). On the other hand, when the bar magnet is the stationary in the ether and the axis of motion through the center of the uniformly moving loop (case 2), there is no induced electric field. According to theory, the Lorentz force exerted by the bar's magnetic field on the free electrons of the wire conductor sets them in motion around the loop, creating a current. In either case, the same observable phenomenon – current in the wire – results from the relative motion of either body, depending only on the relative velocity $|\vec{v}|$ of magnet and loop. The unsatisfactory theoretical situation is summarized in the paper's first sentence: "Maxwell's electrodynamics . . . when applied to moving bodies, leads to asymmetries which do not appear to be inherent in the phenomena".[17]

The example provides an easily visualized demonstration of an already known fact: the Maxwell-Lorentz electrodynamics is not consistent with the Galilean relativity principle valid in classical mechanics. Einstein's diagnosis of why this was so, however, was original: the failure stemmed from a tacit but questionable supposition that the motion of a body with respect to the ether had theoretical significance. Denying this meant that the distinction between electric and magnetic fields could be only a relative one, depending on the state of motion of the observer; as Einstein wrote in 1919, "only the electric and magnetic field taken together could be ascribed some kind of objective reality".[18] It further suggests that electrodynamic bodies do not move with respect to the ether but only relatively to each other. Einstein accordingly stated the ether to be a "superfluous" hypothesis, adding that there was need for a generalization of the relativity principle to include optical and electrodynamic phenomena.

This perceptible lack of diffidence in dropping the ether hypothesis was surely enabled by his contemporaneous light-quantum hypothesis; after all, if light can be deemed to possess particle properties, there is no necessary reason to retain a medium for supporting propagation of waves. But at the heart of the new kinematics is a simple epistemological analysis of the physical meaning of the simultaneity of distant events. In what appears a *non sequitur*, Einstein observed that "insufficient attention" has been paid to the basic procedures by which notions of length, duration, and simultaneity are defined. The determination of time is the most fundamental:

> We have to take into account that all our judgments in which time plays a role are always judgments of simultaneous events. If, for instance, I say, "The train arrives here at 7 o'clock", that means something like this: "The pointing of the small hand of my watch to 7 and the arrival of the train are simultaneous events".[19]

The last sentence, among the most banal passages ever to appear in the pages of a respected scientific journal, has the intent of high-lighting a profound epistemological disparity between judgments concerning simultaneity of events occurring in the vicinity of one another, where the relation of simultaneity can be, as it were, "immediately" apprehended, and judgments regarding simultaneity of events separated at considerable distance. The reasoning goes as follows: A resting clock at a point of space p can unambiguously assign time to events occurring at p or in the immediate vicinity of p. But the time of an event occurring at a point q distant from p cannot be evaluated directly with this clock; "news" of its occurrence at q can reach p most efficiently by means of a ray of light, yet the time of reception of the light ray at p is not the time t_q of the occurrence of the event at q but a time greater than t_q by the time taken by light to travel from q to p. The obvious retort that an observer at p could infer t_q by knowing the one-way velocity of light from q to p is quickly seen to be circular as presupposing t_q is already known. In short, the Newtonian presupposition of time as the independent variable of events, assigned by a universal clock ticking absolute time, lacked evident physical meaning. In its place Einstein argued for a physically unambiguous definition of time, i.e., "from the standpoint of the measuring physicist". An unambiguous assignment of time to events occurring at arbitrarily distant spatial locations required a physical means of comparison; that is, a protocol of clock synchronization.

This proposal, defining time within an inertial system S by synchronizing resting clocks located at different points throughout S by light signaling, is essentially the same protocol for distant clock synchronization suggested by the French mathematician, physicist, and philosopher Henri Poincaré (1854–1912) in 1898.[20] A clock C_1 at rest at A in S can be synchronized with another identically constructed clock C_2 at any distant location B in S by light signaling. An observer emits a light ray at C_1 time t_1 (event e_A) toward B, such that it arrives at B (event e_B) then is immediately reflected back to A, arriving at C_1 time t_2 (event e_C). Then observers at A and B can synchronize their clocks by agreeing that the event e_D at A, occurring at C_1 time $(t_1 + t_2) / 2$, is simultaneous with e_B as measured by C_2 at B. This is equivalent to a stipulation that the travel times of the light ray in each direction are the same (hence that space is isotropic). By following this rule, identically constructed clocks located at all points of S can be synchronized throughout S. The protocol relies on an assumption that light is the fastest possible means of signaling between any two points of space and the absence of any physical method to measure the one-way velocity of a light ray that does not already presuppose a determination of time.

Relativity of simultaneity

The simple and highly intuitive procedure for clock synchronization hardly requires comment. But the simplicity is deceptive if one considers that what has been obtained is a time determination made by clocks at rest within S. According to the light principle, the velocity of light c is the same in S and in S', a system moving uniformly with respect to S. Then exactly the same operations of clock synchronization can be carried out within S' (whose clocks always retain the same separation from one another, and so are "at rest" with respect to one another). And this means there is no *a priori* guarantee that the temporal relations between two events e_1 and e_2 as judged by an observer in S will always be the same as judged by an observer in S'. A simple example shows this (Figure 5.3).

Consider a system S' consisting of three rocket ships R_1, R_2, R_3 traveling with uniform velocity $\bar{v}$ in this order, and in the same direction, maintaining the same constant distance between them. At time t_1', the middle ship R_2 emits a light pulse in all directions. According to the light principle, an observer in R_2 will judge the events e_1 and e_2 (light arriving at ships R_1 and R_3, respectively) to

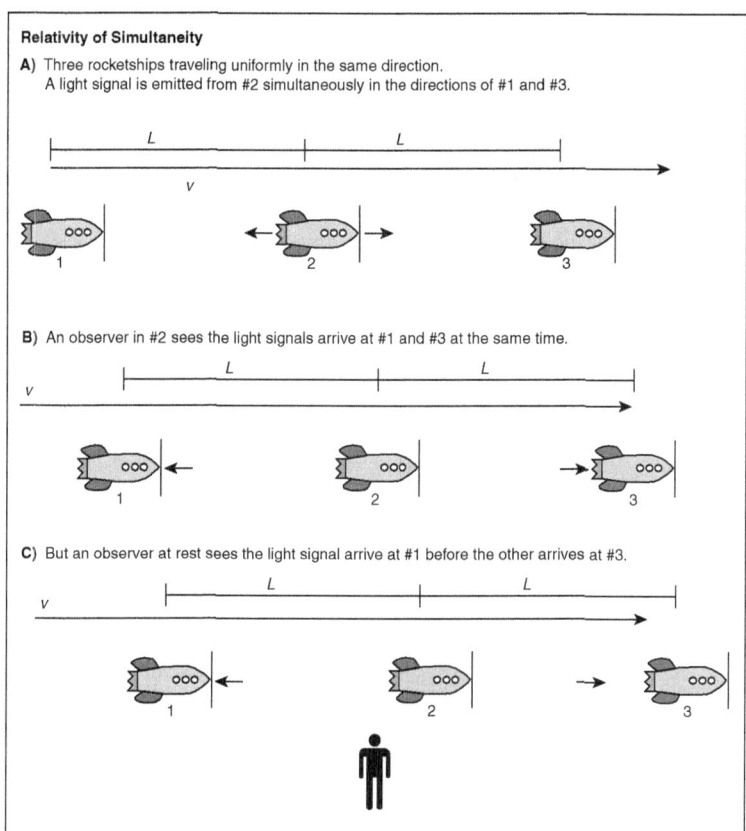

Figure 5.3

be simultaneous. But to an observer at rest (Figure 5.3C, to whom the rocket ships are moving by to the right), e_1 will occur before e_2 since this observer will see R_1 approaching the oncoming light signal with velocity $\bar{v}$ and R_3 as receding with the same velocity from the light signal overtaking it. This is the relativity of simultaneity: two events judged simultaneous by one observer may be judged to occur sequentially by another observer in uniform relative motion with respect to the first.

Time dilation and length contraction

Changing the concept of simultaneity alters other kinematical concepts in its turn. Given the constancy of the velocity of light, consider that a light signal reflected back and forth between mirrors attached to the endpoints of a rigid rod (distance L) comprises an ideal, if impractical, clock. Each round trip between the mirrors defines one "tick". A clock considered at rest in its own rest frame can then be represented this way (see Figure 5.4). If L is the length of the rod,

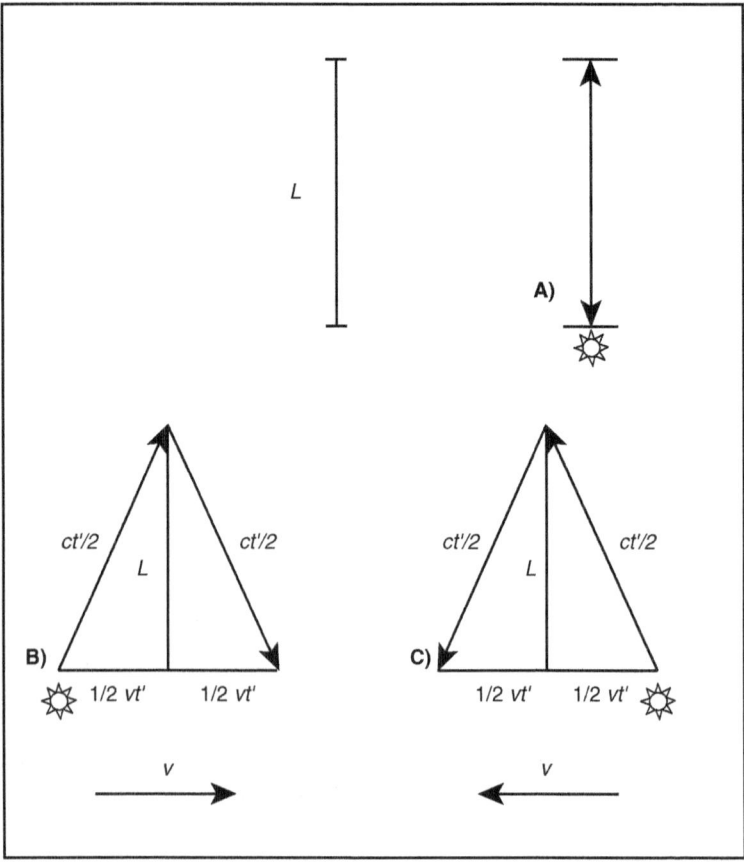

Figure 5.4

then the round-trip time for light to travel to the top mirror and back is $t = 2L/c$. Suppose three such clocks, one considered "at rest" in a laboratory (A), one considered by the laboratory observer to be moving uniformly to the right in a direction perpendicular to the rod's length (B), and one considered by this observer to be moving with the same uniform velocity but to the left (C). Designate the time between ticks of the moving clock t'. Then the distance the light has to travel from the bottom to the top mirror is $ct'/2$ while the rod travels v times the time elapsed on this halfway journey, $t'/2$. One can see that the path of light for a clock considered moving is the hypotenuse of a right triangle whose other sides are L and $\frac{1}{2}vt'$. The Pythagorean theorem can be used to calculate the relation between t and t'. Then $\dfrac{ct'}{2} = \sqrt{L^2 + \left(\dfrac{1}{2}vt'\right)^2}$ and solving for the unknown t',

$$t' = \frac{2L}{c\sqrt{1 - v^2/c^2}}$$

As $t = 2L / c$, the relation between t' and t is

$$t' = \frac{t}{\sqrt{1 - v^2/c^2}} \ .$$

Since $\sqrt{(1 - v^2/c^2}$ is less than 1, t' is greater than t and so to a resting observer the moving clocks appear to tick more slowly. But relativity is symmetrical, and either of B or C can consider themselves "at rest" so that the geometry of their situation is judged to be A. The statement that "moving clocks run slow" often encountered in discussions of relativity always means: as determined by an observer who considers himself or herself 'at rest'. The less misleading formulation is that any inertial observer will determine the rate of a clock moving uniformly with respect to him or her to be less than the rate of a clock that is stationary with respect to him. A clock in a moving frame will be seen to be running slow or "dilated" (according to the Lorentz transformations; see below) whereas the time measured in the clock's own rest frame is the "proper time" of the clock.

In relativity theory, determination of the length of any object amounts to the simultaneous measurement of its initial point and endpoint; that is, a measurement of its length at an instant. The word "simultaneous" cues that measurements of length will depend on the observer's inertial frame of reference. It is easiest to see this using a simple space-time diagram and the primitive language of "events". Here the initial point and the endpoint of a measuring rod at rest are spatially separated events occurring at the same time. A meter stick will be 1 m in length in its own rest frame. But what length would a moving stick have to an observer at rest? According to the relativity of simultaneity, the resting observer, noting the two events when the two ends of the rod pass by, will measure the meter stick as less than one meter in length.

Indeed, the faster the stick travels by, the shorter it appears to be. So in the same way that an observer "at rest" will judge a moving clock to run more slowly than his clock, the relativity of simultaneity leads to the relativity of length. The four-dimensional geometry of space-time, formulated by Polish-German mathematician Minkowski in 1907, enables a graphical representation of the kinematical relations between inertial observers. Figure 5.5 illustrates a space-time diagram in only two dimensions – time, and the one spatial dimension – representing the system S of an observer and meter stick "at rest" (solid lines, the t − x system) as well as another system S′ (dashed lines with primed coordinates) that S regards as

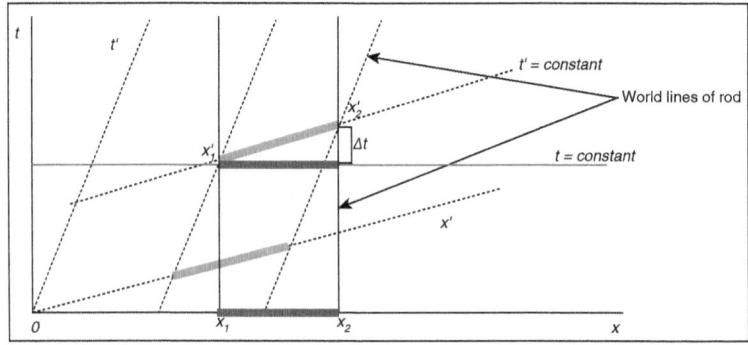

Figure 5.5

meter stick moving uniformly past the observer. The frames have a common origin at O.

Representing four dimensions in two yields inevitable distortion; in particular, for both observers the space and time axes are orthogonal. The two systems obviously disagree on which events occur at the same time. For the observer in S, all events taking place along the x-axis (or on every line parallel to it) are simultaneous; for a moving observer S′, all events on the tilting x′axis (or its parallels) are simultaneous. In other words, observers in S and S′each experience an event at O at the same time $t = t′ = 0$ but disagree on which events are simultaneous with O.

In relativity, the length of an extended object at rest is defined as the distance between its endpoints at a given instant. Consider the measuring rod lying along the x-axis having length $x_2 - x_1$ at $t = 0$. The two vertical red lines in the diagram are the "world lines" in space-time of the endpoints of this rod at rest in S; at each time t, its endpoints are the same distance apart. Similarly in S′, the measuring rod always has the same length $x_2′ - x_1′$. But to an observer in S, the moving meter stick in S′ (the t′-x′ system) will intersect the world lines of the ends of the rod in S at two non-simultaneous instants that occur Δt apart. In the S′ system, these points have the coordinates $(x_1′, t′)$ and $(x_2′, t′)$ with the same t′. There is a general formula first derived by Lorentz to compare these two lengths: where $(x_2 - x_1)$ is L_0 (length of the stick at rest in the S frame) and $(x_2′ - x_1′)$ is L (length of the moving rod as measured by S), and v the velocity of the moving primed system with respect to the rest system, the relation of the two lengths is

$$L = L_0 \sqrt{1 - v^2 / c^2}.$$

In particular, the length of the moving rod L as determined by S is shorter than L_0, its length as measured in the rest system. That this does not appear the case from the diagram is an inevitable distortion of space-time diagrams in two dimensions. Nonetheless, space-time diagrams make it easy to visualize the reciprocity between measured lengths in two inertial frames S and S′. Simply imagine the two frames rotated counterclockwise so that the world lines of the rod in S′are now vertical while maintaining the same angle with S.

Then S will be tilted at the same angle but in the opposite direction as previously S′ was, indicating motion with respect to resting S′ but in the opposite direction.

The "clock paradox"

Equivalence of observers in uniform relative motion coupled with the idea that "moving clocks run slow" led to the following "paradox" first formulated by French physicist Paul Langevin (1872–1946) in 1911. Suppose identical twins Peter and Paul (Figure 5.6) are together at P, each with identically constructed clocks. Bringing his clock with him, Peter takes a high-speed journey (at a considerable fraction of c; notice the light cone at P representing a flash of light) on a rocket ship to a distant location R, then turns around and rejoins Paul who has remained at home. On meeting again at Q, Paul has aged more than Peter.

This result appears paradoxical because according to relativity, each twin may view the other twin as traveling, and so each should judge the other to be younger.

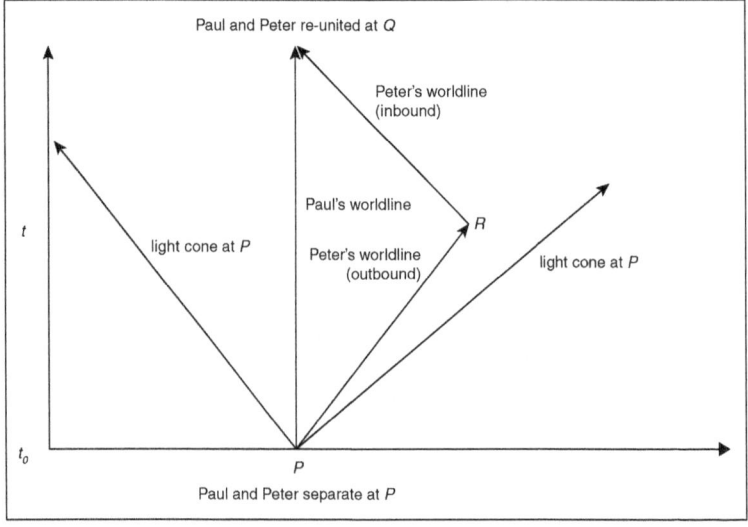

Figure 5.6

A common explanation of the "paradox" is that the contradiction disappears when one considers that Peter has undergone several accelerations and decelerations on his journey and so has occupied several distinct inertial frames, whereas the stay-at-home twin Paul has occupied but one. However, as early as 1913 German physicist Max von Laue (1879–1960) argued that any effect on Peter's clock (and so on Peter) during the periods of acceleration could not account for the asymmetrical age difference between the twins, since stretches of Peter's uniform motion between P and R and between R and Q can be made arbitrarily long as compared to the initial and final accelerations and the turn-around time.[21]

In fact, the resolution of the "paradox" is much simpler. Recalling the notion of "proper time" (time measured by a clock in its own rest frame), the invariant length of a world line (a space-time curve) between two events is, by hypothesis, proportional to the number of ticks of a clock carried along the world line measuring time elapsed between the two events. Since a resting observer has the longest space-time trajectory between two events, it follows that Paul's clock ticks more times than Peter's simply because Paul's world line between P and Q is longer in space-time than Peter's traveling world line. Again in this regard, two-dimensional space-time diagrams are inevitably misleading, but in the four-dimensional geometry of Minkowski space-time, the triangle inequality PR + RQ ≤ PQ is valid, testifying to a departure from Euclidean geometry. Paul's world line is longer.

Lorentz transformations and the space-time interval

In Einstein's kinematical interpretation, the differences between the length and time measurements made by co-moving inertial observers are codified in the Lorentz transformations between space and time coordinates in the different inertial frames of reference. The Lorentz transformations, the mathematical core of special relativity, are linear transformations of the coordinates (x, y, z, t) of a given event to new coordinates (x', y', z', t'), representing those assigned by an observer in a frame of reference moving with constant velocity with respect to the first (unprimed) frame. Although the Lorentz transformations may be written in different ways, depending on the relative (uniform) motions between the primed and the unprimed frames of reference, the simplest example is of a system S and a system S′that S considers moving with uniform velocity v along the x-axis:

$$t' = \gamma\left(t - \frac{vx}{c}\right), \ x' = \gamma\left(x - \frac{vt}{c}\right),$$

$$\gamma = \frac{1}{\sqrt{1 - \frac{v^2}{c^2}}} \ ,$$

with $y' = y$, $z' = z$, and c the speed of light in vacuo. γ is essentially a factor of proportionality between any velocity v less than c and c.

Any theory whose results remain valid under Lorentz transformation is a relativistic theory, while the theory's laws, retaining the same form under such transformation, are called *Lorentz covariant*. The defining property of a Lorentz transformation, first identified by Poincaré but generally attributed to Minkowski, is that it is the unique linear transformation for which an algebraic identity I^2 holds between the quadratic space and time coordinates in the two frames of reference:

$$I^2 = \left(t'^2 - x'^2 - y'^2 - z'^2\right) = \left(t^2 - x^2 - y^2 - z^2\right)$$

The above algebraic expression combining the space and time coordinates on each side of the identity sign is neither a quantity of space nor of time but of space-time. It is called the *space-time interval*, and its significance is apparent: as invariant with respect to the Lorentz transformations, it is a quantity that will be found to be numerically the same by any inertial observer. Thus it enables observers measuring lengths and times in their respective frames to compare their results with those of other inertial observers. Despite the relativity (or frame dependence) of measures of lengths and times (indeed, all physical quantities with the dimensions of lengths or times), the invariance of the interval shows that relativity theory is really misnamed. Not by Einstein. He had proposed the term *invariant theory* before learning that this term had already been appropriated within pure mathematics. The designation relativity theory is due to Max Planck, who sought to emphasize the universal scope of the principle of relativity. Planck's term caught on, and very quickly it was too late for any rebaptism.

Conclusion

Dropping the ether as a "superfluous" hypothesis, Einstein's new kinematics began with a simple epistemological analysis of the physical meaning of the simultaneity of distant events. The analysis engendered a chain of increasingly radical implications. First it entailed that there is no "true" rest frame, so that the velocity of light is the same *in all inertial frames*, a modification of the light principle of ether-based theories of electrodynamics. In consequence, the familiar parallelogram law of addition of velocities of classical mechanics required modification when one of the composed velocities is that of light. Then, Einstein derived the Lorentz transformations from the two explicit postulates, the relativity principle and the light postulate, the latter now making no reference to the ether. And he succeeded in showing that the relativity principle, now based upon the Lorentz transformations, encompassed both mechanics and Maxwellian electrodynamics and optics. But while these transformations have the same mathematical form in Einstein's theory as in those of Lorentz and Poincaré, they have a completely different interpretation. The differences between lengths and durations measured by a "moving" observer and one considered "at rest" – the supposed Lorentz-Fitzgerald dynamical contraction of matter in the direction of its motion through the ether – are explained by Einstein simply as kinematic consequences of the differences between measurements of lengths and durations defined by observers in relative inertial motion. Only the relative velocity of these observers moving uniformly with respect to one another is a valid concept. The resultant "theory of special relativity" is essentially then a revamped kinematics, an account of the behavior of rigid rods and perfect clocks moving uniformly relative to one another in a new arena that, since the geometrical reformulation of special relativity devised by Minkowski in 1907 to be known as "space-time".

Notes

1 Einstein, "On the Electrodynamics of Moving Bodies", *Annalen der Physik* v. 17 (1905), CPAE 2 (1990), Doc. 23, pp. 275–306; p. 277; English Supplement, pp. 140–71, p. 141.
2 Galileo Galilei (1632): "Shut yourself up with some friend in the main cabin below decks on some large ship, and have with you there some flies, butterflies, and other small flying animals. Have a large bowl of water with some fish in it; hang up a bottle that empties drop by drop into a wide vessel beneath it. With the ship standing still, observe carefully how the little animals fly with equal speed to all sides of the cabin. The fish swim indifferently in all directions; the drops fall into the vessel beneath; and in throwing something to your friend, you need throw it no more strongly in one direction than another, the distances being equal; jumping with your feet together, you pass equal spaces in every direction. When you have observed all these things carefully (though there is no doubt when the ship is standing still everything must happen in this way), have the ship proceed with any speed you like, so long as the motion is uniform and not fluctuating this way and that. You will discover not the least change in all the effects names, nor could you tell from any of them whether the ship was moving or standing still". (*Dialogue Concerning the Two Chief World Systems*; Second Day, S. Drake translation, University of California Press, 1953, pp. 186–7).
3 Newton, Isaac. *The Principia. Mathematical Principles of Natural Philosophy*. A New Translation by I. Bernard Cohen and Anne Whitman. Berkeley-Los Angeles-London: University of California Press, 1999, p. 423.
4 "Does the Inertia of a Body Depend on Its Energy Content?", CPAE 2 (1990), Doc. 24, pp. 311–15; p. 314.
5 "On the Relativity Principle and the Conclusions Drawn From It", CPAE 2 (1990), Doc. 47, pp. 432–88; p. 464.
6 Ohanian, Hans C., "Did Einstein Prove E = mc²?", *Studies in History and Philosophy of Modern Physics* v. 40 (2009), pp. 167–73.
7 "Autobiographical Remarks", p. 58; p. 59.
8 Einstein (1953), "H.A. Lorentz, His Creative Genius and His Personality", in G.I. de Haas-Lorentz (eds.), *H.A. Lorentz, Impressions of His Life and Work*, Amsterdam: North Holland Publishers, 1957, pp. 5–9; p. 5.
9 "Über die Möglichkeit einer elektromagnetischen Begündung der Mechanik", *Annalen der Physik* v. 5 (1901), pp. 501–13.
10 Lorentz, Hendrik Antoon, *The Theory of Electrons: A Course of Lectures Delivered in Columbia University, New York, in March and April 1906*. Leipzig, Germany: B.G. Teubner, 1909, pp. 30–1.
11 See Arabatzis, Theodore, *Representing Electrons: A Biographical Approach to Theoretical Entities*. Chicago: University of Chicago Press, 2006.
12 See McCormmack, Russell, "H.A. Lorentz and the Electromagnetic View of Nature", *Isis* v. 61 (1970), pp. 459–97.
13 Lorentz, *The Theory of Electrons*, p. 216.
14 Lorentz, *The Theory of Electrons*, p. 212.

15 Shankland, Robert S., "Conversations with Albert Einstein, *American Journal of Phys-ics* v. 32 (1964), pp. 47–57.

16 *CPAE* 1 (1987), Doc. 52.

17 *CPAE* 2 (1990), Doc. 23.

18 *CPAE* 7 (2002), Doc. 31.

19 *CPAE* 2 (1990), Doc. 23.

20 Poincaré, Henri, "La mesure du temps", *Revue de Métaphysique et de Morale* v. 6, pp. 1–13; translation by G.N. Halsted as "The Measure of Time", in Poincaré, *The Foundations of Science*. New York: The Science Press, 1929, pp. 223–34.

21 Von Laue, Max, "Das Relativitätsprinzip", *Jahrbücher der Philosophie*. Berlin: Mittler & Sohn, 1913. See the discussion in Miller, Arthur I., *Albert Einstein's Special Theory of Relativity: Emergence (1905) and Early Interpretation (1905–1911)*. Reading, MA: Addison-Wesley Publishing Co., 1981, pp. 261–2.

Further reading

Janssen, Michel, "Appendix: Special Relativity", in M. Janssen and C. Lehner (eds.), *The Cambridge Companion to Einstein*. New York: Cambridge University Press, 2014, pp. 455–506.

Maudlin, Tim, *Philosophy of Physics: Space and Time*. Princeton, NJ: Princeton University Press, 2012.

Six
General relativity

"Even if no deviation of light, no perihelion advance and no shift of spectral lines were actually known, the gravitation equations would still be convincing, because they avoid the inertial system (this ghost that acts on everything but on which things do not react)".

Einstein to Max Born, May 12, 1952.[1]

Introduction

The general theory of relativity, Einstein's premier contribution to physics, displaced Newton's law of gravity, for two hundred years the acme of classical mechanics and epitome of an exact law of nature. General relativity remains the cornerstone of the contemporary understanding of gravity, even though it is a classical, not a quantum, field theory. Today, most cosmological models of a homogeneous and isotropic expanding universe are based on an exact solution of Einstein's field equations of gravitation, the Friedmann-Lemaître-Robinson-Walker metric. But general relativity was largely ignored by physicists (though not by astronomers) after the advent of quantum mechanics in 1925–1926; on account of the paucity of empirical data supporting it, general relativity was taught for decades mainly in mathematics departments, as having more mathematical than physical significance. A renaissance occurred in the 1960s with the discovery of the cosmic microwave background (CMB) signature of the "Big Bang" (1964), followed by the theorems of Roger Penrose, Stephen Hawking, and Robert Geroch on the existence of singularities in a wide class of solutions to Einstein's field equations, and John A. Wheeler's coining of the term "black hole" in 1967 (see Chapter 11). Gravitational radiation, a prediction of the theory in 1916, was not observed for nearly one hundred years (in 2015). Yet to those physicists who knew it well, the theory always retained its luster; to Paul Dirac, it was "probably the greatest scientific discovery ever made" while Russian Nobel physicist Lev Landau pronounced it "probably the most beautiful of all existing physical theories".[2]

Born in the throes of World War I, general relativity was communicated to the wider world in London on November 6, 1919 following the theory's confirmation by data gathered by English scientists from a solar eclipse the previous May. The announcement was made at a rare joint meeting of the Royal Society of London and the Royal Astronomical Society, in an atmosphere of intense expectation compared by the philosopher Whitehead to that of a Greek drama. Standing under a portrait of Newton, J.J. Thomson, Nobel laureate and President of the Royal Society, pronounced Einstein's gravitational theory "one of the highest achievements of human thought". Coming just a year after the Armistice of November 11, 1918 ended an unimaginable slaughter in the trenches of Flanders and France, the public found irresistible solace, even romance, in the fact that the scientific establishment of Great Britain showered its highest accolades on a hitherto obscure Berlin physicist whose theory had surpassed the defining achievement of England's greatest scientist. To a world weary of war, revolution, and a raging influenza pandemic, Einstein and his incomprehensible theory quickly became a spectacle possessing seemingly inexhaustible potential to sell journalistic copy. In short order, Einstein acquired worldwide renown and the mystique of iconoclastic genius, accolades he did not seek but was never to lose, even in death.

A tortured path

Although possibly contravening details are lost in the mists of history, Einstein appears to have formulated special relativity in a flash of insight, the theory emerging fully grown suddenly in the first half of 1905, Minerva springing from the brow of Jove. Such was not the case with general relativity; its author, already widely known, left an extended and considerable trail of evidence. Eight years in gestation, the theory was completed in Berlin in late November 1915 in a frenzied race with Göttingen mathematician David Hilbert. Beginning in 1907, his last year working as a patent clerk in Bern, Einstein sought to extend the principle of relativity to non-inertial systems, i.e., systems in non-uniform motion. Persuaded by Ernst Mach's rejection of Newton's absolute space as the agency responsible for the effects of inertia, Einstein attempted to construct a theory of relativity applicable to all motions, hence the name general relativity. Straightforward analogy to the principle of *special* relativity (asserting the equivalence of all inertial frames of reference for the non-gravitational laws of physics) would have it that a principle of *general* relativity correspondingly affirms the equivalence of *all* reference systems (both inertial and non-inertial) for *all* laws of physics. That is, just as velocities are relative to an observer's frame of reference within special relativity, so according to this putative principle, accelerations are to be relative to an observer's frame of reference within general relativity. Yet the analogy cannot be correct, for it is tantamount to ignoring the press of inertial forces familiarly experienced by any rapidly accelerated observer (e.g., riding a roller coaster).

Too-close analogy to special relativity proved misleading in another crucial respect. Recall that a covariance requirement pertains merely to the form of the equations under consideration. In the theory of special relativity, satisfaction of a covariance requirement (Lorentz covariance – the laws of non-gravitational physics retain the same form under Lorentz transformations between inertial frames) implies a relativity principle; namely, that these laws are the same in all inertial frames of reference. After learning Riemannian geometry in Zurich in 1912–1913, Einstein was swayed by the power of analogy into thinking that the requirement of *general* covariance, i.e., that the laws of physics (now including gravitation) must be tensor expressions (see Box 6.1) that retain the same form in all coordinate systems, was mandated by a principle of relativity generalized to accelerated motions. The thought was this: to essentially eliminate any privileged roles for observers or coordinate systems, it was necessary to put all observers and coordinate systems on an equal footing, i.e., to impose the requirement of general covariance. From a contemporary point of view, it is easy to catalogue Einstein's various conflations, of the relativity of inertial mass with relativity of frames of reference, of relativity principles with covariance requirements, and of "observers" (or "reference frames") and "coordinate systems".

Box 6.1   Tensors and general covariance

The concept of vector belongs to linear algebra and should be familiar from elementary physics; vectors represent quantities possessing both magnitude and direction (e.g., force or velocity). A vector quantity is expressed by a set of components (that can be written as a row or column matrix); these are projections of the vector onto basis vectors (for the given coordinate system describing the points of the space) that lie orthogonally along each axis of the space. Since the components depend on the coordinate system, they must change when the coordinate system is changed; to obtain the components in a new coordinate system, the components in the old system are multiplied by a matrix expressing the linear change from one set of basis vectors to another with the same origin. A vector has a single index that indicates how its components linearly transform when changing the coordinate system in which the components are expressed. If the index is "downstairs" (written as a subscript), the components transform like the basis vectors and the vector itself is said to be "covariant". If the index is "upstairs" (written as a superscript), the vector is said to be "contravariant", which simply means that its components transform by a matrix that is inverse to the matrix that transforms the basis vectors.

Tensors are a mathematical generalization of the concept of vector (vectors and scalars are particular cases: vectors being tensors of rank one, scalars of rank zero); they represent more complicated quantities, involving, as it were, several vectors at once. They possess two or more indices; like vectors, these can be downstairs (covariant) or upstairs (contravariant) or, where there are two or more, both ("mixed tensor"). Whether covariant or contravariant, both left-hand and right-hand sides of an equation written in tensor form have the same transformation law with respect to changes of coordinates; the equation will have the same form when transformed to any other set of coordinates (with some technicalities governing the transformation).

A vector A at a given point p of space-time (labeled by four coordinates, x_1,x_2,x_3,x_4) has component A^1 with respect to the axis x_1, component A^2 relative to axis x_2, A^3 relative to x_3, and A^4 relative to x_4. More compactly: the contravariant vector A has four components A^μ ($\mu = 1,2,3,4$). In a second coordinate system, x'_ν (x'_1,x'_2,x'_3,x'_4), the components of vector A at the same point p are represented A^ν ($\nu = 1,2,3,4$). The transformation machinery of vectors and tensors provides an automatic transition from a representation of A in one coordinate system x_μ to another x'_ν (and vice versa), *accounting simultaneously for changes in the coordinate system and in the components of* A. Tensors generically have more components (the fundamental, or metric, tensor $g_{\mu\nu}$ of general relativity has 16, it is a 4 × 4 matrix) but the transformation properties are automatic. Tensorial quantities transform invariantly between all "suitably smooth" space-time coordinate systems; physical laws written solely in their terms have the same form in all of them, hence are said to be *generally covariant*.

To complete the general theory of relativity, Einstein would have to come to a new understanding of the requirement of general covariance. This did not completely sever the faulty implication from covariance condition to relativity principle that he retained more or less to the end of his life, but it did require him to "unlearn" the very lesson that guided him to the theory of special relativity, that by using postulated rigid rods and ubiquitous synchronized resting clocks, the resulting coordinates could be used by themselves to identify events in space and time, i.e., that space-time coordinates had an immediate physical significance. Much later, in 1946, he admitted this message to be the "main reason" for the lengthy gestation period of the general theory of relativity:

> This (the discovery of the principle of equivalence) happened in 1908. Why were another seven years required for the construction of the theory of general relativity? The main reason lies in the fact that it is not so easy to free oneself from the idea that coordinates must have an immediate metrical meaning.[3]

How in fact Einstein freed himself from this idea is one of the more storied episodes in history and philosophy of physics, for it involves entering into the meaning, and philosophical significance, of general covariance, a discussion that continues today.[4]

Milestones on the way

At first, progress was immediate: Sitting at his desk in the patent office in Bern (on January 7, 1907, to be precise), Einstein had what he later described as "the most fortunate thought of my life", that "the gravitational field has only a relative existence, since *for an observer freely falling from the roof of a house no gravitational field exists while he is falling*".[5] If an unfortunate roofer drops a hammer while he falls, the tool descends at the same rate of acceleration (at sea level, 9.8 meters/sec^2) and so remains – relative to him – at rest. Since Galileo it was known that if air resistance is neglected, all bodies regardless of material composition fall in the same way with constant acceleration toward the Earth. But this largely ignored fact suggested to Einstein that a freely falling observer would be justified in assuming that he occupied a mini-inertial system. Alternatively, an observer enclosed within a box (unbeknowst to him, at sea level on Earth) could consider objects released from shoulder height to be inertial bodies whilst he himself was accelerating, the elevator floor pressing against his feet with a force of 9.8 m/sec^2. Just this glimpse of recognition that acceleration (relative to one observer) might cancel freefall (according to another) and so be in some sense equivalent to gravitation, led Einstein to infer that a gravitational field, if only in the vicinity of the freely falling body, has a relative existence. That insight, deemed the *principle of equivalence* (Figure 6.1), proved the

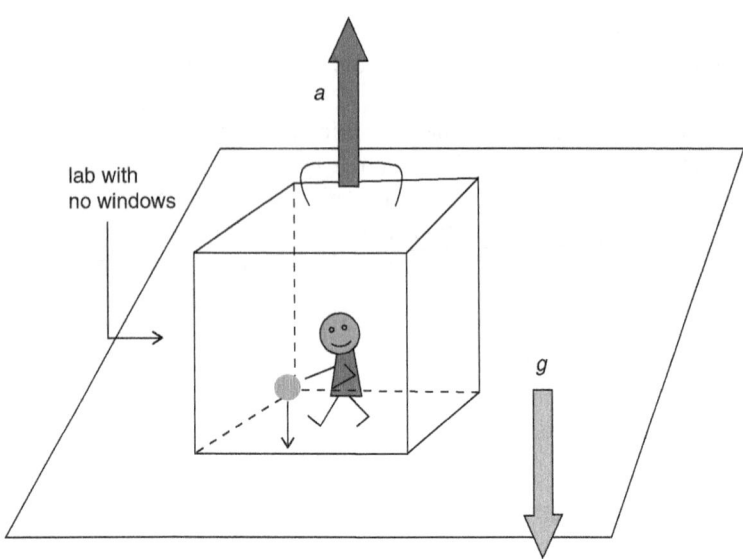

Figure 6.1 Principle of Equivalence 1907: Uniform acceleration a without a g-field is equivalent to being at rest in a homogeneous g-field. Experiments within the box cannot distinguish a from g.

invaluable heuristic on which the development of the new theory would depend. For if the principle of relativity is to be extended to systems whose relative motion is non-uniform, even systems in relative uniform acceleration, then the phenomenon of gravitation must be taken into account.

Other guidelines also were clear from the beginning. Any relativistic theory of gravitation had to be compatible with the abolition of Newtonian absolute time and simultaneity already demanded by special relativity. Moreover, unlike the theoretically instantaneous gravitational attraction between any two bodies in Newtonian theory, special relativity mandated that the effects of gravity propagate no faster than the speed of light. On the other hand, Newtonian planetary astronomy is highly accurate, so some close approximation of Newtonian gravity must be recoverable within the relativistic theory. Indeed, the enormous empirical success of Newtonian gravitational theory deterred nearly everyone else from seeking another.

With but a single known observational anomaly, the tiny 43' of arc per century unaccounted for in the precession of Mercury's orbit, the Newtonian theory had long been the firmest arch supporting physicists' world view. The understandable opinion, that where there is nothing really wrong there is nothing to rectify, was nearly universal. On traveling to Zurich in 1913 to learn whether Einstein would consider a move to Berlin, Max Planck used the opportunity to give Einstein a bit of advice. An initial and enthusiastic supporter of special relativity, Planck frankly urged Einstein against wasting his time on gravitational theory and to turn his talents elsewhere:

> As an older friend I must advise you against it for in the first place you will not succeed; and even if you do succeed, no one will believe you.[6]

A crucial step occurred in 1912 when Einstein came to realize that a relativistic theory of gravitation would require a non-Euclidean geometry for the appropriate mathematical description of the gravitational field. The realization came from a very simple example prompted by the principle of equivalence, extending the principle of relativity from uniformly moving to uniformly accelerated systems. In place of systems accelerating uniformly in a rectilinear direction, Einstein considered a rigid circular disk uniformly rotating in a Minkowski space-time, a non-inertial system in a region free of gravitational fields. The disk is a circle in the x-y plane in an inertial "Galilean system" K, and it is a circle in the x'-y' plane of a system of reference K' rotating with uniform velocity with respect to K, where it is supposed that the origin and z-axis of two systems of reference coincide. The question then is: What is the geometry of the disk? An observer in K, measuring with rods at rest in K, will find that the ratio of the circumference of the disk to its diameter to be π, concluding the geometry of the disk to be Euclidean. But in the context of special relativity, rods at rest in K' determine a quotient different from π. From the standpoint of the observer in K, measuring rods laid out along the periphery of the rotating disk appear to suffer Lorentz contraction, slightly shortening in the direction of their rotational motion, while measuring rods laid out along the diameter are not affected since these are perpendicular to the direction of motion.

Since more rods are then needed to span the circumference while the diameter remains the same, the ratio of the former to the latter is greater than π when both distances are measured within the frame in which the disk is at rest. Einstein concluded that Euclidean geometry has to be abandoned in a relativistic gravitational theory. This led him to the study non-Euclidean geometries, a subject that came to fruition in the second half of the 19th century (see Chapter 7).

In 1913, Einstein, recently arrived at the ETH, collaborated with his friend from school days, Marcel Grossmann, now a mathematics professor also at the ETH, to produce an initial "Outline (Entwurf) of a Relativistic Theory of Gravitation".[7] On leaving Prague in 1912, Einstein immediately sought out Grossmann for mathematical assistance with geometries of curved spaces. Grossmann obliged by introducing Einstein to the most general of these, Riemannian geometry, in its natural mathematical setting, tensor calculus, and to the so-called Riemann (or Riemann-Christoffel) tensor, the most general expression for the curvature of a Riemannian space. The Entwurf has a "mathematical part" (under Grossmann's name) and a "physical part" (under Einstein's); Einstein's main result is to have found a field equation for the gravitational field, showing how the field is generated from its sources, matter and energy. This equation featured an essential aspect of the final field equations formulated two years later in November 1915; it contains a tensor (the Ricci tensor, see below) representing the curvature of space-time, derived from the fundamental curvature tensor of Riemannian geometry. The Entwurf field equation is inherently generally covariant by this fact. Einstein, however, erroneously believed the covariance of the equation had to be restricted (by imposing special coordinate conditions) for two distinct reasons: to arrive as a first approximation at the Newtonian theory of planetary orbits (the Newtonian limit) and additionally to satisfy the conservation laws of energy and momentum. He subsequently concocted a spurious argument (the "hole argument", considered below) purportedly showing that the field equations in any case could not be generally covariant, if they were to be deterministic (in the manner of Maxwell's equations) by enabling univocal prediction of the values of field strengths from the initial data of matter and energy sources. As Einstein also erroneously thought, the principle of general covariance was required by a principle of general relativity, i.e., the extension of the principle of relativity to accelerating frames of reference, this came as a heavy blow. For some two years he believed that a relativistic theory of gravitation could not be generally covariant, i.e., that the field equations could not keep the same form in all coordinate systems.

Breakthrough

In 1915 Hilbert invited Einstein to lecture in Göttingen on his progress, and in late June and early July he did so, entering into extensive discussions with Hilbert. Champion of the axiomatic method, Hilbert may well have seen that general covariance was an achievable goal, ideal from the axiomatic standpoint since in accordance with the spirit of Riemannian geometry it removed any metrical significance from space and time coordinates. In any case, by autumn Einstein was engaged in a "race" with Hilbert, now similarly engaged in the project of finding generally covariant field equations of gravitation.[8] By November, Einstein had returned to the Riemann tensor, giving four presentations to the Prussian Academy of Sciences in Berlin; one at each of the weekly Thursday sessions of the section for mathematics and physics (Nov. 4, 11, 18, and 25).[9] On November 4, he presented a variation of the 1913 Entwurf theory, with a coordinate restriction to ensure energy conservation. On November 11, he introduced into the previous week's theory the hypothesis that the only matter sources of gravitation are electromagnetic in origin, essentially the electromagnetic theory of matter that Hilbert was concurrently attempting to couple with gravitation. In the third presentation on November 18, he used the not-yet completely covariant theory to calculate the observed tiny anomaly in Mercury's orbit by which the orbit did not form a closed ellipse; the result differed from the theoretical Newton value by exactly the observed anomaly, 43 seconds of arc per century.[10] Though not yet the final theory, the calculation relied only on assumptions that would be taken over by the generally covariant theory of the following week, and it showed that space-time is curved in the presence of a strong gravitational field, such as that of the sun. Finally on November 25, Einstein presented the field equations of general relativity in a form equivalent to the one widely employed today. (see Box 6.2). On the left-hand side are the Ricci tensor ($R_{\mu\nu}$) and the Riemann curvature scalar (R), both constructed from the metric tensor ($g_{\mu\nu}$) as are all invariant expressions within Riemannian geometry. On the right-hand side is a coupling term and $T_{\mu\nu}$, the so-called stress-energy-momentum tensor representing the mass and energy sources of the gravitational field in merely phenomenological fashion. The left-hand side represents space-time curvature (which is responsible for the effects of gravitation and inertia) and the right-hand side the matter-energy sources of this curvature. According to Einstein in 1936, the left-hand side is "a palace of fine marble" while the right-hand side is but "a house of cards", since the $T_{\mu\nu}$ is a crude skeleton, the fleshed out more particular form needed for any application depending on the type of matter and the specific interactions chosen for representation within a particular model.[11] Following John Wheeler, it is customary to read the expression from right to left, as "matter-energy tells space-time how to curve", and from left to right, as "space-time tells matter how to move".

Box 6.2

Einstein's equation in components (in the form adopted by Einstein in 1918) is $R_{\mu\nu} - \frac{1}{2}g_{\mu\nu}R = 8\pi T_{\mu\nu}$ in units where $c = G$ (Newton's gravitational constant) $= 1$. The left-hand side represents gravitational curvature of space-time; on the right-hand side, apart from the proportionality constant, stands the energy-momentum tensor representing matter-energy sources of gravitational curvature. The apparent simplicity of the expression is deceptive; in four-dimensional space-time, the two indices μ, ν each run from 1 to 4, and the expression stands for 16 complicated nonlinear partial differential equations (PDEs), 10 of which are independent. The most general space-time curvature tensor, the Riemann tensor constructed from the metric tensor, has four indices (is of "rank four"); however, already in 1912, Einstein believed that in analogy with electromagnetism, the energy-momentum tensor on the right-hand side had to be of rank two. Hence the two terms on the left are of rank two; the first, the Ricci tensor, represents mean curvature, and is obtained from the Riemann tensor by the operation of *contraction* that reduces the rank of a tensor by two. The second term on the left contains the *trace* of the Ricci tensor (the sum of its diagonal components); adding it to the left-hand side is dictated by the requirement of conservation of the energy-momentum tensor on the right. Introducing coordinates on space-time, the Einstein equation is 10 PDEs for the independent component functions of the metric tensor $g_{\mu\nu}$, a function of the arbitrary space-time coordinates $g_{\mu\nu} = g_{\mu\nu}(x^{\sigma})$, $\sigma = 1, 2, 3, 4$.

Connection to measurement and observation is made via the space-time *interval* $ds^2 = g_{\mu\nu}x^{\mu}x^{\nu}$ (with implicit summation over $\mu, \nu = 1, 2, 3, 4$), the square of the invariant "line element" on whose value all observers agree. Intuitively, the metric tensor can be considered as enabling the Pythagorean theorem between nearby space-time events p and q in a way that does not depend on any particular coordinate system.

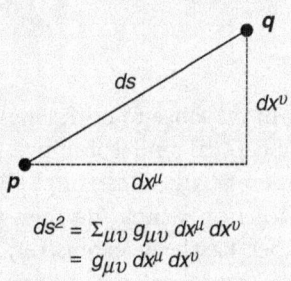

$$ds^2 = \Sigma_{\mu\nu}\, g_{\mu\nu}\, dx^{\mu}\, dx^{\nu}$$
$$= g_{\mu\nu}\, dx^{\mu}\, dx^{\nu}$$

Overview

General relativity's two core ideas are briefly summarized: 1) distributions of matter and energy determine the geometry of space-time encoded in the metric tensor $g_{\mu\nu}$; 2) the trajectory (world line) of a freely falling uncharged "test" particle is a *geodesic* of this geometry, essentially the longest world line (an extremal curve) connecting two space-time events. Such world lines define a "local inertial frame" in general relativity, a small region of locally curved space-time indistinguishable from flat Minkowski space-time. With respect to the freely falling test body, in this region accelerations due to other (non-gravitational) forces may be defined. Taken together, these statements affirm that there is no prior geometry; rather the metric of space-time geometry is itself a *dynamical* inertio-gravitational field accounting for the dependence of spatio-temporal relations between events on the surrounding distributions and motion of matter and energy. In this book we cannot presuppose the requisite mathematical and physical background required for full comprehension of the theory of general relativity. Nonetheless it is possible to come to a qualitative understanding of the theory, and of how it was crafted to pursue the philosophical objective Einstein set for it, of ridding physics of fixed inertial systems, and of any residue of the old Newtonian concept of "absolute space". The theory can be illuminated by reference to the three principles Einstein in 1918 considered undergirding the theory. These are the principle of equivalence, the principle of general relativity (and associated requirement of general covariance), and Mach's principle.

Three principles

In the spring of 1918, in response to critics and others who sought clarification of the conceptual underpinnings of the newly completed theory of general relativity, Einstein published a brief discussion of three assumptions on which, he then claimed, the theory rested.[12] This short paper has been extensively discussed, since in it Einstein conceded, in response to a pointed criticism by Kretschmann, that the assumption of general covariance (as noted above; Einstein believed it was required by the principle of general relativity) possesses by itself no physical content but had a "heuristic significance" if conjoined with a principle of simplicity (see below). The concession lends itself to various interpretations, with the result that disputes over the physical meaning, if any, of the mathematical requirement of general covariance continue to this day.[13] But the paper merits scrutiny for another reason, for it brings out complex interconnections between the philosophical requirements Einstein placed on the theory. These assumptions, each denominated a *principle*, are:

A) *the principle of (general) relativity*: The laws of nature are only statements concerning space-time coincidences: they accordingly find their sole natural expression in generally covariant equations.
B) *the principle of equivalence*: Inertia and gravitation are essentially the same (*sind wesensgleich*).
C) *"Mach's principle"*: The G field (i.e., all $g_{\mu\nu}$ solutions) is *exhaustively* determined by the masses of bodies. Since mass and energy are the same, according to results of the special theory of relativity, and since energy is formally described by the symmetrical energy tensor ($T_{\mu\nu}$), this affirms that the G-field is conditioned and determined by the energy tensor of matter.

The 1918 formulations of A and C are the upshot of recent controversies with critics. But prefacing the list is a frank admission that previous failure to distinguish between A and C, statements ostensibly having little to do with one another, had been a confusion. The relations among the shifting content of all three principles is highly intricate; it is not difficult to find passages in Einstein's writings that deem B to be the foundation of A or that A is a special case of B [A → B], but also that B requires A [B → A]. There are also statements to the effect that what is now distinguished separately as C is an extension or generalization of B [C → B] or that satisfaction of C requires A [C → A]. Scrutiny of earlier writings on the problem of gravitation, both published and in private correspondence, shows that the three principles took on various non-synonymous meanings during the gestational years of the theory from 1911–1915. Inadvertent confusion or lack of clarity regarding the meaning of each was the result, so this paper seems an analytical attempt to demarcate their respective boundaries, and thus to isolate the distinctive requirement each should impose on the completed theory in Einstein's retrospective and presumably re-considered opinion.

Two general but related comments might be made before proceeding to a closer discussion of the three principles. First, a reader of this 1918 paper is bound to be struck by a notable admission that by encapsulating general relativity under the above three principles, the theory has been assessed "as it now appears to me", seemingly suggesting that the theory, or at least its underlying principles, appeared to Einstein somewhat differently in the past. But how can a successful theory appear differently to its own creator within just a few years of its inception? To make sense of this, it is helpful to distinguish two distinct meanings of what might be meant in referring to "the theory", distinguishing philosophical underpinnings from the theory narrowly construed, understood in contemporary terms as the set of all the theory's models – here, models of general relativistic space-times. The latter, e.g., space-time descriptions of particular gravitational fields (such as the sun's) did not fundamentally change between 1916–1918, nor have they since, although today more solutions of Einstein's equations are known and studied. Rather, the changed appearance had to do with Einstein's reappraisal of the theory's motivating underlying assumptions.

The distinction between theory, as formalism, and underlying assumptions recalls Heinrich Hertz's famous characterization of Maxwell's theory of electromagnetism merely as Maxwell's equations,[14] jettisoning the attempts of Maxwell and others who sought to understand the theory's central theoretical structures, i.e., the electric and magnetic fields, in terms of mechanical disturbances propagating through an ethereal substance supposedly required to support these processes. In this regard, Hertz is reported to have quipped that the formalism of the theory is wiser than its creator.[15] Something like this is also true of the equations of general relativity, as it has been subsequently recognized that each of the supposed "core" principles of the theory either does not have, or cannot have, quite the meaning Einstein intended. First up is the principle of equivalence.

The principle of equivalence

Only formulated as a principle in 1912, the principle of equivalence is commonly illustrated by the "elevator" thought experiment used by Einstein, initially in a lecture in Vienna in December 1913. Imagine observers ("physicists awakening from a drugged sleep"), enclosed in a windowless elevator ("chest") who are attempting to determine whether the elevator is at rest on a planet possessing a gravitational field (of a very special homogeneous kind, see below), or is undergoing a uniform acceleration upwards (in the opposite direction) in a region of space far from any planet or other massive body (for all intended purposes, a gravity-free region). Conceptually, the two cases can be distinguished as follows. In a gravitational field, an unsupported test object is observed to freely fall with an acceleration $g = -a$. Its mass will be regarded as weight, measurable in a scale balance. In the other, the test object, seemingly weightless and at rest, is met by the up-rushing floor of the elevator, rising with an acceleration a. Its inertial mass is manifested as resistance to acceleration, determinable by the magnitude of the required uniformly accelerating force. In both cases the object encounters the same location on the floor with the same absolute magnitude of acceleration. While there remains a conceptual distinction, the test object's behavior is empirically indistinguishable in the two cases, indicating that its gravitational mass can be considered as identical to its inertial mass, $m_g = m_i$. This identity states the so-called Galilean, or "weak" principle of equivalence. As gravitational freefall and uniform upward acceleration are both non-inertial motions, the equivalence between them suggests that the principle of relativity extends beyond inertial frames of reference.

"Einstein's elevator" has been employed at times (by Einstein himself[16]) to draw a stronger conclusion: that the laws of physics are the same in a laboratory in a gravitational field and in an accelerating laboratory in "empty space", far from any source of gravitational force. That statement – known as the "strong" principle of equivalence – requires qualification, for it is correct only within the accuracy of measurements (of distances, speeds, angles) that a freely falling observer might make in his "local" region of space-time. Outside this limited region it is not difficult to detect the presence of a true gravitational field by showing the effects of "tidal forces" generated by the non-uniformity of the field. These forces, manifested in the different strengths of the gravitational field at different space-time points, are responsible for the relative accelerations of two test objects placed at a small separation. In an unchanging (static) and completely homogeneous gravitational field, a highly artificial situation almost never to be found in nature, these relative differences – by definition – do not exist. But generically they do, and they can be measured. For example, if two identical rubber balls are placed at horizontal separation high above the Earth and then released, then, on account of the inhomogeneity of the Earth's gravitational field, their separation will decrease as they fall toward the center of the Earth, the greater the initial separation, the greater their relative acceleration toward each other. The non-zero mathematical expression (a tensor, see Box 6.1) linking the vectors of relative acceleration of the two balls to their separation vector is an unmistakable signature of a true gravitational field. As this phenomenon does not appear in an accelerated frame far away from all gravitational sources, the "strong" principle of equivalence is not generally valid. This has led some leading relativity physicists to question the need for the principle of equivalence.[17]

Correctly interpreted, the principle of equivalence essentially consists of four assertions:

1. a body's gravitational mass is equivalent to its inertial mass, $m_g = m_i$
2. an uncharged test particle of negligible mass traverses a "geodesic" of space-time, i.e., the straightest possible curve between two points along which time is maximal (i.e., "proper" time, according an idealized clock accompanying the particle)
3. since according to Galileo gravitational accelerations at a given location are the same in magnitude as well as direction, these accelerations may be regarded as zero with respect to bodies present there subject only to gravity but to no other forces. Then freely falling bodies, maintaining their mutual separation (and so forming "a congruence of timelike curves") physically define a "locally inertial" reference frame, existing only in the limited region surrounding the freely falling bodies, but with respect to which accelerated motions due to other forces may be specified
4. therefore, within a small region surrounding a space-time point-event in an arbitrary gravitational field, it is possible to choose a local inertial system in which the laws of physics have the same form as in the special theory of relativity. Since the special theory of relativity is gravitation-free physics, this means that gravitational force is (locally) an apparent force: its effects

(acceleration) can be transformed away by a suitable choice of space-time coordinates in a restricted region around the freely falling bodies. That indeed is the main point of the elevator thought experiment.

So understood, the principle of equivalence is the *sine qua non* of the theory of general relativity. Equally importantly, it furnishes a necessary basis for the physical interpretation of the purely mathematical (pseudo-) Riemannian geometry of space-time on which the theory is based for it permits measuring rods and clocks to be coordinated with the purely mathematical quantity $g_{\mu\nu}$, the metric tensor.[18] Without such direct linkage to a conceptually simple observable physical process, the Riemannian theory of manifolds arguably has no claim to be descriptive of the phenomena of physical space-time (see Chapter 7). In this regard, the principle of equivalence is fundamental in characterizing the transition between the four-dimensional globally flat space-times of special relativity and the (generically) curved four-dimensional space-times of general relativity. In particular, according to the "for all practical purposes" absence of gravity allowed locally by the principle of equivalence, a local inertial frame may be arbitrarily extended, becoming an inertial frame of special relativity. As a result, the principle, however stated, does not possess rigorous mathematical validity but has the character of a heuristic approximation.

On the other hand, on account of the principle of equivalence, manifestations of a body's inertial behavior (its resistance to acceleration, or its uniform straight-line motion) can no longer be accounted as the effects of its situation with respect to a space (or space-time) of fixed predetermined structure. Rather, this behavior can be considered the result of the body's dynamical interaction with the combined inertial-gravitational field generated by *all surrounding masses and energies*, changing as these change according to the Einstein field equations. That the effects of inertia and gravitation have a common origin and a common explanation as a *dynamical interaction* between given bodies and their material surrounding, is a core Machian idea that motivated the general theory of relativity (see below).

Finally, two subtle but striking predictions can be drawn from the principle of equivalence alone. The first is *gravitational redshift*. In the frame of a freely falling observer, light should behave just as if there were no gravity, i.e., traveling at fixed speed c with unchanging frequency (color, in the visible spectrum). However, frequency is not an intrinsic property of light. This is shown by the Doppler effect due to the relative motion of an observer and the source of a light wave or acoustic wave, most familiarly in the suddenly lowered pitch of the horn of a train engine rushing by. According to the principle of equivalence, if a flashlight pointing upwards on the floor of the enclosed elevator is switched on during the elevator's upward acceleration (in the absence of gravity), the frequency of the light wave arriving at the ceiling will be slightly lower than its frequency on emission because the speed of the ceiling at the reception of the signal is greater than the speed of the floor when the flashlight was switched on. A gravitational redshift is then equivalent to a Doppler shift between two accelerating frames. But gravitational redshift is also the most direct observational test of the curvature of space-time. If one considers clocks as periodic emitters of light (e.g., the spectral lines of atoms), then the spectral lines of atoms near a large mass will exhibit somewhat lower frequencies than the lines emitted by the same kind of atoms elsewhere. In this regard, gravitational redshift is a special case of a more general effect: all processes occurring in a gravitational field surrounding a large mass appear "slowed down", as now will be seen.

Ignoring the motion of the Earth in its orbit, Einstein believed it possible to confirm general relativity by measuring the redshift of light emitted from the surface of the sun when seen from Earth. But measurements did not at all agree with the predicted value of the shift, in large measure because the surface of the sun is a highly complicated and violent gaseous region, though this was not well understood until some decades later. As a result, for many years not a few critics of the theory of general relativity pointed to the failed prediction of gravitational redshift as sufficient reason to reject the theory. The first confirmation of gravitational redshift occurred in the period 1960–1965 (after Einstein's death) in an experiment using the 73.8-foot tower of the Jefferson Physics laboratory at Harvard. R.V. Pound, G.A. Rebka, and J.L. Snider were able to detect the tiny effect expected in the Earth's modest gravity by exploiting a subtle effect of atomic crystals previously demonstrated by Möss-bauer in Heidelberg. Placing an emitting crystal at the top of the tower, and a detecting crystal in the basement and monitoring the detection rates, Pound, Rebka, and Snider were able to displace the emitter with just the right motion relative to the detector that the velocity difference between emitter and detector compensated for the gravitationally induced blueshift of the emitted photon falling in the Earth's gravitational field to the detector. Although the predicted frequency shift was merely two parts in a thousand trillion (10^{13}), the initial experiments produced a result to within 10 percent of predicted value. Subsequent experiments narrowed the gap between prediction and observation to 1 percent.

From the gravitational redshift of light, another implication of the principle of equivalence is immediate: time dilation in a gravi-tational field. This is most clearly seen using atomic clocks based on the oscillations of a source (e.g., cesium atom) having a constant, stable, and well-defined frequency. Suppose that at a given location two such identically built clocks are synchronized so that at one moment they tick at the same rate and read the same time. Then carry one to the top of a skyscraper while placing the other at the bottom, leaving them in place for a few days. On bringing them together again, it will be found that they are no longer synchro-nized, but that the clock from the top of the skyscraper is running faster than the clock left at the bottom; it will have ticked more times in the interval during which they were separated. Both atoms and clocks (whether mechanical, atomic, or biological) perform their functions more slowly where gravitational field strengths are stron-ger. This is a relative effect: the frequency of the light emitted by the atom at the bottom looks normal to an observer there, but appears redshifted to the observer at the top, whose clock is running faster. Recent advances in navigation using Earth-orbiting atomic clocks and highly accurate time-transfer technology routinely take gravita-tional redshift and time-dilation effects into account. The accuracy of Global Positioning System (GPS) devices rests on the difference in rate between satellite and ground clocks as a result of relativistic effects amounting to 39 microseconds per day (46 μs from the grav-itational redshift, and −7 μs from time dilation).

The principle of general relativity

The 1918 formulation of this principle cited above requires considerable explanation. It would seem that some such principle is implicated
in the very name of Einstein's theory. In his first comprehensive review
of relativity theory, written in 1907, Einstein raised the question, "Is it
conceivable that the principle of relativity also holds for systems that
are relatively accelerated with respect to each other?" Such an extension of the principle of relativity appeared possible on the grounds
of the thought experiment concerning the freely falling observer
described above. In 1907, Einstein spoke not of a principle, but of the
assumption (*Annahme*) of equivalence between an inertial system containing a homogeneous gravitational field and uniformly accelerated
system. By 1911, the supposition of equivalence between an inertial
system K in a homogeneous gravitational field and a uniformly accelerated system K' in gravitation-free space came to serve as the basis for
a much stronger claim, of the relativity of accelerated motions:

> (I)f we accept assume that the systems K and K' are exactly
> physically equivalent . . . (then) one can just as little speak
> of the *absolute acceleration* of the system of reference as one can
> speak in the usual theory of relativity of the *absolute velocity* of
> the system.[19]

In more cautious moments, as in a lecture in Vienna in September
1913, Einstein frankly denied that the relativity principle extended
to nonuniform motions.[20] Yet the postulate of general relativity is
given pride of place in the early sections of the first complete exposition of the final theory in April 1916. In §2 ("On the reasons suggesting an extension of the postulate of relativity") of that paper,
Einstein presented a thought experiment hypothetically involving
an otherwise empty universe containing only two fluid spheres in
relative rotation. Observers on each can find by local measurements
which sphere is actually rotating: it has a bulging equator. The explanation of this fact given by Newtonian mechanics, singling out the
"true" inertial frame of the non-rotating sphere as the legitimate
space for the laws of mechanics, is criticized for its appeal to a "factitious cause", an unobservable absolute space, rotations in respect
to which occasion the observed inertial effects. In contrast, Einstein

demanded that any epistemologically satisfactory account of the relative rotations appeal only to observable causes, a restriction that ostensibly appears vintage Machian positivism. But this demand is followed by another having only tenuous relation to epistemological strictures requiring observability.

> The laws of physics must be of such a nature that they apply to systems of reference in any kind of motion. Along this road we arrive at an extension of the postulate of relativity.[21]

A few paragraphs later in §3 ("The space-time continuum. Requirement of general covariance for expressing the equations of the general laws of nature"), the implicit inference of §2 is made explicit: the postulate of general covariance ("the general laws of nature are to be expressed through equations which are valid for all coordinate systems, that is, are covariant with respect to arbitrary substitutions") is implied by a "general postulate of relativity".

> It is clear that a physical theory that satisfies this postulate (of general covariance) will also be suitable for the general postulate of relativity. For the sum of all substitutions in any case includes those which correspond to all relative motions of three-dimensional systems of coordinates.

This is precisely the inference seen above to be false. The principle of general relativity is *not* a generalization of the principle of special relativity in the sense that "all motions are relative" since uniform and accelerated motions are not (outside of highly contrived situations) physically equivalent. But then what is it? From a contemporary perspective, it is essentially the requirement that the laws of nature must be formulated in terms of geometrical relationships between physical quantities, the latter expressed in coordinate-free geometric form (i.e., represented as tensor quantities).

As Hilbert surely knew, and as Erich Kretschmann (a Planck student) would observe in a critical article in 1917, the requirement of general covariance is purely formal: it amounts to no more than a stipulation that the laws of nature are to be stated in tensor form; after all, tensors are expressions that have the same form in all coordinate systems.[22] Both gravitation-free special relativity and

Newtonian gravitational theory can be formulated in a generally covariant manner using only tensor equations. In his 1918 response, Einstein had to agree but countered that Newtonian gravitational theory, if expressed in tensor form, would be considerably more complicated than general relativity. In short, the principle of general covariance together with the constraint of simplicity can have physical significance, at least if it is assumed that the simpler of two theories in generally covariant formulation is the more likely to be true. This kind of defense of general covariance as a physical principle was paraphrased much later in what remains the lengthiest textbook of general relativity, affectionately known (in its day) as "the telephone book":

> Another viewpoint . . . constructs a powerful sieve in the form of a slightly altered and slightly more nebulous principle: "Nature likes theories that are simple when stated in coordinate-free geometric language". According to this principle, Nature must love general relativity and it must hate Newtonian theory. Of all theories ever conceived by physicists, general relativity has the simplest, most elegant geometric foundation (three axioms: [1] there is a metric; [2] the metric is governed by the Einstein field equation . . .; [3] all special relativistic laws are valid in local Lorentz frames of metric). By contrast, what diabolically clever physicist would ever foist on man a theory with such a complicated geometric foundation as Newtonian theory?[23]

Indeed, Newtonian theory in tensor form appears horribly complex when expressed as (now) ten differential field equations for the metric coefficients in a specific coordinate system. Still it turned out to be a reasonably straightforward matter in the 1920s to write down Newtonian theory (and its absolute motions) in coordinate-free geometric form. This is just what careful contemporary comparisons of Einstein's gravitational theory with Newton's do, employing an empirically equivalent geometric version of Newton's theory.[24] Furthermore, and more significantly, Kretschmann observed that formulated in the geometry of Minkowski space-time, the relativity postulate of special relativity, i.e., Lorentz transformations carrying inertial states (geodesic trajectories) into inertial states (geodesic trajectories), is a symmetry of the space-time and so a physical, not a merely formal, postulate. But in general relativity the local iner- tial (geodesic) structure about a freely falling test body depends on surrounding (and variable) mass-energy distributions; there are as many local inertial frames as there are variable distributions of mat- ter with no symmetries between them. In brief, in general relativity there are no (non-trivial) symmetries of space-time. Hence, general covariance, even as a purely formal requirement, cannot be asso- ciated with any relativity of motion principle. On the other hand, while Newtonian gravity *need not* be expressed in tensor form, Ein- steinian gravity *must be* because of the relativity of the gravitational field established by the principle of equivalence. In general relativity, there is no invariant (tensorial) way of capturing the core Newto- nian distinction between gravitation and inertia; in general relativity, there is only a common inertio-gravitational field, the distinction between them being observer-dependent.[25]

Even though he repeated the inference in 1916 and even later on, in order to find the final form of the general theory of relativity in late 1915, Einstein had to sever the connection between relativity principle and covariance condition. In returning to general covari- ance, he effectively had to reject the default assumption that space- time coordinates have an immediate metrical meaning. The path along which he traveled in doing this, extending over more than two years, was for a long time considered by historians of relativity and physicists alike as a gradual realization of a mere mathematical confusion.[26] It is now one of the more storied episodes in history and philosophy of physics since 1979 when physicist and Einstein scholar John Stachel – making essential use of Einstein's correspon- dence in late 1915 and early 1916 with Paul Ehrenfest, H.A. Lorentz, Michele Besso, and others – revealed the full extent of Einstein's struggles with general covariance in 1913–1915.[27] It is again a story of a misleading analogy to the special theory of relativity. There, the coordinate system of any inertial frame constructed using rigid rods and synchronized clocks located at each space-time point pro- vides an immediate metrical meaning to these points. The attempt to carry this out in the context of gravitational theory led Einstein to fall into the clutches of what he called the "hole consideration" (*die Lochbetrachtung*), widely known today as the "hole argument". The argument convinced Einstein that his relativistic gravitational theory *could not be* generally covariant, on pain of a failure of the determinism required of all satisfactory field theories: from given boundary conditions and sources, temporal evolution of the field equations should completely determine field strengths at any point in a given region. Until autumn 1915, he remained persuaded of its faulty conclusion, a bitter pill to swallow given the erroneous association of the principle of general covariance with a generalization of the relativity principle to accelerated systems.

The hole argument and the point coincidence argument

Einstein came under the sway of the hole argument in the summer of 1913; it appears in several papers of 1913–1914. Recall that in the 1913 *Entwurf* theory, Einstein erroneously believed that special coordinate conditions (restricting the theory's covariance) were required in order to recover the Newtonian limit and to satisfy the laws of conservation of energy and momentum. Since he believed in the faulty implication above, this, as he wrote to H.A. Lorentz, was an "ugly dark spot" on the theory.[28] He then invented an argument to purportedly show that this blemish was in any case a necessary one as it was impossible to reconcile the requirement of general covariance with the determinism required of all reasonable physical theories. For the next two years, Einstein chose to restrict the covariance of his theory in order to preserve determinism, arguing, as in this review paper from 1914:

> *Events in the gravitational field cannot be determined uniquely by means of generally covariant differential equations for the gravitational field.* If we demand, therefore, that the course of events in the gravitational field be completely determined by means of the laws that are to be established, then we are obliged to restrict the choice of the coordinate system.[29]

The hole argument is rather intricate; to properly extract Einstein from its clutches requires invoking machinery of modern differential geometry not developed until around 1950.[30] But it can be stated in simplified way. Consider, as Einstein did, a limited region of space-time where no matter is present, a "hole" H, completely surrounded by a matter-filled region. Using the notation of Einstein's equation (Box 6.2) inside H, $T_{\mu\nu} = 0$, whereas in the containing region, $T_{\mu\nu} \neq 0$. If that equation is read from right to left ("matter-energy tells space-time how to curve"), the space-time geometry (the values $g_{\mu\nu}$) inside H are determined by matter-energy sources outside H (where $T_{\mu\nu} \neq 0$). As noted, the $g_{\mu\nu}$ are functions of the space-time coordinates $x^\sigma = x^1, x^2, x^3, x^4$. Einstein presupposed that space-time coordinates suffice to individuate physical events of space-time. After all, within special relativity differences between space-time point-events in every inertial system are understood as marked off by rigid rods of unit length and perfectly synchronized clocks at each point of the system; these differences have an immediate and fixed metrical significance. He then argued that if one requires general covariance of the field equations, then a specific matter distribution outside H does not uniquely determine the $g_{\mu\nu}$ within H.

The argument requires three steps.

Step 1: Everyone will agree that under the arbitrary coordinate transformation $x \rightarrow x'$ permitted by general covariance, if $g_{\mu\nu}(x)$ is a solution at some point of space-time for sources at the point $T_{\mu\nu}(x)$, then so is $g'_{\mu\nu}(x')$ at that same point for sources $T'_{\mu\nu}(x')$.

Step 2: According to general covariance, there is a permissible (nonlinear) coordinate transformation over the entire region that is the identity outside H (i.e., leaving the coordinates of the space-time points in this region unchanged) but changes coordinates at points inside H, going smoothly from one to the other on H's boundaries. Then in particular, $T_{\mu\nu}(x) = T'_{\mu\nu}(x') \neq 0$ inside H.

Step 3: Using this coordinate transformation, Einstein was able to construct (for details, see Note 29) a second solution $g'_{\mu\nu}(x) \neq g_{\mu\nu}(x)$ in the same coordinate system x^σ ($\sigma = 1, 2, 3, 4$) at the same point p inside the hole from the same sources $T_{\mu\nu}(x) = T'_{\mu\nu}(x') = 0$ outside the hole.

Conclusion: the matter distribution outside H does not uniquely determine the $g_{\mu\nu}$ at points within, as should be the case if the theory is deterministic. For under the above coordinate transformation, the *same* sources $T_{\mu\nu}$ will assign different values $g_{\mu\nu}$ to the *same* point within H. It is an apparent failure of determinism.

Einstein's response to the hole argument for more than two years was to abandon the requirement of general covariance since it led to a failure of determinism (in the above sense). Only in late 1915 did Einstein see that the argument's conclusion rested on a questionable philosophical presupposition: that space-time has a physical existence independent of the presence of physical fields (i.e., actual physical events corresponding to values of the metrical $g_{\mu\nu}$ field or other matter fields). In particular, this amounts to the claim that the space-time points (and so those within H, where the solutions differ) have an identity fixed prior to the specification of the metric $g_{\mu\nu}$; that is, prior to solving the field equations. Resolution of the hole conundrum lay in seeing that this assumption is by no means necessary and might be denied. And this is to say that space-time points cannot be individuated, or identified, without the physical properties $g_{\mu\nu}$ specified there by the gravitational field equations. In general relativity, the geometrical $g_{\mu\nu}$ is a dynamical variable to be determined at all points of a four-dimensional manifold. Any physical properties attributed to these points – indeed, their physical reality as space-time point-events – are consequences of that determination. Then the conclusion that there are two physically distinct solutions $g_{\mu\nu}$ at some given point p within H is faulty since it relies on the assumption that the space-time coordinates alone give p an independent existence, that it can be the bearer of two physically distinct values of $g_{\mu\nu}$. Under the coordinate transformation, $g'_{\mu\nu}(x) \neq g_{\mu\nu}(x)$ do differ but they are only different mathematical descriptions of the same physical geometry at p, just different tensorial characterizations (eliminating any special role for coordinate systems) of the same event.

Precisely when in the autumn of 1915 Einstein received this illumination, in many ways the most novel philosophical aspect of general relativity, is not known. In an unpublished account written in 1934, Einstein reports specifically on how he was troubled by having to "unlearn" the lesson of special relativity:

I soon saw that the inclusion of non-linear transformation, as the principle of equivalence demanded, was inevitably fatal to the simple physical interpretation of the coordinates – i.e., that it could no longer be required that coordinate differences should signify direct results of measurement with ideal scales or clocks. I was much bothered by this piece of knowledge.[31]

When did he realize that coordinates in general relativity are simply parameters? An elliptical passage in §3 of the canonical April 1916 presentation of the theory accounts for the specific 1918 formulation of the "principle of general relativity" cited above. This is the so-called "point coincidence argument":

In the general theory of relativity, spatial and temporal magnitudes cannot be defined in such a way that differences of the spatial coordinates can be directly measured by the unit measuring-rod, or differences in the time coordinate by a standard clock. . . . So there is nothing for it but to regard all imaginable systems of coordinates, in principle, are equally justifiable for the description of nature. This amounts to the requirement: *The general laws of nature are to be expressed through equations that are valid for all coordinate systems, i.e., are covariant with respect arbitrary substitutions (generally covariant).* It is clear that a physics that satisfies this postulate will be suitable for the postulate of general relativity.

. . . That this requirement of general covariance, which takes away from space and time the last remnant of physical objectivity (*den letzten Rest physikalischer Gegenständlichkeit*), is a natural requirement, follows from the following reflection. All our space-time verifications (*Konstatierungen*) invariably amount to the determination of temporal-spatial coincidences. For example, if natural occurrences consisted merely in the motion of material points, then ultimately nothing would be observable but the encounters of two or more of these points. Also the results of our measurements are nothing other than the verification of those encounters of the material points of our measuring rods with other material points . . .

The introduction of a reference system serves no other purpose than to facilitate description of such coincidences. . . . Since all our physical experiences can be ultimately reduced to such coincidences, there is no immediate reason to prefer certain coordinate systems to others, that is, we arrive at the requirement of general covariance.[32]

Positivists and empiricists took special notice of this passage that seems to rest the observational basis of physics on the verification of "point coincidences". Max Born read Einstein as identifying "*das physikalische Urphänomen*", the primitive phenomena to which the entire empirical content of physics can be reduced; Phillip Frank went further, assimilating Einstein to Machian phenomenalism:

> Einstein joined Mach, and in his general theory of relativity actually erected an edifice of mechanics in which space and time properly speaking no longer occurred, but only the coincidence of phenomena.[33]

The cryptic remarks on point coincidences are in fact the conclusion of an enthymematic argument by which Einstein extracted himself from the hole argument. Their meaning could hardly be recovered from the cited passage except by the few of Einstein's correspondents or confidants for whom the missing context was understood. They do not intend what they appear to say: that in locating the criterion of what is in principle observable in the coincidence of points (intersections of world lines), only such observable phenomena are real. Rather, in the space-time setting of general relativity, "coincidences" are any physical event, a physical occurrence in a small localized region: the pop of a firecracker, the intersection of the trajectories of particles (i.e., a meeting of the world lines of two bodies), an atom's emission of a photon. All observers will agree whether such an event occurs, while coordinates are just arbitrary labels of the event or, as it is sometimes called, "world point". The totality of such events – past, actual, and future – is then the physical reality of space-time. This is the sense in which the requirement of general covariance removes "the last remnant of physical objectivity" from the old Newtonian idea of an absolute fixed background structure to space. Relations of connection and succession suffice to establish the essential meaning of space and time as forms of the coexistence, connection, and order of events.[34]

With the hidden context of the hole argument restored, the 1918 pronouncement,

> *the principle of (general) relativity*: The laws of nature are only state-ments concerning space-time coincidences: they accordingly find their sole natural expression in generally covariant equations,[35]

can be understood as giving rhetorical emphasis to the now well-understood fact that coordinates have no direct metrical mean-ing, that they are only a means of labeling the points of a four-dimensional manifold in such a way that differential calculus can be performed on a mathematical representation of space-time. The physical meaning of general covariance is that only space-time quantities that legitimately appear in the equations of physics are the metric tensor $g_{\mu\nu}$ and quantities derivable from it.[36] Mere points of the underlying mathematical manifold then have no physical meaning. In the absence of fields with their physical sources, man-ifold points are not *points of space-time*; i.e., they have only mathemat-ical, not physical, meaning.[37] Late in his life, Einstein made several attempts to elucidate this point of view. These efforts to clarify the significance of the principle of general covariance notably cluster around Einstein's reliance on the principle as a heuristic guide for the futile unified field theory program.[38] The fundamental meaning of general covariance is then expressed by the statement that there can be no principled distinction between the structure of space-time and its "contents"; in brief, "no metric, no space-time". In this way, space and time by themselves have indeed lost "the last remnant of physical objectivity". In this broadened heuristic sense, the *meaning* of general covariance encompasses the purely formal *requirement* of general covariance (requiring use of tensor expressions and free-dom to make "arbitrary", including nonlinear, transformations of the space-time coordinates). The heuristic meaning sets a program-matic agenda within which the distasteful notion of a fixed inertial system does not appear; it has the latter as an implication. Any theory in which the space-time metric is dynamical must be (like general relativity) formulated in a generally covariant way.

Late in life Einstein thought the matter important enough to underscore once again. Declining, on grounds of illness, an invita-tion to attend the 50th anniversary conference of relativity at Bern in 1955, he noted that for him, the "most essential thing" about the general theory of relativity is its *ambition* to remove from physical theory, once and for all, the notion of a privileged frame of reference, background space-time structures that act (in the explanation of the inertial behavior of a body) but which in turn are not acted upon.[39] A core motivation underlying the unsuccessful unified field theory program dominating the last three decades of Einstein's life was the recognition that the theory of general relativity, as a theory of the gravitational interaction alone, could not adequately accomplish this goal but that a "theory of the total field" encompassing the other matter/energy fields might. It is general covariance in this wider meaning (sometimes also called, "the general principle of relativity") that served as a "limiting heuristic principle" in the search for such a unified theory.[40]

Mach's principle

The final fundamental principle on Einstein's 1918 list while bearing Mach's name is informally paraphrased as "a generalization of Mach's demand that inertia be explained by the interaction of bodies". The demand on the kind of explanation considered satisfactory leaves wide open the nitty-gritty details of how it might be carried out. In particular, an explanation of this kind requires showing how other masses influence the value of a given body's inertial mass, as well as how they exercise their influence on each local segment of a body's inertial trajectory. Mach himself provided only the murkiest idea of how the forces that Newton sought to explain in terms of acceleration relative to absolute space might be better explained in terms of accelerative motions with respect to other masses. Nonetheless "Mach's principle" pays tribute to the inspiration Einstein received from the familiar passages in Mach's *Mechanics* criticizing Newton's purported demonstration of an effect of absolute motion, and hence the existence of absolute motions, in the famous bucket experiment.[41] Mach's tacit appeal to an "interaction of bodies" in accounting for the manifestations and effects of inertia – the straight-line uniform motion of a body on which no net forces act, the body's resistance to acceleration, the appearance of centrifugal forces in accelerating bodies, etc. – comprised, in Einstein's terms, at least a template for the "relativization inertia", a phrase not found in Mach. In general relativity the problem takes the form of attempting to show that the matter content of the universe should determine a body's freely falling local inertial frame. Einstein's difficulties in trying to carry out such a program within general relativity occasioned a shift in explanatory emphasis on "an interaction of bodies" to the purely negative injunction that no physical properties be ascribed to "empty space".

Mach deemed it necessary to purge non-empirical metaphysical elements from the foundation of mechanics as established by Newton. The essential flaws lay in Newton's unsatisfactory definition of mass as "quantity of matter" and his statement of the first law, the law of inertia. Mach reformulated the definition of mass as a ratio of masses, so that the second law appears as a mere definition and the third law, the law of action and reaction, as a consequence of the new definition of mass. To Mach, Newton's mechanics entirely rested on the first law. Mach recognized that Newton's tacit appeal to absolute space in the statement of the law of inertia was occasioned by doubts whether any given fixed star is truly or only apparently at rest. As "we have knowledge only of *relative* spaces and motions",[42] the first law is then to be understood so that uniform motion is with respect to the observable distant stars. If perchance these are discovered to be not at rest, then the fixed reference system presupposed by the law of inertia remains "still to be found". Similarly, Mach sought to explain the appearance of inertial effects in accelerating systems, such as water rising up the sides of Newton's rotating bucket, as due to the influence of observable cosmic masses. In particular, Mach hinted that a body's inertia is not a resistance to acceleration *per se* but to acceleration with respect to the distant cosmic masses. Mach's own concrete suggestions regarding how the principles of mechanics might be reformulated so that inertial forces (as in Newton's rotating bucket) appear in rotations with respect to other masses were left perhaps intentionally vague, with the consequence that they were largely ignored. Much more important was Mach's assertion that it is unnecessary to refer the law of inertia to absolute space. This resonated with Einstein who – according to the "principle of general relativity" – sought to eliminate altogether the notion of inertial frame. Such an elimination points to a "relativization of inertia" in the following sense: without a distinguished class of reference frames, Einstein could conjecture *à la* Mach in 1912 that "the *entire* inertia of a mass point is an effect of the existence of all other masses, resting on a kind of interaction (*Wechselwirkung*) with the latter".[43]

Immediately on publication of the first attempt to formulate a relativistic theory of gravitation, the Entwurf theory completed in Zurich in June 1913 with the collaboration of Grossmann, Einstein posted an offprint to Mach on June 25, acknowledging in an accompanying letter the salutary influence of the Mach's "brilliant (*genialen*) investigations on the foundations of mechanics".[44] Boasting that the Entwurf theory "establishes with necessity that *inertia* has its origin in a kind of *interaction* (*Wechselwirkung*) of bodies, entirely in the sense of your considerations on Newton's bucket experiment", Einstein listed several suggestive results of the theory (described below) that yield only partial realization of the goal of a complete "relativization of inertia". A further postcard to Mach, written in the second half of December that year, contains a declaration that might have been penned by Mach himself, "For me it is absurd to ascribe physical properties to empty space".[45] Einstein must have known that old age, a stroke that paralyzed his right side, and failing health had not diminished Mach's anti-metaphysical ardor. Just a year before, writing in the Preface to the seventh German edition of his *Mechanics*, Mach had described himself as an old man "struck down by a grave malady (who) shall not cause any more revolutions". Nonetheless he could not resist one last expression of scorn for "the monstrous conceptions of absolute space and absolute time". Though Einstein and Mach were not personally close, and may have only met once or at most twice, Einstein clearly felt that it was important to keep Mach apprised of progress on carrying out what Einstein considered Mach's agenda.

Einstein's most comprehensive early treatment of the problem of "relativization of inertia" appeared in a September 1913 lecture "On the Present State of the Problem of Gravitation" delivered, fittingly, in Mach's Vienna.[46] Section 9, bearing the heading "On the Relativity of Inertia", describes more fully the results mentioned in the letter to Mach. Einstein asserts that according to the Entwurf theory, he can demonstrate three effects suggesting a "relativitization of inertia": first, the inertia of a body can be shown to vary with the proximity of neighboring masses; second, he can show that there is a precession of the plane of oscillation of a pendulum contained within a rotating mass shell (see Chapter 11); finally, he argues that a rotating mass shell generates a so-called Coriolis force, whereby objects actually traveling in a straight line appear to an observer rotating with the shell to be deflected. The three Machian predictions are repeated in Einstein's 1921 Princeton lectures, after the completion of general relativity.[47] However, the main Machian result of this 1913 paper is presentation of a two-termed expression for the energy of a slowly moving mass point in a Newtonian gravitational field. Einstein argued that according to the first term, the energy of the body diminishes if the surrounding masses are augmented, corresponding to a greater inertial mass of the point. He then elaborated:

> This result is of the highest theoretical interest. For if the inertia of a body can be *raised* by accumulation of masses in its region, then we can hardly avoid accepting that the inertia of a point is *conditioned* (*bedingt*) by the other masses. Inertia appears thus conditioned through a kind of interaction (*Wechselwirkung*) of the accelerating mass point with all other mass points.
>
> The result appears completely satisfactory if one considers the following. To talk of the motion, therefore also the acceleration, of a body A in itself has no meaning. One can speak only of motion, or acceleration, of a body A relative to other bodies B, C, etc. What holds for acceleration in kinematic terms should also obtain for the inertial resistance of a body opposing acceleration. It is *a priori* to be expected, even if it is not exactly necessary, that this inertial resistance is nothing else than a resistance of the body A to relative acceleration considered with respect to the totality of all other bodies B, C, etc. It is well known that E. Mach first advocated this point of view in his history of Mechanics with all acuteness and clarity, so that here I can simply refer to his remarks. . . . I will refer to the conception sketched above as the "hypothesis of relativity of inertia".
>
> In order to avoid misunderstandings, it must be said again that just as little as Mach, I am of the opinion that the relativity of inertia corresponds to a logical necessity. However, a theory in which the relativity of inertia is established is more satisfying than the theory familiar to us today, since in the latter the inertial system is introduced, whose state of motion is, on the one hand, not conditioned by the states of observable objects, therefore is caused by nothing accessible to perception, but on the other hand, should be determinative for the behavior of material points.[48]

In this lengthy exposition, noticeable leeway is given to the distinction between an interaction with other masses *determining* a given body's inertia, as opposed to merely *influencing* or *conditioning* it. As subsequent events will show, this will prove to be a necessary qualification. The extended discussion reveals also a distinction between a purely "kinematic" principle of relativity of motion – only the *relative* motion of a body with respect to other bodies is observable; the motion of a body with respect to absolute space is unobservable – and a "dynamical" principle of relativity of motion – the *inertial motion* of a body is influenced by surrounding masses. While relationist accounts of motion from Leibniz onward criticized the unobservability of absolute motion, it is clear that Einstein considered Mach's unique contribution to relationist/absolutist debate to have posed the explanatory requirement of a functional dependence (causal, see below) account of inertia as a criterion to be met by any satisfactory account of mechanics.

Nonetheless, a Machian insistence on "observable empirical fact" becomes the centerpiece of an argument in §2 of the above-mentioned first full synthetic exposition of general relativity in April 1916. Entitled "On the Reasons That Suggest an Extension of the Relativity Postulate", Einstein presented a thought experiment involving a relative rotation. He considered two fluid spheres "hovering" in space at a constant but considerable distance from one another and at such a great distance from all other bodies that external influences can be ignored. On each is an observer who judges his own sphere at rest but the other sphere to be rotating with constant angular velocity about the line joining the bodies. However, using local measures, one body is found not to be a sphere but an ellipsoid of rotation, i.e., it bulges along its equator as does the Earth. A variant of Newton's familiar thought experiment of two rotating globes, Einstein's purpose is to identify an "inherent epistemological defect" in the Newtonian account of the situation that "perhaps for the first time was pointed out by Ernst Mach". The defect is this: the Newtonian explanation – that the bulging body is "truly" rotating, i.e., rotating with respect to the preferred inertial frame of absolute space – appeals to a "factitious cause", absolute space, as the reason for the observable difference between the two bodies. This charge is coupled with a highly restrictive codicil governing causality, namely, that application of the "law of causality" to experience has meaning only if it pertains to *observable* causes and effects.[49] Einstein insisted that an observable effect must have an observable cause. But this demand unfortunately sidetracks, even distorts, the salient epistemological issue.

Mach himself regarded the very use of the concepts of cause and effect in physics to be merely picturesque description, a holdover from everyday experience, but liable to give rise to metaphysical fetishism (see Chapter 8). In its place, he urged substitution of the language of functional dependence of (observable) quantities on one another, e.g., the variation of position with differential increments of time as expressed by the equations of mechanics. In a 1912 paper "On the Notion of Cause", well-known to philosophers, Bertrand Russell, following Mach, had urged just this understanding of functional dependence as replacement for talk of causation in physics.[50] Secondly, these observational constraints on causal attributions were read by subsequent philosophers of science as providing yet further confirmation that Einstein's philosophy and methodology of physics fully conformed to logical positivist strictures, which indeed had been fashioned on this understanding of Einstein. But consistency with the overall non-positivist character of general relativity, the archetype of a "principle theory", suggests that criticism of the Newtonian invocation of absolute space as appeal to a "factitious cause" is really a methodological indictment that the alleged cause is not independently discoverable. As such, absolute space is an unexplained explainer invoked only, and conveniently, in the Newtonian story about absolute motion. If, independently of that familiar story, absolute space played an explanatory role in accounting for other mechanical phenomena, its causal standing in the explanation of the observable difference between the two spheres would no longer merit the epithet "factitious", despite the fact that absolute space remains unobservable. However, it plays no other such role in Newtonian mechanics.

This criticism of absolute space can be pushed yet further, from methodology to metaphysics. The dismissal of the Newtonian appeal to absolute space as a "factitious cause", though packaged in positivist window-dressing for publication, is a complaint that absolute space is an artificial, contrived, and unnatural posit. And this is because absolute space has the metaphysical attributes of absolute substance, i.e., something standing outside the nexus of causal relations because it acts, but is not in turn acted upon. From the incipient beginnings of general relativity to the end of his life, Einstein returned again and again to lodge this objection to absolute space. For example, in his Princeton lectures of 1921:

> It is contrary to scientific understanding (*zu dem wissenschaftlichen Verstande*) to posit a thing that in fact acts but is not acted upon.[51]

And in a 1954 letter, a year before his death:

> I see the most essential thing in the overcoming of the inertial system, a thing that acts upon all processes, but undergoes no reaction. This concept is in principle no better than that of the center of the universe in Aristotelian physics.[52]

In recent times, Einstein's objection to unidirectional causal agency in physics has been understood to mean that in general relativity there can be no non-dynamical objects, though there continues to be disagreement upon precisely what counts as dynamical and as an absolute, or non-dynamical, object. The broad metaphysical prohibition against the one-way causal isolation of substance long antedates these discussions.

Mach's principle and relativistic cosmology

During the long months of struggle between 1913 and the completion of general relativity in November 1915, Einstein sustained his faith that such a theory was possible with the idea that a body's inertia was due to its interaction with all other matter in the universe. The Dutch astronomer Willem de Sitter proved the sharpest critic of Einstein's Machian ambitions with general relativity. To de Sitter, Einstein confessed that Mach's conception was "psychologically important because it gave me courage to continue work on the problem when I absolutely could not find covariant field equations".[53] Indeed, the 1918 formulation of "Mach's principle" above originated in a dispute with de Sitter about conflicting cosmological models. In discussions with Einstein in late 1916 in Leiden, de Sitter emphasized that determination of the inertial-gravitational field requires reference not only to distant masses but also to boundary conditions. Since these were generally assumed to be fixed "at infinity" with the fixed, non-dynamical values of the metric of special relativity, they are effectively *a priori* unilateral causal agents. To the extent that they play a role in the explanation of motions in general relativity (e.g., that a freely falling test particle moves on a geodesic), those motions are accordingly absolute in the sense forbidden by Mach.

Having completed the theory of general relativity, Einstein sought to eliminate such "un-Machian" solutions to the field equations of gravitation. In doing so, he created relativistic cosmology in 1917 in a paper declaring:

> In a consistent theory of relativity, there can be no inertia *relative to "space"*, but only inertia of masses *relative to one another*. . . . If I remove a mass sufficiently far in space away from all other masses in the world, then its inertia must drop to zero.[54]

Initially considering the cosmological problem, Einstein assumed a universe infinite in both space and time. Its mass was assumed finite and essentially static, with no large-scale motions between cosmic bodies; the relative distances between stars and "nebulae" (galaxies) remained fixed. These assumptions presumably were made to render cosmology a scientific and not a theological question, and so to avoid the problem of origin. As de Sitter emphasized, solutions of the gravitational field equations in this model required fixing boundary conditions at spatial infinity. For Machian reasons (mass-energy completely determines inertia, see below), Einstein assumed that space (the spatial components of the metric tensor) should vanish at infinity. However, he was unable to find any such solutions, even approximately.

Einstein then effectively "abolished infinity".[55] Under the above requirement on a "consistent theory of relativity" Einstein projected the first relativistic cosmological model, a spatially closed cylindrical world positively curved in the three spatial dimensions hence spatially finite, but indefinitely extended in the time dimension. It is also a quasi-static universe with a uniform distribution of stars. He immediately recognized, however, that such a universe would collapse under the mutual gravitational attraction of cosmic masses. In order to ensure the model's stability, he inserted into his field equation a new "cosmological term" that would be effectively zero at scales smaller than the solar system but at larger scales acted as a repulsive force (a "negative pressure") to exactly balance against gravitational pull:

$$R_{\mu\nu} + \frac{1}{2} g_{\mu\nu} R + \Lambda g_{\mu\nu} = 8\pi T_{\mu\nu}.$$

While it detracts from the simplicity of the equation, the Λ term ('weighted' by the metric $g_{\mu\nu}$) is the only type of term that could be added in a way consistent with local conservation of energy and momentum of $T_{\mu\nu}$.[56] The new term sought to ensure that at the scale of galaxies, there are no secular (non-periodic) motions. Opposing gravitational attraction with a tendency to increase the spatial separation between objects, Einstein deemed the term

> only necessary for the purpose of making possible a quasi-static distribution of matter, as required by the fact of the small velocities of the stars.[57]

Yet if one considers the matter for a moment, it is easy to see that the cosmological term can provide at most only an extremely fragile balance against gravitational collapse. In order to do its job, it has to be "fine-tuned" to the detailed specifics of actual distances between stars and galaxies. And since the stars do have small relative velocities resulting in changing local condensations of matter, to maintain balance, the cosmological term must continually re-adapt to the varying gravitational pull between them. In short, the Einstein universe is highly unstable even for tiny perturbations in the distribution of cosmic masses, a result that some years later would be shown rigorously by Lemaître and Eddington.

Within just two weeks, Dutch astronomer Willem de Sitter produced an exact solution of the now-modified field equations for a matter-free universe, inconveniently demonstrating just how elusive is the goal of a complete relativization of inertia. De Sitter's universe contains just the repulsive energy density of empty space represented in the new cosmological term. The inertial-gravitational field in such a world, having no material sources, is essentially indistinguishable from absolute space; accordingly, from a Machian point of view, it is completely unacceptable. Einstein could only hold his nose while hoping to find some flaw in de Sitter's argument. In this he was unsuccessful, and an ensuing correspondence with de Sitter centered precisely on the question of whether the theory of general relativity satisfied any version of a Machian requirement to relativize inertia. In a letter to de Sitter in March 1917, Einstein made is his philosophical position fully explicit:

> In my opinion, it would be unsatisfactory if there existed a world without matter. Rather the $g_{\mu\nu}$ field should be *fully determined by matter and not able to exist without it*.[58]

This is just the formulation baptized as "Mach's Principle" in 1918.

De Sitter's model universe turned out to be non-static and expanding, though this was not clear for some time (today, de Sitter's model is regarded as a standard early universe vacuum solution to Einstein's field equations). In 1922 Russian mathematician Alexander Friedmann showed that a closed universe of roughly uniform density will inevitably expand to a maximum dimension, then contract in complete gravitational collapse. The most natural solutions of Einstein's unmodified field equations resulted in a dynamic (expanding or contracting) universe, not a static one. Without knowledge of Friedmann's solutions, Belgian Jesuit priest Georges Lemaître rediscovered them and in 1927 formulated a general relativistic model of an expanding universe to accommodate new astronomical data showing an apparent radial velocity of distant galaxies in the line of sight from Earth. Lemaître's model, which crucially contained the cosmological term, was a "fireworks theory" (Eddington),[59] and a precursor of today's hot big bang models. The universe has a beginning, a very small highly compact mass (an initial singularity) and subsequently expands due to radiation pressure. In March 1929 American Edwin Hubble reported data from the 100-inch telescope at Mt. Wilson in California indicating a roughly linear relation between velocities and distance, as measured by redshifts from receding galaxies. Theory and observation had coalesced, agreeing that the universe is described by general relativity and is expanding.

From a philosophical point of view, Einstein was far more comfortable with the notion of an eternal, static, bounded universe. His initial reactions to the possibility of dynamic universes by Friedmann, then Lemaître, were accordingly critical; with Friedmann, he was (wrongly) convinced a mathematical error had crept in, and while he could find no errors in Lemaître, he wrote that "from the physical point of view it was "*tout à fait abominable*".[60] Later on, after Hubble's measurements (together with his assistant Humason) of stellar redshifts convincingly demonstrated the recession of stellar systems from one another, Einstein had to capitulate. On February 5, 1931, the *New York Times* reported that, during a current visit to Caltech, conversations with Hubble and with physicist Richard C. Tolman had convinced Einstein that his original 1917 cosmology, assuming a static and uniform closed universe, "would have to be modified in accordance with more recent data".[61] Since the term had been introduced to ensure the static, fixed character of the universe, Einstein is reported (by George Gamow) to have called its introduction into the field equations of gravitation his "biggest blunder". Whether he in fact said this is unlikely.[62] Today, however, the so-called cosmological constant is a fundamental link, and glaring problem, between quantum field theory and general relativity (see Chapter 11).

The Schwarzschild solutions

Early in January 1916, soon after completion of general relativity the previous November, German physicist and astronomer Karl Schwarzschild (1873–1916), serving as an artillery officer on the Russian front, sent Einstein the first exact solution of the new gravitational field equations. Schwarzschild's solution pertained to the vacuum space-time exterior to the surface of a star or other massive object, idealized as a static (time-independent), non-rotating spherically symmetric body. The so-called Schwarzschild exterior solution describes how the gravitational field of a compact object such as the sun determines the paths of particles and light rays in its vicinity; as with Newton's law, the influence of the localized source of gravity falls off with distance. While congratulating Schwarzschild on the simplicity and elegance of the solution, Einstein was dismayed by its apparent anti-Machian character, for it could be taken to be a solution of his field equations corresponding to the presence of a single body in an otherwise empty universe. On the other hand, it also defined what is now called the gravitational or *Schwarzschild radius* of the body: where r is the body's radius, this is $r = 2GM/c^2$ (where G is the Newton gravitational constant and M is the body's mass; here we follow custom and use units where $G = c = 1$, so $r = 2M$). For a star of the sun's mass, this is about 3 km, whereas the Sun's actual radius is 696,000 km. At the time, and for years afterward, it was generally thought that no actual body would ever become so compressed as to lie within its Schwarzschild radius.[63] Einstein reasoned similarly that a clock kept at this radius would cease to tick (as the g_{44} component vanishes) and so both light rays and material particles would take an infinitely long time (in "coordinate time") to reach it when coming from outside. He therefore considered that a sphere of radius $r = 2M$ to be a place where the gravitational field is singular, and hence unphysical. Gradually (over more than forty years) it become clear that the supposed "singularity" at $r = 2M$ is merely a coordinate effect (similar to the singularity of polar coordinates at the origin), due to a breakdown of the Schwarzschild time coordinate here. Not until the 1960s was it generally accepted that should a massive body collapse under gravitation (e.g., a star that has exhausted its fuel) to less than the Schwarzschild radius, the escape velocity from the body becomes equal to the speed of light, and so the object becomes a "black hole" (a term coined by John Wheeler in 1968). From a general relativistic perspective, the Schwarzschild radius is indeed physical; it defines the radius of the horizon of a Schwarzschild (non-rotating) black hole, the surrounding space-time region into which things may enter from without but from which nothing – not even light – can escape.

That light could not escape the critical circumference of a massive compact "dark star" was not at all a new notion. Though without the concept of space-time curvature, and still assuming the Newtonian corpuscular theory of light, that light could not escape the gravitational pull of a very massive small star had been theoretically considered already in Newtonian gravity in the late 18th century.[64] A few days later in January, however, Schwarzschild sent Einstein an exact computation of the space-time geometry extending the solution to also within a massive spherical body, idealized as a incompressible fluid with a definite radius r. In addition to the Schwarzschild radius, this "interior" solution had the peculiar property that there was a "singularity" at the center ($r = 0$) where the space-time curvature became infinite.[65] While Einstein had to agreed that the exterior Schwarzschild solution indeed showed that the more compact the star, the greater the curvature of space-time around it, as well as the larger the gravitational redshift of light from its surface, on account of the singularity at $r = 0$, Einstein completely rejected the interior Schwarzschild solution as illogical (on grounds that the concept of an incompressible fluid is not compatible with relativity theory: elastic waves propagating in it would have to travel at infinite velocity). As late as 1939, he wrote a paper whose principal conclusion was to show that "the 'Schwarzschild singularities' do not exist in physical reality".[66] Despite Einstein's antipathy to singularities, general relativity indeed predicts that an $r = 0$ singularity exists at the center of a black hole where the density and curvature become infinite, and the known laws of physics break down. Some contemporary relativists assume that such singularities must be "hidden" behind a black hole horizon, and so cannot influence what happens outside, where the normal laws of physics apply (see Chapter 11).

The return of the ether

Many were taken aback when Einstein, in a 1920 lecture at Leiden, rehabilitated the old concept in a discussion of the problem of space. In the presence of the august Lorentz, who never surrendered his belief that an etherial medium is explanatorily preferred in electrodynamics, Einstein declared the "denial of the ether is not necessarily required by the special principle of relativity".[67] A surprising admission, surely, since fifteen years before, in his epochal paper on the theory of special relativity, Einstein had scornfully deemed the ether a "superfluous" theoretical posit. Yet, as he went on to explain both here and in a similar lecture four years later, the general theory of relativity has given a new and distinct sense to the term, bereft of every mechanical and kinematical property ever accorded it.

   In Leiden Einstein had urged that "to deny the ether is ultimately to assume that empty space has no physical properties whatever".[68] According to this criterion, Newton might just as well have used the term "ether" for he accorded some kind of physical reality to absolute space, in his explanation of the presence of inertial effects in rotating bodies. In this sense one can speak of an "ether of mechanics", but it is justifiably called "absolute" since it plays only the asymmetrical role of causal actor, the nexus of causation is not closed. Einstein returned to this point in 1924:

> If Newton called the space of physics "absolute" he was thinking of yet another property of that which we call "ether". Each physical object influences and in general is influenced by others. The latter is not true of the ether of Newtonian mechanics.[69]

Nor is it true of Lorentz's ether. Lorentz posited an ether to explain the transmission of force and radiation across empty space. But his immobile ether is a substance that dynamically acts (contracting lengths and dilating times of bodies and clocks in motion), but is not acted upon by "ponderable matter". As Poincaré forcefully objected, Lorentz's ether failed to obey Newton's third law of action and reaction. Such a failure similarly afflicts the Minkowski metric of special relativity that stands aloof from matter and energy, and so from this point of view the claim to have completely abolished the ether must be seen as over-reaching. In the aftermath of general relativity, Einstein spoke of the "ether of special relativity" which is "absolute, because its influence on inertia and light propagation was thought to be independent of physical influence of any kind".[70] At the end of his life, he even characterized Minkowski space-time as "a four-dimensional analogue of H.A. Lorentz's three-dimensional ether", for failing to eliminate "the *a priori* existence of "empty space".[71]

Summary

In the new conception afforded by general relativity, brought painfully home when Einstein extracted himself from the clutches of the hole argument, empty space-time is not really "empty" but "occupied" or, rather, constituted by the metric inertial-gravitational field. The field is not a quiescent container but has the properties of a dynamical system, interacting with other systems. The shift of emphasis allows the achievement of general relativity, stated in terms of Einstein's cosmological model, to be characterized in terms of the ether:

The ether of the theory of general relativity ... is not "absolute",
but is determined in its locally variable properties by ponderable
matter. This determination is complete if the universe is closed
and spatially finite. The fact that the theory of general relativity
has no preferred space-time coordinates that stand in a determinate relation to the metric is more a characteristic of the mathematical form of the theory than of its physical content.[72]

Einstein's rehabilitation of ether terminology stems from the continued attempt to underscore the Machian ambitions of the theory
of general relativity. It is not Machian à la lettre since general relativity
"excludes direct distant action",[73] perhaps a supposition that Mach
actually put forward as a putative dynamical explanation of inertia
that may not be justified. It is highly implausible that Mach advanced
or even intended to suggest an explanation of inertia in terms of
the mysterious action at a distance of cosmic masses. More charita-
bly, Einstein may be seen as claiming that general relativity gives a
field-theoretic implementation of an attempted explanation of inertia along Machian lines. That Mach has been Einstein's inspiration for
the change in viewpoint regarding the ether is clear, so much so, that
the ether of general relativity is also called "Mach's ether".

But this conception of the ether to which we are led by Mach's
way of thinking differs essentially from the ether as conceived
by Newton, by Fresnel, and by Lorentz. Mach's ether not only
conditions the behavior of inertial masses but is also conditioned in its state by them.[74]

The ether stands in place of "empty space" in Einstein's conception of general relativity. Mach's idée fixe, that acceleration should be
defined relative to a frame of reference determined by the configuration of the entire universe and so eliminating the concept "absolute
space" from physics, proved to be the guiding and perhaps strongest
motivation in Einstein's pursuit of a generalized theory of relativity.

Notes

1 *Albert Einstein-Max Born Briefwechsel 1916–1955.* München: Nymphenburger Verlagshandlung, 1969, p. 251; *The Born-Einstein Letters,* translated by Irene Born. New
   York: Walker and Co., 1971, p. 192.

2 Dirac, Paul, "Methods in Theoretical Physics", in *From a Life in Physics: Evening Lectures at the International Centre for Theoretical Physics*. Trieste, Italy, June 1968. *The IAEA Bulletin*, Special Supplement, p. 24; Landau, Lev and Evgeny Lifshitz, *The Classical Theory of Fields*. Fourth Revised English Edition. Translated by Morton Hamermesh. Oxford, UK and New York: Pergamon Press, 1975, p. 228.

3 "Autobiographical Notes", pp. 66–7.

4 E.g., Norton, John D., "Did Einstein Stumble? The Debate Over General Covariance", *Erkenntnis* v. 42 (1995), pp. 223–45.

5 "Fundamental Idea and Methods of the Theory of Relativity, Presented in Their Development" ("after 22 January 1920"), reprinted in *CPAE* 7 (2002), Doc. 31, pp. 245–81; p. 265.

6 Remark related by Einstein to a research assistant in Princeton, Ernst G. Straus; reported in a letter of Straus to A. Pais, October 1979, quoted in Pais, Abraham, *"Subtle Is the Lord ..." The Science and the Life of Albert Einstein*. New York: Oxford University Press, 1982, p. 239.

7 Einstein, Albert and M. Grossmann, *Outline of a Generalized Theory of Relativity and of a Theory of Gravitation*. Leipzig: B. Teubner Verlag, 1913; as reprinted in *CPAE* 4 (1995), Doc. 13, pp. 302–44.

8 In Göttingen on November 20, and so five days before Einstein's final presentation, Hilbert presented an axiomatic but highly schematic derivation of the Einstein field equations using a variational method. One of Hilbert's two axioms is the stipulation of a generally invariant (i.e., generally covariant) "world function", the object that subjected to variations by "Hamilton's principle", yields the sought-for field equations. The other axiom concerns the form of the world function, coupling its gravitational part to a specific source term $T_{\mu v}$ (rather than the unspecified $T_{\mu v}$ of Einstein); this is a nonlinear (and of course, non-quantum) generalization of Maxwell's electromagnetic theory due to German physicist Gustav Mie that in 1915 purported to be a complete theory of matter (i.e., matter comprised solely of gravitation and electromagnetism). In this sense, Hilbert's ambition was broader than Einstein's; rather than only a gravitational theory, Hilbert proposed a "theory of everything". The importance of general covariance is underscored by its axiomatic status while the gravitational part of Hilbert's "world function" is today known as the *Hilbert-Einstein action*. Hilbert always maintained that general relativity was Einstein's theory; see K.A. Brading and T.A. Ryckman, "Hilbert's 'Foundations of Physics': Gravitation and Electromagnetism within the Axiomatic Method", *Studies in History and Philosophy of Modern Physics* v. 39 (2008), pp. 102–53.

9 "On the General Theory of Relativity" (November 4, 1915); "On the General Theory of Relativity (Addendum)" (November 11, 1915); "Explanation of the Perihelion Motion of Mercury from the General Theory of Relativity" (November 18, 1915); "The Field Equations of Gravitation" (November 25, 1915), as reprinted in *CPAE* 6 (1996), Docs. 21, 22, 24, 25.

10 Observations in the 19th century showed that Mercury's perihelion precesses (the orbits do not close at the point nearest the sun) at a rate of 574 arcseconds (0.159 degree; 1 degree is 3,600 arcseconds) per Earth century. By taking into account the orbits of the other planets, Newtonian mechanics (as computed

Urbain Le Verrier, in 1859, and refined by Simon Newcomb in 1895) could account for 531 seconds of arc per century. Mercury's orbit is, however, a highly nonlinear problem involving the gravitational equations of the Sun and at least five other planets; no exact Newtonian solutions are known and approximations are required. The residual 43 arcseconds was derived in Einstein's paper of November 18. Le Verrier believed the discrepancy due to an unknown small planet closer to the Sun, which he termed *Vulcan*. For a concise modern treatment of Einstein's result, see Robert Wald, *General Relativity*. Chicago and London: University of Chicago Press, 1984, pp. 142–3.

11 Einstein, "Physik und Realität", *The Journal of the Franklin Institute* v. 221, no. 3 (March 1936), pp. 313–47; p. 335; English translation, pp. 349–82. Translation reprinted in Einstein, *Ideas and Opinions*. New York: Crown Publishers, 1954, pp. 290–323, p. 311. In particular, there may be a distinct $T_{\mu\nu}$ for every conceivable matter Lagrangian. To make the situation manageable, one seeks to formulate a set of generic conditions that all "reasonable" $T_{\mu\nu}$ must satisfy, for example, energy densities must always be positive".

12 "On the Foundations of the General Theory of Relativity", *Annalen der Physik* (1918); reprinted in *CPAE* 7 (2002), Doc. 4, pp. 33–6.

13 See Norton, John, "General Covariance and the Foundations of General Relativity: Eight Decades of Dispute", *Reports on Progress in Physics* v. 56 (1993), pp. 791–858.

14 Hertz, Heinrich, "Introduction", in D.E. Jones (trans.), *Electric Waves, Being Researches on the Propagation of Electric Action with Finite Velocity Through Space*. London and New York: Macmillan and Co., 1893, p. 21.

15 "One cannot escape the feeling that these mathematical formulae have an independent existence and an intelligence of their own, that they are wiser than we are, wiser even than their discovers, that we get more out of them than was originally put into them". Quoted in Freeman Dyson, "Mathematics in the Physical Sciences", in National Research Council's COSRIMS (ed.), *The Mathematical Sciences*. Cambridge, MA: MIT Press, 1969, pp. 97–115; p. 99.

16 Einstein, Albert and Leopold Infeld, *The Evolution of Physics: The Growth of Ideas from Early Concepts to Relativity and Quanta*. New York: Simon and Schuster, 1938, pp. 226–35.

17 See, e.g., Eddington, Arthur S., *The Mathematical Theory of Relativity*. Second edition. Cambridge, UK: Cambridge University Press, 1924, §17.

18 In particular, the space-time equivalent of "distance" between two point-events p and q is given by the space-time interval $ds^2 = \sum_{\mu,\nu=1}^{4} g_{\mu\nu}dx^{\mu}dx^{\nu}$, an expression for the Pythagorean theorem holding only for "nearby" space-time events (see Box 6.2).

19 "On the Influence of Gravitation on the Propagation of Light" (1911), in *CPAE* v. 3 (1993), Doc. 23, pp. 485–97; p. 487.

20 "On the Present State of the Problem of Gravitation", *Physikalische Zeitschrift* Bd. 14 (1913), pp. 1249–62, p. 1254; reprinted in *CPAE* 4 (1995), Doc. 17, pp. 486–503, p. 492: "Abstrakt gesprochen: Es gibt kein Relativitätsprinzip der ungleichförmigen Bewegung".

21 "The Foundation of the General Theory of Relativity" (*Die Grundlage der allgemeinen Relativitätstheorie*), CPAE 6 (1996), Doc. 30, pp. 283–339; p. 287. Emphasis added.

22 For discussion, see Janssen, Michel, "'No Success Like Failure . . .': Einstein's Quest for General Relativity", in Michel Janssen and Christoph Lehner (eds.), *The Cambridge Companion to Einstein*. New York: Cambridge University Press, 2014, pp. 167–227; pp. 186–7.

23 Misner, Charles W., Kip S. Thorne, and John A. Wheeler, *Gravitation*. New York: W.H. Freeman and Co., 1973, pp. 302–3.

24 See Friedman, Michael, *Foundations of Space-Time Theories: Relativistic Physics and Philosophy of Science*. Princeton, NJ: Princeton University Press, 1983.

25 This is seen in representing the gravitational field strengths by Christoffel symbols. Since these expressions are not tensors, inertio-gravitational forces attributed by one observer to gravitation may be regarded by another observer as due to inertia. In general relativity, there is no invariant way of distinguishing the two.

26 For example, Banesh Hoffmann, one of Einstein's research assistants in Princeton in the 1930s, and later Professor of Mathematics at Queen's College in New York, commented,

> As for the principle of general covariance, Einstein's belief that it expressed the relativity of all motion was erroneous. . . . Worse, as was quickly pointed out, the principle of general covariance is, in a sense, devoid of content since practically *any* physical theory expressible mathematically can be put into tensor form.
>
> *Albert Einstein: Creator and Rebel*. New York: Viking Press, 1972, p. 127

27 Stachel, John, "Einstein's Search for General Covariance, 1912–1915", in D. Howard and J. Stachel (eds.), *Einstein and the History of General Relativity* (Einstein Studies v. 1), Basel, Boston, and Berlin: Birkhäuser, 1989, pp. 63–100. This paper is based on the written version of a talk circulated since 1980.

28 Einstein, letter to H.A. Lorentz, August 16, 1913; CPAE 5 (1993), Doc. 470, pp. 352–3.

29 "The Formal Foundation of the General Theory of Relativity", reprinted in CPAE 6 (1996), English translation supplement, Doc. 9, pp. 72–130; p. 110 (emphasis added).

30 In particular, Einstein used the fact that there is a 1:1 correspondence between "passive" and "active" diffeomorphisms; a coordinate transformation at a given point p in the hole is used to construct a tensor field at p that is a "carry-along" from another point q in the hole with the same coordinates as p but in another chart. The result is two metrics $g_{\mu\nu}(x)$ and $g'_{\mu\nu}(x)$ at the same point p in the same coordinate system x^σ ($\sigma = 1, 2, 3, 4$) from the same sources $T_{\mu\nu}$ outside the hole, an apparent failure of determinism.

31 "*Einiges über die Entstehung der allgemeinen Relativitätstheorie*", notes published in C. Seelig (ed.), Mein Weltbild. Amsterdam: Querido Verlag, 1934, pp. 134–8; p. 135; translation in *Ideas and Opinions*, 1954, pp. 285–90; p. 288.

32 "The Foundation of the General Theory of Relativity", *Annalen der Physik* Bd. 49 (1916), pp. 769–822; reprinted in *CPAE* 6 (1996), Doc. 30, pp. 293–339; pp. 290–2.

33 Born, Max, *Die Relativitätstheorie Einsteins und ihre physikalischen Grundlagen*. Berlin: J. Springer, 1920, p. 223; Phillip Frank, "Die Bedeutung der physikalischen Erkenntnistheorie Machs für die Geisteleben der Gegenwart", *Die Naturwissenschaften* Bd. 5 (1917), pp. 65–71; reprinted in Frank, *Modern Science and Its Philosophy*. Cambridge, MA: Harvard University Press, p. 73.

34 See Ryckman, Thomas, "'P(oint)-C(oincidence) Thinking': The Ironical Attachment of Logical Empiricism to General Relativity (and Some Lingering Consequences)", *Studies in History and Philosophy of Science* v. 23 (1992), pp. 471–97.

35 Einstein, "On the Foundations of the General Theory of Relativity", *Annalen der Physik* (1918); *CPAE* 7 (2002), English translation supplement, Doc. 4, pp. 33–36, p. 33.

36 The formulation of Wald Robert, *General Relativity*, Chicago: University of Chicago Press, 1984, p. 68.

37 See Stachel (1989), note 26 and John Norton, "How Einstein Found His Field Equations, 1912–1915", also in D. Howard and J. Stachel (eds.), *Einstein and the History of General Relativity* (Einstein Studies v. 1), Basel, Boston, and Berlin: Birkhäuser, 1989, pp. 101–59.

38 The clearest one is "Relativity Theory and the Problem of Space", Appendix V. (1952). *Relativity: The Special and the General Theory: A Popular Exposition*. New York: Crown Publishers, 1961, pp. 135–57; p. 152: "On the basis of the general theory of relativity . . . space as opposed to 'what fills space' . . . has no separate existence. . . . If we imagine the gravitational field, i.e., the functions $g_{\mu\nu}$ to be removed, there does not remain a space of the type (of special relativity), but absolutely nothing, and also not 'topological space'. For the functions $g_{\mu\nu}$ describe not only the field, but at the same time also the topological and metrical structural properties of the manifold. . . . There is no such thing as an empty space; i.e., a space without field. Space-time does not claim existence on its own, but only as a structural quality of the field".

39 Einstein to Andre Mercier, November 9, 1953 (EA 41–884).

40 On the *pro tem* character of the general theory of relativity, see "Autobiographical Notes", p. 75: "Not for a moment, of course, did I doubt that this formulation (the field equations) was only a makeshift in order to give the principal of general relativity a preliminary closed expression. Certainly, it was not anything more than a theory of the gravitational field that was rather artificially isolated from a total field of yet unknown structure".

41 Mach, Ernst, *The Science of Mechanics: A Critical and Historical Account of Its Development*. First German edition, 1883. Translated by T.J. McCormack with revisions through the ninth (1933) German edition. LaSalle, IL: Open Court Publishing Co., 1960, pp. 276–88.

42 *Ibid.*, p. 283.

43 Einstein, "Is There a Gravitational Effect Which Is Analogous to Electrodynamic Induction?" (1912), as reprinted in *CPAE* 4 (1995), Doc. 7, pp. 174–9; p. 178.

44 Einstein, letter to Mach, June 25, 1913, *CPAE* 5 (1993), Doc. 448, pp. 531–2.

45 Einstein, postcard to Mach, December 1913, CPAE 5 (1993), Doc. 495, pp. 584–5.

46 "On the Present State of the Problem of Gravitation", Physikalische Zeitschrift Bd. 14 (1913), pp. 1249–62; reprinted in CPAE 4 (1995), Doc. 17, pp. 486–503.

47 "Four Lectures on the Theory of Relativity Held at Princeton University in May 1921", CPAE 7 (2002), Doc. 71, pp. 496–577; p. 563; translation in Einstein, The Meaning of Relativity. Fifth edition. Princeton: Princeton University Press, 1956, p. 100.

48 Ibid., pp. 498–9.

49 "The Foundation of the General Theory of Relativity", Annalen der Physik Bd. 49 (1916), pp. 769–822; reprinted in CPAE 6 (1996), Doc. 30, pp. 293–339; pp. 286–88.

50 Russell, Bertrand, "On the Notion of Cause", Proceedings of the Aristotelian Society v. 13 (1912–13), pp. 1–25.

51 "Four Lectures on the Theory of Relativity Held at Princeton University in May 1921", CPAE 7 (2002), Doc. 71, pp. 496–577; p. 535; translation in Einstein, The Meaning of Relativity. Fifth edition. Princeton, NJ: Princeton University Press, 1956, pp. 55–6.

52 Einstein, letter to Georg Jaffe, January 19 1954; cited by J. Stachel, "What a Physicist Can Learn From the Discovery of General Relativity", in R. Ruffini (ed.), Proceedings of the Fourth Marcel Grossmann Meeting on General Relativity. Amsterdam: Elsevier Science Publishers, 1986, pp. 1857–62; p. 1858.

53 Einstein, letter to de Sitter, November 4, 1916 CPAE 8, Doc. 273, pp. 359–61.

54 "Cosmological Considerations on the General Theory of Relativity", reprinted in CPAE 6 (1996), Doc. 43, pp. 540–52; p. 544.

55 Eddington, Arthur S., The Expanding Universe. Cambridge, UK: Cambridge University Press, 1933, p. 21.

56 $\Lambda = \dfrac{8\pi G\rho}{c^2}$.

where G is Newton's gravitational constant and ρ is the energy density of empty space; see Chapter 11.

57 "Cosmological Considerations on the General Theory of Relativity", reprinted in CPAE 6 (1996), Doc. 43, pp. 540–52; p. 551.

58 Einstein, letter to de Sitter, 24 March 1917, CPAE 8, Doc. 317, pp. 421–23; p. 422.

59 Eddington, Arthur S. The Expanding Universe. Cambridge, UK: Cambridge University Press, 1933, p. 59.

60 As quoted in Helge S. Kragh, Conceptions of the Cosmos: From Myths to the Accelerating Universe: A History of Cosmology. New York: Oxford University Press, 2007, p. 141.

61 The New York Times, February 5, 1931, p. 17.

62 See Weinstein, Galina, "George Gamow and Albert Einstein: Did Einstein Say the Cosmological Constant Was the 'Biggest Blunder' He Ever Made in Life?", Ms. Ben Gurion University, October 3, 2013.

63 Contemplating in 1926 the hypothetical example of a star with a density of 61,000 gm/cm³ (the mass of the sun within a radius much less than Uranus),

A.S. Eddington's verdict may be taken as authoritative: "I think it has generally been considered proper to add the conclusion 'which is absurd'". *The Internal Constitution of the Stars.* Cambridge, UK: Cambridge University Press, 1926, p. 171.

64 See Israel, Werner, "Dark Stars: The Evolution of an Idea", in Stephen Hawking and Werner Israel (eds.), *300 Years of Gravitation.* Cambridge, UK and New York: Cambridge University Press, 1987, pp. 199–276.

65 More carefully, since the components of the Riemann curvature tensor are coordinate-dependent but its scalars are coordinate-independent, the scalars formed from the Riemann curvature tensor become infinite. In particular, the so-called Kretschmann scalar $K = R_{\mu\nu\sigma\tau} R^{\mu\nu\sigma\tau}$ blows up as the singularity at $r = 0$ is approached. However, a singularity is better defined as a space-time point beyond which a geodesic is not well behaved. See John Earman, *Bangs, Crunches, Whimpers, and Shrieks: Singularities and Acausalities in Relativistic Spacetimes.* New York: Oxford University Press, 1995.

66 Einsten, Albert"On a Stationary System with Spherical Symmetry Consisting of Many Gravitating Masses", *Annals of Mathematics*, Second Series v. 40, no. 4 (October 1939), pp. 922–36; p. 936.

67 "Ether and the Theory of Relativity", reprinted in *CPAE* 7 (2002), Doc. 38, 305–23; p. 314.

68 *Ibid.*, p. 316.

69 "Über den Äther", *Schweizerische naturforschende Gesellschaft Verhanflungen* v. 105 (1924), pp. 85–93; translated in Simon Saunders and Harvey R. Brown (eds.), *The Philosophy of Vacuum.* Oxford, UK: Clarendon Press, 1991, pp. 13–20; p. 15.

70 *Ibid.*, p. 17.

71 "Relativity and the Problem of Space", Appendix V. (1952). *Relativity: The Special and the General Theory: A Popular Exposition*, pp. 135–57; p. 151.

72 "Über den Äther", *Schweizerische naturforschende Gesellschaft Verhanflungen*; as translated by Simon Saunders in Harvey R. Brown and Simon Saunders (eds.), *The Philosophy of Vacuum*, Oxford, UK: Clarendon Press, 1991, pp. 13–17.

73 *Ibid.*, p. 20.

74 "Ether and the Theory of Relativity", reprinted in *CPAE* 7 (2002), Doc. 38, 305–23; p. 316.

Further reading

Barbour, Julian and Herbert Pfister (eds.), *Mach's Principle: From Newton's Bucket to Quantum Gravity.* Boston-Basel-Berlin: Birkhäuser, 1995. (Einstein Studies, v. 6)

Janssen, Michel, "'No Success Like Failure . . .': Einstein's Quest for General Relativity 1907–1920", in M. Janssen and C. Lehner (eds.), *The Cambridge Companion to Einstein.* New York: Cambridge University Press, 2014, pp. 167–227.

Stachel, John, *Einstein From 'B' to 'Z'.* Boston-Basel-Berlin: Birkhäuser, 2002. (Einstein Studies, v. 9).

Part III
Geometry and philosophy

Seven
Geometry and experience

From Cairo in mid-December 1919, Swedish mathematician Gösta Mittag-Leffler wrote to Einstein inviting his contribution to a special issue of the journal *Acta Mathematica* commemorating Henri Poincaré. The French savant – widely acknowledged to be the leading mathematician of his era, as well as an innovative mathematical physicist and influential philosopher of science – had died at the height of his career unexpectedly after minor surgery in 1912. Publication of the memorial issue languished on account of WW I. Still, the more-than-four-year delay furnished ample time for Mittag-Leffler to garner essays from Europe's leading scientific luminaries, among them H.A. Lorentz and Max Planck. Now he wished to include one from the suddenly famous Albert Einstein. In the five weeks since the announcement in London of the solar eclipse observations confirming his relativistic theory of gravity, a media-created mania made Einstein's face globally recognized while creating a myth of unfathomable genius that would burden the rest of his life. Not surprisingly, Mittag-Leffler suggested Einstein contribute an essay on those ideas of Poincaré of particular significance in the development of the general theory of relativity, notably "concerning the relations between space, matter, and time". For unknown reasons, the letter reached Einstein in Berlin only in early April 1920. Nonetheless Einstein promptly replied, agreeing to contribute if allotted sufficient time to write the essay. Then, in late July, he rescinded his promise, pleading that other responsibilities had prevented him from undertaking a needed review of all Poincaré's writings relevant to the question of geometry and experience ("*zu der Frage Geometrie und Erfahrung*").[1] Even so, a seed was planted for renewed reflection on Poincaré. That seed bore fruit in "Geometry and Experience" (*"Geometrie und Erfahrung"*), given as a public address on January 27, 1921 at the Prussian Academy of Sciences on its annual founder's day (*Friedrichstag*) celebration.[2]

Einstein versus Poincaré

Einstein's lecture ostensibly revisits the dispute over the foundations of geometry between geometric empiricism and geometric conventionalism, set now in the new context provided by the general theory of relativity. Published in the proceedings of the Prussian Academy on February 3, an expanded version of "Geometry and Experience" appeared as a separate booklet and was sold out within weeks. Within the year it was translated into French, English, Russian, Italian, and Polish. It would become one of the widely reprinted and influential texts essays in 20th-century philosophy of science, one of three Einstein texts included in the 1953 anthology of papers in philosophy of science edited by Herbert Feigl and May Brodbeck, the first such collection to be published in the USA. By 1929 it had been deemed an essential part of the "Scientific World Conception" (*Wissenschaftliche Weltauffassung*) in the manifesto of the Vienna Circle, written (mostly) by Otto Neurath, Rudolf Carnap, and Herbert Feigl. In a summary section,[3] the logical empiricists epitomized the philosophical significance of "Geometry and Experience" in four points:

1. the need for a clear distinction between pure mathematical (axiomatic) geometry and applied geometry, a branch of physics;
2. the definition of the latter as "practical geometry" characterized by the possible positions of a rigid body in space;
3. an insistence that the definition of rigid body has an "empirical basis", namely, "the preservation of coincidences (equality of spatial intervals)";

and finally the conclusion

4. that from the standpoint of "practical geometry", the question regarding the spatial structure of the universe (concerning both metrical structure as well as finite or infinite extension) has an empirical answer.

Fundamental tenets of logical empiricism are indicated in this brief and, as will be seen, rather tendentious *précis*. Above all, there is an implied rejection of any metaphysics of space. In declaring the question concerning the geometry of physical space to have an unambiguous empirical answer, Einstein was understood not only to invalidate Poincaré's geometric conventionalism but also to ride the positivist hobby-horse of renouncing metaphysics, the ideological cornerstone of the "Scientific World Conception". In particular, Einstein had shown the error of any *a priori* foundation for geometry, in particular the Kantian doctrine that Euclid's postulates express the "necessity and universality" of the form of outer intuition and so are synthetic *a priori* conditions of possible experience. To logical empiricism, the collapse of this bastion of the *a priori* in the theory of space signaled more generally the triumph of empiricism over idealist and metaphysical philosophies; Einstein's "clear formulations brought order into a field where confusion often prevailed".[4] Furthermore, his sharp distinction between empirical "practical" physical geometry and purely formal axiomatic geometry went in tandem with the logical empiricist account of pure mathematics as grounded ultimately in logic. Inspired by Whitehead and Russell's *Principia Mathematica*, logical empiricism considered mathematical statements to be reducible to logical statements and purely mathematical truths to be a species of logical truth. As logical truths are paradigmatically analytic, true in virtue of the meanings of the terms they contain, the account of mathematics as logic supported the core logical empiricist thesis that any meaningful statement is either analytic or is an *a posteriori* synthetic statement, confirmable or refutable by experience. In this way, the logical empiricists would point to "Geometry and Experience" as an illustrious precursor of their dictum that the synthetic *a priori* statements of metaphysics were literally meaningless. Two decades later, the sharp dichotomy between analytic and synthetic statements was attacked as one of the "dogmas" of empiricism in Quine's famous 1951 critique of Carnap.[5]

Less than apparent in the above clipped summary is just how the empiricist conception of the geometry of physical space (-time) ostensibly presented by Einstein actually responds to, and defeats, Poincaré's geometric conventionalism. Seeing this requires unpacking what was stated with considerable compression in 3). "Practical geometry" explicitly rests on Einstein's concept of the "practically rigid body". Now the rigid body is a problematic concept (a "child of sorrow" — *Schmerzenskind*, according to Einstein) already in special relativistic physics. Imagine a very forceful tug on one end of a long completely rigid measuring rod. Since a rigid body has only rigid motions, the fixed spacing between the molecules of the rod will transmit the impulsive force with an indefinitely large (and so possibly superluminal) velocity to the distal end. Nonetheless 3) states that the concept has an "empirical basis" in that observed coincidences between the endpoints of two rigid bodies are preserved when the bodies are translated in space. That statement entails that whenever the endpoints of two "practically rigid" measuring rods are found to coincide in one region of space, they always will be found to do so when the rods are brought together in any other region. Is this really an empirical statement?

A brief reflection should convince that it is not, at least not in any straightforward sense. And so logical empiricism emphasized the "empirical basis" of Einstein's definition of a rigid body in 3) required a *stipulation* of "preservation of coincidence". This states that if two bodies (e.g. rods), whose endpoints are in coincidence at one time at location *A*, are then separated and translated to a distant location B, their endpoints again will be found in coincidence when compared at B at a later time. The definitional nature of this statement is readily appreciated if one considers that the bodies may travel from *A* to B at different velocities along distinct, possibly circuitous paths. That the application of pure mathematics in natural science requires stipulations of this sort, here investing a physical object (measuring rod) with the meaning of "practically rigid body" (and of "equality of spatial intervals") would be enshrined as a central facet of logical empiricism's account of scientific methodology. The necessary first step in the application of any formal mathematical theory (e.g., pure axiomatic geometry) to empirical phenomena required similar stipulations or "coordinative definitions" associating certain concepts or relations of mathematics with observable objects or processes. That empirical determination of the geometry of physical space rested upon postulation of definitional linkages between formal geometric concepts ("distance", "straight line") and physical objects ("measuring rod", "light ray") became the logical empiricists' paradigmatic example of how formal expressions of a mathematized theory acquire empirical meaning in science. Rudolf Carnap in 1927 provided an early, and certainly most graphic, illustration of the significance of the methodology of coordinative definitions. Only through implementing definitions coordinating formal concepts with concrete empirical objects does "the blood of empirical reality" enter through these touch points to flow upward into the most diffuse veins of the hitherto empty theory-schema.[6]

Einstein's lecture does indeed suggest this – in a way. Yet "Geometry and Experience" is not really concerned with the methodological issues emphasized by logical empiricism. Its message is considerably more tempered, endorsing geometric empiricism after a fashion but merely in the guise of a *pro tem* strategy. Practically rigid measuring rods and ideal clocks play, at least provisionally, an epistemologically privileged but in principle logically objectionable role in the general theory of relativity. It is no coincidence that in January 1921, the privileged role of rods and clocks in Einstein's general theory of relativity had become a live issue of contention between Einstein and mathematician Hermann Weyl, a friend and colleague who sought, though in a mathematically speculative way, to reconstruct general relativity without it. What the logical empiricists did not, nor, for the most part, would not mention is the existence of this controversy, the threat it posed to their methodology of coordinative definitions and to the ensuing conception of empiricism in physical science more generally. Considerable stage setting, requiring a detour into the late 19th century debate about the foundations of geometry, will give a clear understanding of the issues at stake. Einstein's paper is only superficially a pointed intervention on behalf of geometric empiricism in its storied confrontation with geometric conventionalism à la Poincaré. Instead "Geometry and Experience" appropriates that earlier debate, a conflict in any case now outmoded in the different context opened up by the variably curved space-times of general relativity, effectively carrying over the no-longer-suitable terms of the earlier discussion into a new, and considerably more intricate, setting. Interestingly, rather than insisting upon what is novel about the geometry of general relativistic space-times, territory firmly occupied in 1921 by Weyl, Einstein chose to largely mute the controversy, emphasizing the *pro tem* benefits of a pragmatic justification of the "practical geometry" of rods and clocks while admitting that Weyl (though in the persona of Poincaré) is correct, in principle ("*sub specie aeterni*"). In the end, what is really at issue in the new situation – unless one knows the backstory – is only dimly perceptible. Like the masked actor in a Nōh play, Einstein relates an illusive and largely symbolic drama, seasoned with stylized elegance (and memorable quotes), directed at an audience that may or may not be able to read between and behind the lines of ritualized presentation.

19th-century geometric empiricism and conventionalism

Einstein's title echoes the fifth chapter ("*Experiénce et géométrie*") of Poincaré's widely read *La Science et l'hypothèsis* (1902), an essay famously concluding, "Whichever way we look at it, it is impossible to discover in geometric empiricism a rational meaning".[7] Geometric empiricism is a 19th-century doctrine, prompted by the mathematical discovery of non-Euclidean geometries earlier in the century and the resulting challenge they presented to the traditional assumption of Euclidean geometry of physical space. Perhaps its most notable proponent was C.F. Gauss (1777–1855), "prince of mathematicians", whose geometric discoveries, as well as those in many other areas of mathematics, were far in advance of his era. As early as 1816, Gauss formulated a non-Euclidean geometry of constant negative curvature, subsequently named after Bolyai and Lobachevsky, its independent discoverers in the late 1820s; fearing a scandal ("the clamor of the Boetians", according to Gauss), he did not publish. On the other hand, in his capacity as director of the Göttingen observatory, Gauss devised a precise cartographic survey of the German principality of Hannover. Employing an instrument of his own design to focus and reflect rays of sunlight between three distant mountain peaks (Inselberg, Brocken, Hohenhagen), Gauss used the method of triangulation to pinpoint locations within this great triangle with unprecedented precision. As a matter of fact, historians disagree whether Gauss actually intended to test the Euclidean assumption that the interior angles between the triangular mountain peaks must sum to 180 degrees. Nonetheless, in a paper of 1827 he reported that this was indeed the case while intimating that any empirical test of Euclidean geometry would require triangulations over much greater, indeed stellar, distances. Critical of the Kantian attribution of necessity and universality to the Euclidean axioms as the form of outer intuition, Gauss classified geometry not with arithmetic (which Gauss assumed *a priori*) but as an empirical science akin to mechanics.

At the time of Poincaré's essay, around the turn of the century, geometric empiricism was primarily associated (not entirely without qualification, as will be seen) with another of 19th-century Germany's most notable scientists, physicist and physiologist Hermann von Helmholtz (1821–1894). Helmholtz's 1866 essay "On the Factual Foundations of Geometry" was his direct response to the posthumous publication of Bernhard Riemann's 1854 *Habilitationsschrift* "On the Hypotheses that Lie at the Foundations of Geometry".[8] One of the landmarks of modern mathematics, Riemann's lecture was delivered in Göttingen before the aged Gauss himself who reportedly praised it highly. Riemann outlined a vastly broad conception of geometry, generalizing the concept of space to what he termed "multiplicities" or "manifolds" (*Mannigfaltigkeiten*) of n dimensions. In such manifolds, afterwards termed *Riemannian*, distances and angles between points are defined by extending an 1828 theorem of Gauss that determined the intrinsic curvature of a two-dimensional surface (i.e., without regard to an ambient space in which the surface is embedded). Riemann generalized to manifolds Gauss's differential expression showing, e.g., how distances between finitely separated points on a surface can be measured along a path connecting them by summing up all the small differences in coordinates between adjacent points (e.g., between x and x + dx) on that path. Riemann's principal assumption required the validity of the Pythagorean theorem only within the infinitesimal region of any point of the manifold, so that the squared distance ds^2 between two nearby points is equal to the sum of their squared coordinate differences, i.e., $ds^2 = \sum_i dx_i^2$. Riemannian manifolds are accordingly said to be "locally Euclidean", or in Riemann's terms, possess "flatness in their smallest parts". However, Riemann suggested that the flat Euclidean infinitesimal regions might be connected up over finite regions in a manner that may give rise to non-Pythagorean (and so, non-Euclidean) distance relations. He even allowed the possibility that any deviation from flatness (i.e., curvature) in such manifolds might vary with position. Still, he recognized that if solid bodies could freely move around "without distension" (altering their size or shape), this could only be the case in manifolds of uniform (constant or zero) curvature – as on the surface of a sphere or, in the latter case, Euclidean space.

In response to Riemann, Helmholtz sought to show that Riemann's fundamental hypothesis, namely, that the metrical relations of a manifold are characterized by the above generalized (differential) form of the Pythagorean Theorem, is no mere postulate but might be derived from facts summarizing our experience of measurements with rigid bodies and paths of light rays. In particular, Helmholtz argued that Riemann's hypothesis could be derived from observations of the arbitrary continuous motions of rigid bodies throughout space. The physical regularities manifested by observed relations of congruence between translated rigid bodies reveal that space is homogeneous, satisfying a condition of "free mobility". Observed satisfaction of free mobility, i.e., "(t)he independence of the congruence of rigid point-systems from place, location, and the system's relative rotation" is then "the fact on which geometry is grounded".[9] What Riemann stated as a *possibility*, Helmholtz recognized as *fact*. On the other hand, Helmholtz acknowledged the notion of a perfectly rigid extended body to be an idealization, exceeding the bounds of experience and not an actual existent. This admission enabled retention of something akin to a transcendental Kantian theory of space in that the free motions of these ideal bodies are expressions of the necessary form of spatial experience, i.e., the space of external intuition.

In general, free mobility should permit construction in space of all figures licensed by the axioms of the geometry. But which axioms are these? Initially Helmholtz concluded they must be those of Euclidean geometry. However Italian geometer Eugenio Beltrami (1835–1899) quickly pointed out that the condition of free mobility does not uniquely single out Euclidean geometry since it obtains also – as Riemann had foreseen – in spaces of constant but (unlike Euclidean space) non-zero curvature. In a footnote added prior to publication, Helmholtz conceded Beltrami's objection. In effect this meant that the familiar association ("coordination") of paths of non-refracted light rays to Euclidean straight lines is not obligatory. Rather, in spaces of constant negative curvature (hyperbolic, or Lobachevskian geometry) or those of constant positive curvature (elliptic, or Riemannian spherical geometry), although rays of light may be deemed straight lines, they are not Euclidean straight lines – in the former, the number of parallels to a given straight line through a given point is infinite; in the latter, zero. Yet in virtue of satisfying the condition of free mobility, all these homogeneous spaces of constant curvature are imaginable, i.e., in conformity with spatial intuition, and so permit the familiar practices of measurement and geometric construction. Helmholtz summarized his position in a telling aphorism: "Space can be transcendental without the axioms being so". Accordingly, he concluded that free mobility remains a necessary presupposition of measurement and so an *a priori* condition of spatiality or spatial intuition, whereas it remains an empirical question which geometry of constant curvature accurately characterizes the motions of extended rigid figures in space.

Poincaré's geometric conventionalism targets Helmholtz's contention that the geometrical axioms of space rest upon presumed facts regarding the free mobility of rigid bodies. Poincaré flatly insisted that the extended rigid body does not exist in reality and in geometry (though not in mechanics) is an impermissible and unphysical idealization. To the concept of distance, for example, there corresponds no perfectly rigid solid body, impervious to deformation through stress, temperature gradients, or indeed, gravitational force. Even so, granting the existence of rigid bodies for the sake of argument, he denied the premise that experiments with such bodies are capable of univocally picking out the geometric axioms characterizing physical space. Rather such experiments can provide information only regarding the mutual relation of these bodies to one another. Finally, following Norwegian mathematician Sophus Lie (1842–1899), Poincaré emphasized that the totality of free continuous motions of presumed rigid bodies has the structure of a mathematical group, now called a Lie group. To say that these motions form a group means that 1) to every translation of a rigid body in one direction there corresponds an inverse operation (a displacement of the same magnitude in the exact opposite direction), 2) that the translations of the body associatively compose (add) together, and that 3) there is an identity, a "motion" that leaves the body *in situ*. The rigid motions themselves can then be given an explicitly mathematical characterization in terms of the transformations of a particular Lie group acting on a space. In turn, geometrical notions measured in space (e.g., distances and angles) are magnitudes invariant under the action of the group on the space. Geometry accordingly becomes the study of a continuous (Lie) group of motions, the general notion of which Poincaré believed, essentially for reasons of natural selection, to be unconscious and pre-existing innately within us, associated with and informing the motions of our bodies in space.

In the 1880s Lie proved (the so-called Helmholtz-Lie theorem) what Riemann and Beltrami conjectured earlier, that the condition of free mobility is consistent with three distinct groups, corresponding to the three geometries of constant curvature: Euclidean (zero curvature), hyperbolic (Lobachevskian or negative curvature) and elliptical (Riemannian spherical, or positive curvature). Poincaré interpreted Lie's result to mean that experience alone was unable to distinguish between these; hence, there should be no compulsion to choose one as solely correct. Although experiments made with bodies supposed rigid informs the choice of geometrical axioms for physical space, they do not and cannot determine that choice. Rather, if choice need be made, it is made on grounds of convenience.

> Our choice is therefore not imposed by experience. It is simply guided by experience. But it remains free; we choose this geometry rather than that geometry, not because it is more *true*, but because it is more *convenient*.[10]

As there can be no firm empirical basis for the claim that a single geometry is true or false of physical space, geometry is not to be regarded as an experimental science. On the other hand, Poincaré thought that Euclidean geometry would always be selected on the ground of simplicity; mathematically its simplicity resides in the fact that the group of Euclidean rigid motions alone contains a proper normal subgroup comprising both free translations and rotations about a fixed point. But as one cannot say in general that a particular geometry (say, Euclidean) is true of physical space while another (say, that of Lobachevsky) is false, geometric empiricism can have no clear meaning.

How then does a conventional geometry relate to physics? Poincaré argued that if, in astronomy, the path of a light ray is taken to realize the geometrical notion of "straight line", observation by itself cannot determine a unique geometry of space. He illustrated this claim with an example pertaining to measurement of the parallax of distant stars, i.e., the angular difference in the apparent position of a star as seen from opposite ends of the Earth's orbit of the sun, six months apart. On account of the vastness of stellar distances in comparison to the size of the Earth's orbit, stellar parallax are exceedingly small and difficult to measure; only in 1838 did Friedrich Bessel succeed in measuring the first stellar parallax (the star 61 Cygni). Assuming a Euclidean geometry of physical space, these values are in general quite small, below a certain threshold. Conceivably, however a parallax might be observed to be above this limit (if the geometry of space is hyperbolic or Lobachevskian) or to be negative (if it is the elliptical geometry of Riemann). Supposing observation reveals either of these non-Euclidean values, Poincaré insisted it is still possible to retain Euclidean geometry as the geometry of space by modifying the laws of physics (geometrical optics), namely, that light does not exactly propagate in a Euclidean straight line. More generally, the ideal basic concepts of geometry (e.g., "straight line") lack any physical meaning in isolation but can be variably interpreted, depending on the content of the assumed physical laws. For this reason, he concluded, "Euclidean geometry has nothing to fear from new experiments".[11]

Somewhat surprisingly, neither Poincaré nor Helmholtz took into account the full generality of geometries permitted by Riemann's theory of manifolds as viable candidates for the geometry of physical space; they restricted attention to spaces of constant curvature wherein alone free mobility is possible. But generic Riemannian geometries extend far beyond the geometries of constant curvature; in particular, they may also possess variable curvature – curvature varying from region to region, and even (smoothly) from point to point. These geometries describe spaces where rigid bodies in principle cannot exist, let alone possess free mobility. The far-sighted Riemann even anticipated the dynamical character of space-time in general relativity, conjecturing that should space be found to have variable curvature, matter might well be the cause. Apparently such ideas were too radical for both Helmholtz and Poincaré; Poincaré in particular took note of them only to deny that the Riemannian geometries of variable curvature could ever be anything more than mathematical curiosities since they cannot be characterized by deductive (synthetic) axiom systems akin to Euclid's, that is, they "*ne pourraient donc jamais être que purement analytiques*" ("could never be other than purely analytic").[12] By 1921, such conservatism regarding the possible geometries of physical space had been swept away by the advent of general relativity, a theory of variably curved space-times, and so the parameters of the dispute between geometric empiricism and conventionalism had been fundamentally transformed. This was the new situation into which Einstein sought to intervene with "Geometry and Experience".

"Geometry and Experience" 1921

Whereas Poincaré deemed geometry to be the study of the formal properties of a particular continuous group, not an experimental science concerned with imperfect physical realizations of geometrical notions, Einstein began his essay by drawing an apparently related distinction between "axiomatic" and "practical" – i.e., applied – geometry, summarizing the difference in a pithy, oft-quoted comparison:

> As far as the laws of mathematics refer to reality, they are not certain; and as far as they are certain, they do not refer to reality.[13]

The declaration expresses the modern axiomatic conception of pure mathematics, exemplified in Hilbert's 1899 *Foundations of Geometry* (*Grundlagen der Geometrie*), a work continually augmented through its seventh (1930) edition and translated into many languages.[14] Hilbert famously considered Euclidean geometry as pertaining to "three systems of things, called points, . . . lines, and . . . planes", primitive terms whose meanings are established purely formally by five kinds of axioms that determine what can be said about them. The axiom groups (incidence, order, congruence, parallels, and continuity) of geometry, as Einstein put it, are "free creations of the human mind", constrained only by the meta-theoretical requirements of their completeness, mutual consistency and independence. From the outset, the primitive terms of geometry have no reference to the physical world or to the contents of spatial intuition but possess only a contextual meaning bestowed by their occurrence in the axioms and subsequent theorems. This indirect manner of specifying the meaning of primitive terms of a theory had been advanced nearly a century earlier by the French geometer Gergonne, who coined the term "implicit definition" in a paper of 1818. Hilbert gave Gergonne's method its first completely rigorous implementation. Remarkably, in "Geometry and Experience" Einstein does not at all mention Hilbert but approvingly cites Schlick's 1918 treatise on general epistemology for its "highly apt" characterization of axioms as "implicit definitions" of the meanings of primitive concepts (§7 of Schlick's book pointedly refers to Hilbert's *Grundlagen*).[15]

"Practical geometry" on the other hand is the geometry of the practicing physicist, the application of formal axiomatic geometry to the physical world. Most familiar as the Euclidean geometry of surveying and measurement, it arises in an obvious way from an assumption that solid bodies, e.g., measuring rods, relate to one another as do the ideal rigid bodies of Euclidean geometry. Statements of axiomatic geometry are then interpretable as assertions regarding the relations of physical bodies and Euclidean practical geometry is accordingly a physical science, indeed, "the most ancient branch of physics", its assertions at best empirical truths, having approximate validity. To Poincaré's objection that no actual physical bodies are rigid, Einstein countered that experimental practice is able to determine the physical state of a measuring bodies to sufficient accuracy to render their fiduciary metrical behavior with respect to one another free of ambiguity. The practice of spatial measurement is rooted is in this pragmatic assumption and so "all linear measurement in physics (including 'geodetic and astronomical measurement') is practical geometry in this sense".

Having set out the two opposing conceptions of geometry, Einstein made a case for geometric empiricism in the general theory of relativity. The metrical relations of space-time are not a matter of convention but can be empirically established as Euclidean or non-Euclidean using the instruments of practical geometry. In particular, he observed that the (conceptual) availability of practical geometry, i.e., "the possibilities of relative situation" (*Lagerungsmöglich-keiten*) of practical rigid bodies, had been decisive in taking the momentous step to his theory of gravitation. Briefly alluding to the "rotating disk" thought experiment of 1912 (discussed in Chapter 6), Einstein noted that without its heuristic message, concerning the Lorentz contraction (and so, non-Euclidean behavior) of rigid rods laid along the circumference of a system uniformly rotating with respect to an inertial system, he never should have hit upon the idea that led to the general theory of relativity. As he explained, consideration of the uniformly rotating disk enabled him to see that the geometry of accelerated systems, and so gravitational fields, must be non-Euclidean.[16] The crucial link was provided by the principle of equivalence: the geometry of a gravitational field (a non-inertial system) could not be Euclidean. And in fact the geometry of generic space-times (for arbitrary distributions of matter-energy) in the general theory of relativity is a non-Euclidean Riemannian geometry of variable curvature.

Nonetheless, the general theory of relativity mandated a generalization of Euclidean practical geometry. In the Riemannian context, measuring rods can be considered "practically rigid" only in "infinitesimal" (not extended) regions. The concept of an "infinitesimal rigid rod" arises by considering the notion of a "line segment", exemplified in the extension bounded by two nearby marks on solid body. Line segments permit comparison in the obvious way by bringing together two solid bodies, on each of which is a bounded line segment. Segments are said to be equal just in case the boundaries of one segment exactly coincide with the boundaries of the other. On the plainly self-evident nature of this claim, another is piggy-backed that the unwary reader may regard as equally self-evident:

> If two line segments are found to be equal at one time and at some place, they are equal always and everywhere.[17]

Practically rigid rods (of "infinitesimal" extension) physically implement this supposition about equality of line segments. But any assumption of permanent congruence is contentious, though Einstein pointed out that Riemannian practical geometry follows Euclidean practical geometry in adopting it.

A practical geometry suitable for measurements in the theory of relativity must include not only "practically rigid" infinitesimal rods but also "perfect" or ideal clocks. Abstractly considered, in general relativity a clock is simply a periodic physical process by which numbers are assigned to events on the world line of the clock in such a manner that the number of "ticks" of the clock between two events on its world line is directly proportional to the extension of world line between the events. To say that such clocks are "perfect" is to make an assumption analogous to that above for equality of line segments. Namely, if two clocks are found to tick at the same rate at a common initial location but then are separated, they will be found to tick at the same rate when they are brought together again. Einstein asserted "a convincing experimental proof" of this assumption (let us call it constant synchrony) is found in the characteristic frequencies of "natural clocks", i.e., the discrete pattern of spectral lines uniquely characterizing a chemical element when heated to incandescence, no matter where the element is. The spectrum of hydrogen is, for example, the same in the laboratory as when observed in the light from distant stars (with adjustment for redshifts occasioned by the expansion of the universe, confirmed only some years later in the 1920s).

As underscored by the logical empiricists, with the supposition of "practically rigid bodies" (and "perfect clocks"), the metrical relations of physical space (space-time) are not a matter of conventional stipulation but can be empirically determined from measurements of distances, angles, and durations; in this way, a clear decision can be reached regarding the particular character of the geometry of space-time. The known empirical tests of general relativity rest upon the assumptions of Riemannian practical geometry, in particular the association of the line element ds with measurements made using "practically rigid bodies" and "perfect clocks". These implements serve as a bridge coordinating the phenomena of gravitational mechanics to a purely mathematical non-Euclidean Riemannian geometry. Einstein summarized the general viewpoint in this way:

> To be able to make such assertions (concerning the behavior of real objects), geometry must be stripped of its merely formal-logical character by assigning to the empty conceptual schemata of

axiomatic geometry objects of reality that are capable of being experienced.[18]

As noted above, logical empiricists regarded this conclusion a canonical expression of their own methodology concerning how the mathematical structures of a physical theory acquire empirical meaning.

Sub specie aeterni

Having argued for an empiricist conception of the geometry of space-time resting upon the "practical geometry" of rods and clocks he implicitly (but elsewhere explicitly) associated with Helmholtz, Einstein proceeded to consider the alternative resting on an objection that actually rigid bodies (even infinitesimal ones) and perfect clocks do not exist while pure, or axiomatic, geometry by itself affirms nothing about the behavior of actual bodies. According to this "more general conception, which characterizes Poincaré's standpoint", only the combined system (G+P) of geometry (G) plus physical laws (P) may admit of empirical test. Then G, a particular geometry of space-time, may be conventionally chosen as well as parts of P; however, the remainder of physical laws P (above all, the laws of optics) must be such to ensure that the total system (G+P) is brought into agreement with experience. Somewhat surprisingly in view of the just presented case for empiricism, Einstein conceded that only choice of a total system (G+P) could be considered correct *sub specie aeterni*.[19] The logical empiricists chose to simply ignore this apparent concession to conventionalism. What is going on? How can Einstein give back with his left hand what his right hand has taken away? Is admission of the correctness *sub specie aeterni* of the French mathematician's standpoint on the conventionalist relation between geometry and experience another manifestation of Einstein's unscrupulous philosophical opportunism?

So many have thought. The puzzle is amplified by philosophers who maintain that the (G+P) formula is better identified with the conventionalism of Poincaré's contemporary, Pierre Duhem (1861–1916) than that of Poincaré himself.[20] Duhem famously argued that a physical theory comprises an entire group of hypotheses in which no single hypothesis considered by itself has observational consequences of its own. A direct implication of this holist "non-separability thesis" is that empirical confirmation or disconfirmation (falsification) pertains only to the theory – an entire collection of hypotheses – as a whole. More precisely, the empirical test of a physical theory implicates the class of hypotheses with an unsettling consequence that, should observation fail to agree with predictions derived from the theory, the only immediate conclusion to be drawn is that at least one hypothesis in the collection requires modification, though said experiment does not tell which. As is the case with other theoretical physicists of the period, there is textual evidence that Einstein held such a conformational holism regarding the test of a physical theory. In fact, there was broad recognition of this thesis, which is almost obvious once physical theories are considered similar to formal axiomatic systems. Moreover, Einstein mentioned Poincaré not Duhem.[21] Of greater relevance is the fact that Einstein evinced a critically negative attitude, most forcefully in his April 1918 tribute to Planck, toward Duhem's concomitant thesis of underdetermination of theory by empirical evidence (see Chapter 9). But perhaps above all, it is significant that in "Geometry and Experience" Einstein does not state that the (G+P) holism is correct (as presumably would be the case if he had in mind a Duhemian non-separability thesis pertaining to empirical and semantic content), but only that it is correct *sub specie aeterni*. That crucial modifying phrase suggests that Einstein's invocation of Poincaré's name (not Duhem's), is both intended and fully appropriate. For the salient issue with Poincaré concerns the standing of the assumptions of practical geometry that, Einstein admitted, do not survive scrutiny *sub specie aeterni*. Einstein's choice of the Latin expression is revealing, almost certainly an allusion to Spinoza, with whom the phrases *sub specie aeterni* ("under the aspect of the eternal") or *sub specie aeternitatis* ("under the aspect of eternity") are familiarly associated. As he had been also for Schopenhauer, Nietzsche, and Wittgenstein, all of whom similarly borrowed the Latin phrase for their own purposes, Spinoza was an intellectual hero of Einstein.[22]

Now Einstein's concession that only the sum G+P admits of empirical test *sub specie aeterni* is surely not an appeal to the infallible intuition of rational insight. Nor is it an endorsement of Poincaré's own position regarding a conventional choice of geometry.

But it is an explicit admission that the physical assumptions (P) of "practical geometry", posits concerning the behavior of measuring instruments considered fiduciary (permanent congruence, constant synchrony), are impermissible *sub specie aeterni*. The surrounding text shows two reasons for this. First, Poincaré's staunch unwillingness to consider any physical object as "geometrical", i.e., as ideal, is in principle correct. Such an admission might be thought to be the voice of logical conscience but it has, as will be seen, far-reaching implications for the foundations of physics. Secondly, Einstein has another target in his sights, contrasting the existing standpoint of physical theory with a distantly future, presumably far more complete, point of view. The imperfect state of physical knowledge at the time (P) did not (and does not today) furnish a sufficiently detailed microphysical understanding of bodies to permit derivation of such macroscopic properties as solidity and rigidity. More precisely, the current state of physical knowledge did not permit derivation of the above two assumptions of practical geometry. This is the voice of a pragmatic thinker, expressing a readiness to tolerate (as Weyl, see below, did not) a conceptual blemish in an otherwise aesthetically pleasing and (most importantly) empirically successful theory. The two reasons are clearly distinguished in this passage from "Geometry and Experience":

> The concept of the measuring body as well as in the theory of relativity also the coordinate concept of a measure clock find no exact corresponding object in the real world. It is also clear that the rigid body and the clock do not play the role of irreducible elements in the conceptual edifice of physics, but that of composite structures, which must not play any independent part in the construction of theoretical physics. However, it is my conviction (*Überzeugung*) that in the present state of development of theoretical physics they must still be employed as independent concepts, for we are still quite far from such secure knowledge of atomic physics (*Atomistik*) to be able to provide their exact theoretical construction.[23]

The first sentence simply concedes Poincaré's point that geometrical notions have no exact counterparts or physical realizations. As

Einstein well knew, there are reasons of principle why there can be no extended rigid bodies already in the special theory of relativity. The invariance of the speed of light, restricting the velocity of possible physical influences, in effect meant that no actual physical body could be perfectly rigid. The second sentence affirms a consideration of more relevance to the general theory of relativity: to accord rods and clocks the status of "irreducible elements" is to place them outside, and so independent of, the laws of gravitation pertaining to all other physical objects, as well as of dynamical laws of matter that might explain the behavior of such complicated structures. Yet the last sentence of the passage affirms a "conviction" that nonetheless in the existing imperfect state of theoretical knowledge, it is still reasonable to consider rods and clocks as *de facto* independent of these laws. A conviction is typically a matter of faith or belief; later on it even becomes a "sin". What in "the present state of development of theoretical physics" elicits statement of such a conviction? The answer appears several paragraphs further along, where Einstein offers a justification for the concept of an ideal clock in the theory of relativity:

> If two ideal clocks are going at the same rate at any time and at any place (being then in immediate proximity to each other), they will always go at the same rate, no matter where and when they are compared with each other at one place. If this law were not valid for natural clocks, the characteristic frequencies of individual atoms of the same chemical element would not be in such exact agreement as experience demonstrates. The existence of sharp spectral lines is a convincing (*überzeugenden*) experimental proof of the above-mentioned basic principle of practical geometry. On this ultimately rests the fact that we can meaningfully speak of a metric of the four-dimensional space-time continuum in the sense of Riemann.[24]

The "conviction" allowing *de facto* independence of ideal clocks from fundamental theory is then licensed by the "convincing experimental proof" concerning the sharp spectral lines of the chemical elements. The target of this particular argument is not Poincaré but the mathematician Hermann Weyl with whom, in 1921, Einstein was engaged in an ongoing disagreement.

Weyl's theory of gravitation and electromagnetism

In the spring of 1918, mathematician Hermann Weyl raised an objection of principle to the pragmatic assumptions of Einstein's practical geometry. Weyl's geometrical theory of gravitation and electromagnetism essentially rests upon a generalization of the Riemannian geometry of the theory of general relativity. It is highly mathematically sophisticated, termed "the most beautiful of all theories" by the physicist Paul Dirac many years later. But its fundamental idea is both simple and relevant to the argument of "Geometry and Experience". Recall that vectors have two properties, magnitude and direction. Consider two points p and q at finite separation in a space or manifold. The metric of a Riemannian geometry (and so of general relativity) permits direct comparison of magnitudes of two vectors A at p and B at q but not with respect to directions. In 1917 Italian geometer Tullio Levi-Civita (1873–1941) analytically showed how directional comparisons can be made between vectors at finitely separated points in a Riemannian space. This can be visualized as the "parallel transport" of a fiduciary vector C parallel to A at p along a continuous path from p to q so that at each successive point along the path, C remains parallel to itself at the previous point. At q the angle difference, if any, between B and the transported standard C can be determined. In a curved manifold the result of such direction comparisons will depend on the path taken from p to q. But the fact is that in Riemannian geometry, comparisons of magnitude and direction are treated quite differently. To eliminate this last vestige of "action at a distance" from field physics, Weyl constructed a non-Riemannian geometry to remove the asymmetry: both magnitude and direction comparisons require a "transported" standard. The result, to Weyl's surprise, was a metric for space-time geometry with degrees of freedom that incorporated not only gravitation (Einstein's field equations of general relativity) but also electromagnetism. As these forces were the only fundamental interactions known in 1918, Weyl could claim that according to his theory all physics is at base a manifestation of his space-time geometry.

In a bit more detail, Weyl constructed his theory to expressly deny the Einstein-Riemann assumption about the permissibility of distant comparisons of line segments and durations, equivalent to the

existence of "practically rigid" infinitesimal rods and "ideal clocks".
In Weyl's geometry, congruence – the primary metrical concept – is
"purely infinitesimal" in that magnitudes (lengths) can be immediately compared only at neighboring points but not "at a distance".
Length comparisons between p and any point q at finite separation
are not assumed as in Einstein-Riemann but must be constructed.
A comparison of lengths at finitely separated points begins with a
comparison at a given point p and a neighboring one p′ "infinitesi-
mally adjacent" to p (more precisely, in the linear tangent space surrounding p). Weyl extended Levi-Civita's idea of "parallel transport"
pertaining to directional comparison to a length comparison whereby
a standard of length (normalized unit vector) is "transported" from
p to p′. ("Parallel transport" only figuratively involves moving a vector; the operation is defined purely analytically.) A length comparison between a vector at p and one at any distant point q is established
by transporting the fiduciary standard point by point along a path
p′, p″, p‴, . . ., q. Since there are (infinitely) many distinct paths
between p and q, this procedure is clearly path-dependent, that is,
the "distance" between p and q is said to be "non-integrable". In
Weyl's "purely infinitesimal" standpoint, the twin assumptions of
Einstein's Riemannian practical geometry, pertaining to equality of
lengths and of durations independently of place, are no longer permissible postulates. If, as in general relativity they are assumed, they
must be understood as empirical facts to be explained by solving
the combined field equations for gravitation and matter. In more
generic space-times according to Weyl's geometry, two infinitesimal
measuring rods congruent at one space-time point p subsequently
transported by different paths to another point q at a finite distance
from the first are no longer congruent. This is just the denial of Einstein's assumption of "practically rigid" measuring rods.

Einstein versus Weyl

Although Einstein immediately accorded Weyl's theory the accolade
of "an achievement of genius of the first rank", from its inception
in early 1918, he inveighed against it, both in public and in private
correspondence. Arranging for the initial publication of Weyl's theory in the *Proceedings of the Prussian Academy of Sciences*, Einstein appended

a short comment stating his belief that the theory could not be in agreement with observation. For according to Weyl's theory, measuring rods and clocks (or the radii of atoms and their spectral frequencies) are not independent of their position in space and time but rather depend, Einstein argued, on their "prehistory", more precisely, on the strength of the electromagnetic fields through which their world lines had passed. So if two chemical atoms together at one location are then separated, travel different paths through spacetime and are subsequently reunited, they should display different line spectra if one but not the other passed through a strong electromagnetic field. But astronomical observation does not detect this difference; the spectra of all the chemical elements are sharp and do not vary in light sources distributed throughout space (though, as discovered by Hubble in the 1920s, there are Doppler-like shifts in frequency of light from stars moving away from or towards the Earth due to the expanding universe). This then is the objection, not to Poicaré, but to Weyl, reiterated in "Geometry and Experience":

> The above assumption for line elements must also hold good for intervals of clock time in the theory of relativity. Consequently this assumption may be formulated as follows: If two ideal clocks are going at the same rate at any time and at any place (being then in immediate proximity to each other), they will always go at the same rate, no matter where and when they are compared with each other at one place. If this law were not valid for natural clocks, the characteristic frequencies of individual atoms of the same chemical element would not be in such exact agreement as experience demonstrates. The existence of sharp spectral lines is a convincing experimental proof of the above-mentioned basic principle of practical geometry.[25]

Weyl and others did not find this objection convincing; Weyl held on to his theory until in the early 1920s he became convinced that a continuum-based theory, such as his, could not account for atomic phenomena.[26]

In September 1920, just a few months before "Geometry and Experience", at the 86th annual meeting of German scientists at Bad Nauheim a resort spa near Munich, Einstein confronted Weyl's theory in public, but with a distinctly different argument. The context may provide a reason for this. The meeting became infamous for the intransigent opposition to the theory of relativity expressed by certain well-placed German physicists, above all, Nobel prize winner Philip Lenard. Lenard and Nobel prize winner Johannes Stark, both experimentalists, objected in particular to general relativity, a highly abstract theory employing advanced mathematics unfamiliar to most physicists.[27] With this witches' broth brewing (it came to full boil only after the Nazis seized power in 1933), Einstein in 1920 was understandably sensitive to the fact that there were then but three posed empirical tests of general relativity, only two of which had been satisfactorily met. Despite the dramatic confirming solar eclipse observations of 1919, Einstein was understandably concerned to show that the general theory of relativity had clear ties to observation (through the coordination of the metric tensor to "practically rigid bodies" and "perfect clocks"). Accordingly, he had very plausible expedient reasons for the halfway house *pro tem* defense of the use of "practically rigid rods".

Responding to Weyl's presentation at Bad Nauheim, Einstein reiterated that general relativity is empirically based on measuring-rod geometry. Weyl's demand that the reliability of rods and clocks must be explained through derivation as solutions to a combined theory of gravitation and matter (electromagnetism) would rob general relativity "of its most solid empirical support and possibilities of confirmation":

> Temporal-spatial intervals are physically defined with the help of measuring rods and clocks. If I consider two (such) structures, then their equality is empirically independent of their prehistory. Upon this rests the possibility of coordinating a number ds to two neighboring world points. Insofar as the Weyl theory renounces this empirically-founded coordination, it robs the theory (general relativity) of its most solid empirical support and possibilities of confirmation.[28]

This is a different kind of objection to Weyl's theory, placing more emphasis on the empirical character of general relativity than on confronting Weyl's theory. Reverting to an injunction Einstein will later make in Oxford, (see Chapter 10) that to know anything about the method of theoretical physics one should examine the theoretician's deeds and not listen to his words, a more ambivalent position emerges. Despite concerns to defend the use of rods and clocks as necessary to the empirical ties of the general theory of relativity, Einstein would not be constrained by concerns of empiricist methodology in attempting to advance physical theory beyond general relativity. Already in 1923, in one of the first of his many proposals for a unified field theory, the ties of theory to experience are not even considered. In a report on this theory, which starts from a non-metrical (affine) basis, he essentially affirms that observational concerns are completely subordinated to the overriding goal of attaining a theory of greatest mathematical simplicity, a theme that will become increasingly prominent in his later years.

> The search for mathematical laws that shall correspond to the laws of nature . . . resolves itself into the solution of the question: What are the most natural formal conditions that can be imposed upon an affine relation?[29]

Moreover, in his "Nobel lecture" that same year, he mentioned the "deficiencies of method" in tying relativity theory to observation through the posits of rigid rods and clocks, contrasting it with the "logically purer method" of Levi-Civita, Weyl, and Eddington (who had been spurred by Weyl's theory to present his own geometrical unification of electromagnetism and gravitation in 1921):

> Certainly it would be logically more correct to begin with the whole of the laws and to apply the "stipulation of meaning" to this whole first, that is, to put the unambiguous relation to the world of experience last instead of already fulfilling it in an imperfect form for an artificially isolated part, namely, the space-time metric. We are not, however, sufficiently advanced in our knowledge of nature's elementary laws to adopt this more perfect method without going out of our depth. . . . we shall see that in the most recent studies there is an attempt, based on ideas by Levi-Civita, Weyl, and Eddington, to implement that logically purer method.[30]

Weyl was not merely being unreasonable in sticking to his guns. Einstein clearly recognized that a "logically purer" field physics denying the last vestiges of "action at a distance" must derive, not postulate, the regularities of "practically rigid" measuring rods and "ideal clocks" from underlying field dynamics. Despite his objections, Einstein had to concede that Weyl had touched upon an explanatory sore point and a logical inconsistency in the story Einstein felt compelled to tell about the connection of geometry to experience. *Sub specie aeterni* something similar to Weyl's dynamical explanation is a more principled story.

Reprise: Constructive vs. principle theories

Einstein's long trek to general relativity was spurred by the denial that an irreducibly non-dynamical entity might exist, an "absolute space" that acted in producing inertia-gravitational effects but was not in turn acted upon. Yet rigid rods and perfect clocks, independent of the dynamical laws of physics, are similarly absolute. The similarity becomes palpable in a moment's reflection: Ideal measuring rods and perfect clocks satisfying the permanent congruence and constant synchrony assumptions yield absolute (though inertial frame-dependent) measures of lengths and durations. In the special theory of relativity, rods and clocks are absolute: Lorentz contraction of practically rigid rods and time dilation of ideal clocks in a moving frame (as determined from practically rigid rods and ideal clocks in a rest frame) are regarded as merely kinematical effects, pertaining not to dynamical changes in these bodies but to the relative motion of reference frames of observers. The permanent congruence and constant synchrony assumptions are obviously satisfied in the inertial frame in which both sets of instruments are initially together at rest, and the one in which they are subsequently rejoined. So the special theory of relativity accordingly treats rods and clocks as ideal bodies independent of the dynamical laws of physics.

General relativity, to the contrary, must also consider rods and clocks in accelerated frames; this is essentially the reason motivating Einstein's generalization of Euclidean practical geometry of rods and clocks to a Riemannian practical geometry of infinitesimal rigid rods and clocks ticking (generating time) only along their own world line. But in the variably curved space-times of general relativity, the above assumptions guaranteeing the ideal behavior of instruments measuring space and time are considerably more problematic. One can think about it this way. Imagine the microstructure of a solid (albeit "infinitesimal") rod to be essentially a lattice of atoms, each at constant spacing from its neighbors. Suppose the body to freely fall horizontally through a strong gravitational field. As discussed in Chapter 6, the tidal forces of that field will act differentially along the length of the rod, with the result that particles along its length will traverse non-parallel freefall trajectories. The rod will deform as the rigid connections between the atoms give way. Similarly, in a gravitational field, clocks run more slowly, the stronger the field strengths, the slower the rate of the clock. But how can infinitesimal measuring rods and perfect clocks be immune from the tidal forces of gravitation? And yet they must be if they are stipulated to satisfy the two permanence assumptions.

Recall Einstein's 1919 distinction between theories of principle and constructive theories (see Introduction) and the superior explanatory force of the latter. The voice of logical conscience in "Geometry and Experience" allowing that "the solid body and the clock do not in the conceptual edifice of physics play the role of irreducible elements, but that of composite structures, which must not play any independent part in theoretical physics" affirms not only the physical inconsistency of exempting certain bodies (rods and clocks) from the dynamical laws of microphysics. It also accords recognition to the superior understanding and explanatory depth that a constructive account of the behavior of these bodies in the theories of relativity could provide, especially within the general theory since the permanence assumptions flagrantly conflict with fact that gravitation acts on all matter. General relativity, a principle theory, does not have the resources to account for the structure of matter; it lacks an explicit theory of matter, an acknowledged gap Einstein unsuccessfully sought to fill over a three-decades-long pursuit of a unified field theory. Correspondingly, there is occasional admission of the unsatisfactory treatment in general relativity of rods and clocks as "theoretically self-sufficient entities":

Strictly speaking measuring rods and clocks would have to be represented as solutions of the basic equations (objects consisting of moving atomic configurations), not, as it were, as theoretically self-sufficient entities.[31]

Yet immediately following this voice of conscience there is another, that of the pragmatic physicist who has chosen "practical geometry". It appears both in 1921:

But it is my conviction that in the present stage of development of theoretical physics these concepts (i.e., rods and clocks) must still be invoked as independent concepts, for we are still far from possessing such certain knowledge of theoretical principles as to be able to give exact theoretical constructions of such constructs (solid bodies and clocks).[32]

and in 1946:

However, the procedure justifies itself because it was clear from the very beginning that the postulates of the theory are not strong enough to deduce from them sufficiently complete equations . . . in order to base upon such a foundation a theory of measuring rods and clocks . . . But one must not legalize the mentioned sin so far as to imagine that intervals are physical entities of a special type, intrinsically different from other variables ("reducing physics to geometry", etc.).[33]

Recently it has been argued that these passages, as well as the admission that Poincaré is correct *sub specie aeterni*, are to be understood as expressions of Einstein's "unease" with both theories of relativity as mere "theories of principle".[34] The above remarks then are interpreted as expressing Einstein's preference for the explanatory superiority of an as yet non-existent constructive account of the macroscopic practical rigidity of rods and practical efficacy of clocks in terms of their underlying microphysical dynamics of the constituent atoms and molecules. The claim has an initial plausibility, particularly in view of Einstein's several statements about the logical inadmissibility of explicitly assuming, as does practical geometry, rods and clocks to be objects independent of the rest of physics.

But Poincaré may be deemed correct *sub specie aeterni* for any entirely different reason. A clue lies in the statement quoted above when Einstein agreed in principle with Poincaré's insistence that no actual body is capable of physically realizing the ideal notions of geometry. Inasmuch as geometry is a formal mathematical discipline, there is no pre-axiomatic understanding of geometrical primitives. Geometry in itself makes no physical assertions; in this conception, metrical concepts are postulated and geometrical primitives (straight line, etc.) are implicitly defined within abstract axiomatic structures. But of course physics (both kinematics and dynamics) presupposes such geometrical notions. So geometry is logically prior to physics. This logical hierarchy of disciplines is of particular relevance here since empirical evidence for any dynamical constructive account of the observed behavior of rods and clocks will make use of geometrical notions. More precisely, since evidence for the dynamical theory in question will be stated in terms of measurable quantities having the dimensions of a length and/or a time, a dynamical or constructive account of rods and clocks must distinguish degrees of freedom for metrical concepts such as length and duration. Poincaré may be ultimately correct for a fundamentally logical reason (What presupposes what?), and not simply as a concession to the potential explanatory superiority of constructive/dynamical treatments of the behavior of rods and clocks, an understanding that in any case does not exist even today.

A fundamental length?

At this juncture an old issue reappears; already entertained by Riemann in 1854, it was subsequently completely ignored by Helmholtz and Poincaré. It concerns the possibility that space at the smallest scales may be discrete rather than continuous. Riemann raised the matter as he considered the question of the applicability of geometry "in the infinitely small":

> Now it seems that the empirical notions on which the metric determinations of space are based, the concept of solid body and

that of a light ray, lose their validity in the infinitely small; it is therefore quite definitely conceivable that the metric relations of space in the infinitely small do not conform to the hypotheses of geometry; and in fact one ought to assume this as soon as it permits a simpler way of explaining phenomena.[35]

Riemann's point concerns the validity of the usual assumption that space is a continuum even down to the scale of "the immeasurably small", or whether the continuous character of space, assumed by the differential equations of physics, may be only a course-grained approximation, a kind of averaging over an ultimately primitive discrete structure. Yet on the assumption that space is a continuum rest the hypotheses on which the geometry of Riemannian manifolds (including the group-theoretical characterization of rigid body continuous motions) as well as all of classical physics. But if space is fundamentally comprised of discrete parts, then the concept of a spatial continuum, as Mach averred, can be only an economical idea. Certainly if the nature of space (or space-time) were found to be discrete at the smallest scale, it would be an empirical discovery with momentous implications for the relation of geometry to physics. Riemann noted the problem in this way, terming "something else" just the hypotheses or postulates that he placed at the foundation of geometry of manifolds:

> In a discrete manifold the principle of metric relations is already contained in the concept of the manifold, but in a continuous one it must come from something else.[36]

To say "the principle of metric relations is already contained in the concept of a discrete manifold" means that metrical notions (such as distance, length, area, volume) ultimately rest on counting discrete units of space. And if space (and time) is fundamentally comprised of "atoms", then it is possible to counter the Poincaré objection that no actual physical bodies or objects could serve as physical realizations of the chrono-geometrical concepts of length or duration. In a discrete space-time such notions "already contained in the concept of a discrete manifold", in the sense that they physically correspond to the primitive elements of space and time themselves.

Should this be the case, the criterion of *sub specie aeterni* correctness requires modification. Recall that for Poincaré the dual core of that elevated perspective affirms both the logical priority of geometry to dynamics and the "logically purer method" of "put(ting) the unambiguous relation (of the theory) to the world of experience last instead of already fulfilling it in an imperfect form for an artificially isolated part, namely, the space-time metric". In turn, the logical fact that any dynamics presupposes geometrical notions gives rise to a necessary consistency requirement on any constructive account of rods and clocks. In particular, chrono-geometrical notions – length, area, volume, duration, period – presupposed by the empirical measures that could verify or confirm the derived predictions of the constructive theory cannot be taken from "something else" but must be "already contained in the concept of the manifold". But this can be the case only if space and time are fundamentally discrete.

In sum, any dynamical construction of the universal behavior of rods and clocks must begin in the "immeasurably small" by positing a fundamentally universal unit length and, corresponding to this, a fundamentally universal duration, say, the time for light to travel this length. Ultimately a constructive theory of microphysics, to explain the practical rigidity of rods and the practical reliability of ideal clocks, cannot presuppose any particular geometry (as Poincaré presupposed Euclidean geometry on grounds of simplicity) but must rest upon notions of a fundamental length and fundamental time from which the geometry of space-time is built up. In the context of a fundamental length, the quantity $(G+P)$ that alone renders space-time geometry empirically testable has been transformed. The quantity is no longer a sum of two distinct terms, one of which may be chosen conventionally adjusting the other as needed, but a fundamental, and non-conventional, unity that is both.

In 1942, some twenty years after "Geometry and Experience", Einstein explicitly underscored this consistency requirement on constructive theories in a letter to W.F.G. Swann, a British physicist working in the USA who had proposed a quantum dynamical account of rods and clocks in the special theory of relativity:

If one does NOT introduce rods and clocks as independent objects into the theory, then one has to have a structural theory in which a length is fundamental, which then leads to the existence of solutions in which that length plays a determinant (constitutive) role, so that a continuous sequence of similar solutions no longer exists.[37]

Einstein was well aware that the quantum theory had only a problematic relation to the space-time continuum, and for that reason a "purely algebraic physics" might be adopted.[38] However, should there be a fundamental microphysical length (as was independently proposed around that time by Heisenberg in an attempt to circumvent the infinities appearing in solutions of the equations of quantum electrodynamics), the continuum-based physics of the general theory of relativity, and all of Einstein's attempts to build a unified theory of fields by generalizing upon it, cannot be considered fundamental. Einstein explicitly recognized this possibility in the conclusion of the last of his many letters to his old friend Michele Besso on August 10, 1954:

> I certainly admit it is perfectly possible that physics cannot be based upon the notion of the field, that is to say, on continuous elements. But then practically nothing would remain of all my scaffolding – including the theory of gravitation – and also of existing physics.[39]

In fact physics cannot be based upon Einstein's notion of a field. This judgment is now orthodoxy in contemporary physics as the result of the dialectical sequence of physical theories from Einstein to the present. Relativity theory (including all classical physics) presupposes a spatial-temporal continuum; classical physics led to quantum physics; quantum physics probed into nether regions of higher and higher energies and smaller and smaller scales, led to theoretical postulate of the fundamental discreteness of space and time to which the geometry of continuous space-time may be a low energy approximation. Today both string theories as well as theories of quantum gravity view space-time at the smallest physically possible scales – of Planck length (10^{-33} m) and Planck time (10^{-43} sec.), (the time taken for light to travel a Planck length) – as something discrete, perhaps a "quantum foam". What remains of Einstein's vision of a unified fundamental theory is perhaps only his tempered belief in the method of mathematical speculation (see Chapter 10).

Summary

Einstein's admission in 1921 of the sub specie aeterni correctness of Poincaré's point of view is not a concession to conventionalism, that any geometry may be adopted as a matter of convenience only. It is rather a recognition of the inevitable holist character of any theory that, as a matter of principle, is capable of explaining its own measuring appliances, and so its ties to observation. Poincaré's position is valued for an unwillingness to consider actual physical objects as "geometrical", that is, as ideal bodies, and thus as independent of the dynamical laws to which all matter is subject. This is the perspective of a "consistent field theory". Without a completed theory of matter, however, the adoption of "practical geometry" is a pragmatic way forward to such a theory.

Notes

1 Letter of Gösta Mittag-Leffler to Einstein, dated December 16, 1919 (CPAE v. 7, Doc. 218, pp. 308–9; Einstein's reply to Mittag-Leffler in Stockholm, a one page typed letter copy, noted in "Calendar 1920", entry for April 12; CPAE, vol. 7, p. 611.

2 "Geometrie und Erfahrung", CPAE 7 (2002), Doc. 52, pp. 383–402; English translation supplement, pp. 208–22, also in Pesic, Peter, Beyond Geometry. New York: Dover Publications, 2007, pp. 147–57.

3 As reprinted in Stadler, Friedrich, and Thomas Uebel (eds.) (2012): Wissenschaftliche Weltauffassung. Der Wiener Kreis. Reprint of the first edition on behalf of the Institute Vienna Circle on the occasion of its 20th anniversary. Wien, New York: Springer. Originally published by Arthur Wolf Verlag, Wien, 1929, p. 108.

4 Frank, Phillip, Einstein: His Life and Times. New York: Alfred Knopf, 1947; fourth printing January 1953, p. 177.

5 Quine, Willard Van Orman, "Two Dogmas of Empiricism", The Philosophical Review v. 60, no. 1 (January 1951), pp. 20–43.

6 Carnap, Rudolf, "Eigentliche und uneigentliche Begriffe", Symposion Bd. I, Heft 4 (1927), 355–74, p. 373.

7 Poincaré, Henri, La Science et L'Hypothèse. Paris: Flammarion, 1968, p. 101; translation by G.B. Halsted, The Foundations of Science, p. 86.

8 Pesic, Beyond Geometry, contains translations of Helmholtz (1866) pp. 47–52, and Riemann (1854), pp. 23–40.

9 Ibid., 2007, p. 50.

10 Poincaré, "On the Foundations of Geometry", The Monist (1898), as reprinted in Pesic, Beyond Geometry, pp. 117–46; p. 145.

11 Poincaré, La Science et L'Hypothèse, p. 96; translation by G.B. Halsted, The Foundations of Science, p. 81.

12 Ibid., p. 63.

13 CPAE 7 (2002), Doc. 52, pp. 385–6; Pesic, Beyond Geometry, p. 147.

14 Hilbert, David, Grundlagen der Geometrie: Siebente Auflage. Leipzig und Berlin: B.G. Teubner, 1930.

15 Schlick, Moritz, Allgemeine Erkenntnistheorie. Berlin: J. Springer, 1918.

16 CPAE 7 (2002), Doc. 52, p. 388; Pesic op. cit. (2007), pp. 148–9.

17 Ibid., p. 391; Pesic, Beyond Geometry, p. 150.

18 Ibid., pp. 387–88; Pesic, Beyond Geometry, p. 148.

19 Ibid., p 390; Pesic, Beyond Geometry, p. 149.

20 Most recently, Howard, Don, "Einstein and Duhem", Synthese v. 83 (1990), pp. 363–84.

21 There is only indirect evidence that Einstein actually read Duhem's well-known discussion of the non-separability thesis in Aim and Structure of Physical Theory; see Howard, "Einstein and Duhem", pp. 368–9.

22 During a stay at Leiden less than three months earlier, on the occasion of delivering his inaugural lecture "Ether and the Theory of Relativity", Einstein used the opportunity to visit the house in which Spinoza lived from 1660–1663 in nearby Rijnsburg, signing the visitors' book on November 2, 1920. Sometime in 1920, quite possibly triggered by the emotion of that visit, he composed a poem Zu Spinozas Ethik whose first line is "How much do I love that noble man". (See van Delft, Dirk, "Albert Einstein in Leiden", Physics Today, April 2006, pp. 57–62).

A glance at Spinoza's Ethics, completed in Latin by 1675 but only posthumously published (first in Dutch) after Spinoza's death in 1677, may illuminate the significance or purpose the qualifying Latin phrase may have had for Einstein. Spinoza sharply distinguished between knowledge based on sense experience (things as they appear to a person from a particular perspective at a particular time) and rational knowledge that reveals the essences of things, according to the nature of reason to regard things as necessary, not as contingent. Sensory knowledge arises "from fortuitous experience" ("experientia vaga"). As sensations correspond both to the nature of external bodies and also to the subjective conditions of the perceiver, sensory experience alone can lead to falsehood and error. Discursive inferential knowledge furnished by reason stands in sharp contrast to sensory knowledge. This rational knowledge is capable of providing adequate ideas of things; the model for such knowledge appears to have been knowledge of necessary truths of mathematics. As applied to nature, rational knowledge must aspire to encompass the causal nexus that necessarily makes a thing what it is. For to have an adequate idea of a thing is to perceive the necessity inherent in nature: not just to know that a thing is, but how it is and why it could not be other than what it is. In later years, Einstein will characterize the intellectual hubris of the theoretical physicist in just this way. Nonetheless, adequate knowledge acquired in this manner requires both concepts and inference and, in this regard, is subordinate to the third and highest level of knowledge, direct intuitive knowledge ("*scientia intuitiva*"). This type of knowledge yields the adequate idea of a thing conveyed by an immediate, intuitive grasp of how the thing follows necessarily from the attributes of God. All other knowledge is judged an inferior approximation to this superior form exemplifying "the nature of Reason to perceive things under a certain aspect of eternity" ("*sub quondam specie aeternitatis*"), and so without relation to time. It is the mind's greatest striving to attain this manner of conception of things, aspiring to an ideal of understanding that is also the mind's greatest virtue (*Ethics* 5 p. 25). For to the extent the mind has knowledge of itself and of body under a species of eternity, it has knowledge of God (*Ethics* 5 p. 30), and so knowledge of the eternal necessity of God's essence. At this point, however, we must recall that in the original Latin text (but not the Dutch translation) of the *Ethics*, Spinoza heretically stated (*Ethics* 4 Preface): "That eternal and infinite being we call God, or Nature (*Deus, sive Natura*), acts from the same necessity from which he exists". And, in a string of propositions towards the end of the *Ethics*, passages that surely appealed to Einstein, Spinoza contrasted the man who is led by opinion or by passions (a "slave") with the "free man . . . who lives according to the dictate of reason alone" (*Ethics* 4 p. 67). The free man is one who becomes progressively detached from the transient interests of the individual in a particular environment and, in proportion to the increase of his knowledge, seeks to view all things *sub specie aeternitatis* ("under the aspect of eternity"). Through contemplation of the necessary system of nature made possible by this third kind of knowledge, such a man can attain the greatest happiness (*beatitudo*) (*Ethics* 5, p. 27).

23 "Geometry and Experience", CPAE 7 (2002), English translation supplement, Doc. 52, pp. 208–22; p. 213; Pesic, *Beyond Geometry*, pp. 147–57; pp. 149–50.

24 Ibid., pp. 391–2; Pesic, *Beyond Geometry*, p. 150.

25 Ibid., pp. 391–2; Pesic, *Beyond Geometry*, p. 150.

26 Weyl, Hermann, "Gravitation und Elektrizität", *Preußischen Akademie der Wissenschaften* (Berlin) *Sitzungsberichte. Physikalisch-Mathatische Klasse*, pp. 465–78; 478–80; translation in L. O'Raifeartaigh, *The Dawning of Gauge Theory*. Princeton Series in Physics. Princeton, NJ: Princeton University Press, 1997, pp. 24–37. Einstein, "*Nachtrag*", Ibid., p. 478, reprinted in CPAE 7 (2002), Doc. 8, p. 61. On the episode and its consequences, see Thomas Ryckman, *The Reign of Relativity: Philosophy in Physics 1915–1925*. New York: Oxford University Press, 2005.

27 Following Lenard and Stark (who would become members of the Nazi party), anti-Semites regarded relativity theory a violation of Aryan "healthy common sense" and a prime example of an abstract formal contamination inflicted on physics by the non-Aryan mind.

28 "Discussions of Lectures in Bad Nauheim", *Physikalische Zeitschrift* v. 21 (1920), p. 651, as reprinted in CPAE 7 (2002), Doc. 46, p. 352.

29 Einstein, "The Theory of the Affine Field", *Nature* v. 112 (1923), pp. 448–9; p. 448.

30 Einstein, "Fundamental Ideas and Problems of the Theory of Relativity", in *Nobel Lectures in Physics, 1901–1921*. Amsterdam, London, and New York: Elsevier Publishing Co., 1967, pp. 482–90; pp. 483–4.

31 "Autobiographical Remarks", p. 59; p. 61.

32 CPAE 7 (2002), Doc. 52, p. 390; Pesic, *Beyond Geometry*, pp. 149–50.

33 "Autobiographical Notes", p. 58; p. 59.

34 Brown, Harvey R., *Physical Relativity: Space-Time Structure from a Dynamical Perspective*. Oxford, UK: Clarendon Press, 2005.

35 Riemann, Bernard, "On the Hypotheses That Lie at the Foundation of Geometry", as translated in Pesic, *Beyond Geometry*, pp. 23–40; p. 32.

36 Pesic, *Beyond Geometry*, p. 33.

37 Einstein to W.F.G. Swann, January 24, 1942 (EA 20–624); the original German and translation in Amit Hagar, *Discrete or Continuous? The Quest for Fundamental Length in Modern Physics*. New York: Cambridge University Press, 2014, pp. 153–5.

38 See Stachel, John, "The Other Einstein: Einstein Contra Field Theory", *Science in Context* v. 6 (1993), pp. 275–90.

39 Speziali, Pierre (ed.), *Albert Einstein Michele Besso Correspondance 1903–1955*. Paris: Hermann, 1979, Doc. 210, pp. 305–7; p. 307.

Further reading

Brown, Harvey R., *Physical Relativity: Space-Time Structure from a Dynamical Point of View*. New York: Oxford University Press, 2005.

Hagar, Amit, *Discrete or Continuous? The Quest for Fundamental Length in Modern Physics*. New York: Cambridge University Press, 2014.

Ryckman, Thomas, *The Reign of Relativity: Philosophy in Physics 1915–1925*. New York: Oxford University Press, 2005.

Weyl, Hermann, *The Philosophy of Mathematics and Natural Science*. Reprint of the 1949 edition. Princeton, NJ: Princeton University Press, 2009.

Eight
Philosophy of science – realism

"For most of us, the paradigm Realist is Einstein".[1]

Introduction

With Einstein's criticism of quantum mechanics in mind, both physicists and philosophers of science have long regarded Einstein as a
paradigm of the realist conception of scientific theories. Of course
the meaning of the term "realism", like many others in philosophy,
has distinct context and time-sensitive connotations. One result is
that the notion of a realist conception of scientific theories changed
considerably during the course of the 20th century, largely under
pressure of philosophical dialectic with quite different and distinctive non- or anti-realist viewpoints. In this chapter, and in the next
two, a case is made that Einstein is legitimately considered a *realist*,
though not a *scientific realist* in the contemporary sense, largely because
of constructivist and empiricist tendencies that remain anathema to
scientific realism. Aspects of realism are present in Einstein's earliest
contributions to physical theory on "the molecular-kinetic theory of
heat" and the 1905 account of Brownian motion. Later confrontations over quantum mechanics, and a broader disagreement with an
overt positivism and instrumentalism of many quantum physicists,
led to heightened emphasis on realism. It became increasingly clear
just how much of a breach quantum mechanics demanded with
what Einstein presumed to be a largely uncontroversial previous
conception of physical theory.

The changing face of scientific realism

As understood today, scientific realism is broadly the view that current best theories in the "mature" sciences are true, or at least "approximately true", descriptions of both observable and unobservable aspects of a mind-independent nature. Though emphases may be differently placed, contemporary scientific realists widely subscribe to three subordinate theses. In the detailed expositions of Psillos (1999) and Chakravartty (2011),[2] these are:

1. a *metaphysical thesis*, that the external world has a definite structure existing independently of mind or consciousness;
2. a *semantic thesis*, that theoretical terms have definite factual reference, while the true statements in which they occur are to be construed as "corresponding to" parts of this definite structure; and
3. an *epistemic thesis*, that a reliable methodology licenses inferences from the predictive success of theories in the mature sciences to the existence of the entities and processes to which these theories refer, and to the approximate truth (in the correspondence sense) of theoretical claims about them.

As with most longstanding philosophical theses, this particular packaging reflects both contemporary debate and philosophical developments. Bas van Fraassen influentially argued that scientific realism crucially involves the thesis that the aim of science is to develop theories that are true, in the above correspondence sense.[3] Steven Leeds (*op. cit.*) on the other hand, points out that the "correspondence theory" of truth, and so thesis 2), only became a key component of scientific realism after 1970, presumably impelled by adoption of Tarski-Davidson theories of reference and truth within analytic philosophy of logic and language. Just as "realism" has distinct and changing connotations, *a fortiori* this is the case with expressions of realist commitments. It is all too easy to overlook the rather obvious hermeneutic warning that the significance of terms and assertions from an earlier era, with their largely implicit context, should not, at least without argument, be simply identified with word-for-word equivalents occurring in contemporary discussions.

The semantic apparatus of reference and truth characteristic of contemporary scientific realism presupposes both empirical success as well as relative stability of theory. For the two revolutionary developments of 20th-century physics, relativity theory and quantum mechanics, these conditions were clearly met in the first decades following WW II, a period coinciding with a gradual liberalization of positivist currents in philosophy of science. Cracks began to appear in the reigning hegemony of complementarity and the "Copenhagen Interpretation" of quantum mechanics; Bohm's ontological version of quantum mechanics dates from 1952. But up through the 1960s, the furthest reaches of quantum theory in high-energy physics yielded little more than a confusing and imperfect taxonomy of elementary particles, congenial to an instrumentalist view of physical theories. In the philosophical literature, it is difficult to find an expression of the contemporary doctrine of scientific realism prior to the work of Grover Maxwell around 1960. Employing the new tool of Tarski-style semantics, Maxwell promoted a "radically realistic interpretation of theories" seeking to extend semantic notions of truth and reference to statements and terms pertaining to unobservable or theoretical entities.[4] Psillos recognizes this paper as the initial expression of contemporary scientific realism. However, the first book in English in which the term scientific realism prominently appears in the title, J.J.C. Smart's Philosophy and Scientific Realism (1963), targets not the semantic double standard of the observable/unobservable dichotomy that animated Maxwell, but a considerably wider collection of foes, inter alia the later Wittgenstein, phenomenalism, biblical literalists, anti-mechanist dualists, idealism of all varieties, and perhaps above all Kant's "anti-Copernican counter-revolution . . . his metaphysics, putting (man) back in the centre again".[5] Resistance to putting "man back in the centre" appeared also in reaction to the mind-dependence or subjectivity of Thomas Kuhn's account of mature science using the core notion of "paradigm" in Structure of Scientific Revolutions (1962). Kuhn's account of theory-change, caricaturized by some as "mob rule", provided further impetus to scientific realism, initiating a new round of debate between realism and antirealism that continues into the 21st century. In the light of these later developments, it is plausible that putative realists in the first half of the 20th century subscribed only to a version of thesis 1), but hardly to 2) or to 3).

Realisms

Even the above cursory overview indicates that both the term "scientific realism" and the doctrine are of fairly recent vintage in the philosophical literature. What about the term "realism" itself? Its usual metaphysical sense is captured in the following Ur-intuition, recently described by Kit Fine:

> One might think of the world and of the propositions by which the world is described as each having its own intrinsic structure; and a proposition will then describe how things are in themselves when its structure corresponds to the structure of the world.[6]

But realism's original philosophical meaning is otherwise, standing in opposition to the doctrine of nominalism, a dispute stemming from medieval philosophy concerning the existence (or not) of abstract universals and their relation to the concrete individuals, or particulars, that instantiate them.

Only in the course of the 19th century did the term "realism" come to acquire a wholly distinct meaning within a new context of discussion. One source of change was a revival of direct, or causal, theories of perception, directing attention from the quality or character of an individual's percept or idea to the object itself, as the external cause of perception. Thomas Reid's critique of Lockean and Humean representationalist theories of ideas was an initial counter to empiricism's epistemological idealism (that we know or perceive objects only via our representations of them). Realism also would come to signify opposition to ontological idealism (denial of the existence of objects external to mind, as in Berkeley, Fichte, Schelling, and, in some readings, Hegel). By the 1850s, idealist metaphysics was widely regarded as an embarrassment by a burgeoning technological age accustomed to regular advances in the sciences.

One response was a disregard for professional philosophy as irrelevant to their concerns by many, if not most, scientists. Post-Hegelian German philosophy itself was in transition from idealism to various forms of naturalism or materialism, positivism, and to an emerging school ("Back to Kant") of neo-Kantian thinkers, some urging a return to Kant's empirical realism. Even so, as can be seen by a glance at John Dewey's 1902 entry for "Realism" in what was then a standard reference work, Baldwin's *Dictionary of Philosophy and Psychology*, the traditional anti-nominalist significance of the term within philosophy retained precedence. Noting that "the term has two important meanings in philosophy, wholly distinct from each other", Dewey's article is almost entirely a discussion of the realism-nominalism controversy. Dewey notwithstanding, the contemporaneous "New Realism" of the young Cambridge philosophers Bertrand Russell, G.E. Moore, and G.F. Stout, targeted equally both traditional nominalism as well as the idealism of the British Hegelians, such as F.H. Bradley and Bernard Bosanquet.

Realism: Boltzmann and Planck

At the beginning of the 20th century, a "new theory of knowledge" (*die neuere Erkenntnistheorie*) had arisen "out of the soil of the natural sciences", and so independently of traditional philosophy.[7] An influential trend in this development was the purely instrumentalist conception of physical theory promoted by Gustav Kirchhoff followed by emerging realist responses to it. In 1876 Kirchhoff famously stated the task of mechanics as that of describing "the motions occurring in Nature *completely* and *in the simplest way*".[8] Kirchhoff's descriptivism is better known to most philosophers of science today through the version advanced by Pierre Duhem, lauding Kirchhoff and Mach as "more modest and more far-sighted" physicists who

> recognized that physical theory is not an explanation, but a simplified and orderly representation grouping laws according to a classification which grows more and more complete.[9]

Mach had expressed similar views, even "more radical" (as he put it in his *Mechanik* of 1883), earlier in 1872. Mach elevated Kirchhoff's descriptivism in mechanics to a stricture on science as a whole; in an 1894 lecture in Vienna, against those who maintained that "description leaves unsatisfied the *requirement of causality* (*Kausalitäts-bedürfnis*)", Mach expressed his hope that "the science of the future will discard the idea of cause and effect, as being formally obscure" as well as his belief that "these ideas contain a strong tincture of fetishism".[10] Causal explanations of phenomena, if not cashed out as regularities between groups of sensations, are metaphysical ballast that science doesn't need and adopts only on pain of mischief. The legitimate task of the scientist is to provide the simplest and most economical description of phenomena, employing the language of functional dependences between sensations in place of the metaphysical notions of cause and effect.

Mach's anti-metaphysical positivism met with resistance from many scientists who refused to surrender what might be termed a "common-sense realism" regarding a physical world familiar to the experience of all individuals but whose existence and nature is independent of all who experience (or think). Among certain theoretically inclined physicists, such a view was accompanied by a more expansive conception of physical theory than descriptivism permitted. To its most prominent turn-of-the-century advocates, Boltzmann and then Planck, the augmented common-sense realism largely took the form of defending the use of hypotheses, such as atomism, in testable and empirically confirmed theories. Hypotheses allowed theory the resources of autonomous development, and in doing so, underscored the explanatory benefits of theoretical concepts that go beyond and supplement experience. Whether, and in what sense, Boltzmann was a realist about atoms is a matter of some delicacy; he surely believed in kinetic theory and that kinetic theory presupposes atoms and molecules. However, he died in 1906, just after Einstein's 1905 paper on Brownian motion that would play a pivotal role in atomism's ultimate vindication. That paper derived the hypothetical mechanism producing invisible molecular motions whose consequences might even be observed; Jean Perrin's experiments in Paris (in 1908) confirmed Einstein's theory and the existence of physical atoms was ever afterwards taken for granted. Einstein referred to his early realist motivations much later, in 1946:

> My principal aim . . . was to find facts that would guarantee as much as possible the existence of atoms of definite size. . . . The agreement of these considerations with experience together with Planck's determination of the true molecular size from the law of radiation (for high temperatures) convinced the skeptics,

who were quite numerous at that time (Ostwald, Mach), of the reality of atoms.[11]

Many contemporary philosophers of science view the confirmation of Einstein's (and independently in 1906, Polish physicist Marian Smoluchowski's) molecular hypotheses explaining the mechanism of Brownian motion as vindicating realism over positivism, and there is a sense in which this is correct. But it should be recognized that atomism's triumph is not one of realism regarding scientific theories but of the justifiable posit of unobserved physical entities in the explanation of observations. This fallible entity realism is also the basis of Boltzmann's atomism. It presupposes a conception of physical theory as a deductively complex structure that is capable of deriving the phenomena of observation from hypotheses regarding unobserved theoretical entities and underlying processes. This is a view Einstein also shared, as will be discussed in Chapter 9.

Planck had consistently argued against Boltzmann's statistical interpretation of the second law and accordingly rejected the hypothesis of atomism that interpretation presupposed. But he was converted to atomism during the course of his investigations of blackbody radiation. With the zeal of a new convert, in a notable lecture in Leiden on December 9, 1908, Planck attacked the leading anti-atomist Mach and the latter's positivist conception of physical theory. The lecture was Planck's philosophical debut, a first intervention in ongoing discussions by physicists and others, such as Poincaré, over the nature and character of physical theory. Planck's thesis is prominently stated in the lecture's title "The Unity of the Physical World Image" ("*Die Einheit des physicalischen Weltbild*"). The issue of atomism was now peripheral and largely passé; Planck simply pronounced atoms to be "as real as the heavenly bodies".[12] Rather the lecture's significance concerns Planck's shift of realist focus from the existence of atoms to the definition of the task of the theoretical physicist. This was, *contra* Mach, the search for a unifying physical world image (*Weltbild*). The lecture was extensively discussed, not least because Planck reprinted it no less than four times in various collections of philosophically themed essays. In fact, the main themes of 1908 appear again and again in later essays, extending almost up to his death in 1947.

One suspects that the source of Planck's ire was not so much directed at Mach's philosophy as a perceived pernicious influence on the younger generation of physicists. Under the banner of positivist anti-metaphysics, Mach had urged that the task of physical theory is simply to provide the most "economical" description of empirical connections between sensations (that is, observable quantities), having no further epistemological, and certainly no metaphysical, significance. To Planck, such an emaciated conception of the role of physical theory, apparently having to do only with the ordering of sensations, was unbearably anthropomorphic. Against this, Planck proclaimed the goal of the theoretical physicist to be the construction of a de-anthropomorphized, complete, consistent, and unified physical world image (*physikalischen Weltbild*) of such permanency that it can be taken to be a *definition* of physical reality:

> The constant unified *Weltbild* is . . . the fixed goal to which actual (*wirkliche*) natural science progressively approaches through all its changes. In physics we even may justifiably claim that our contemporary *Weltbild* contains certain traits that will never disappear through any revolution, either in Nature or in human thought. This constancy, independent of any human, more generally any intellectual individuality, is now plainly that which we call the real (*das Reale*).[13]

The term *Weltbild* merits special emphasis. The next chapter looks into the origin of the term *Bild* in the context of late 19th-century discussions of the constructive character and nature of physical theory. It will be sufficient here to note that an image (*Bild*) is distinct from what it portrays or represents; the possibility of comparing image and object, and establishing their difference, is normally presupposed. Yet this is not the case, at least regarding the *Weltbild* of the future, the projected unified complete representation of physical reality. Herein lies its suitability in establishing the very meaning of physical reality:

> Is there any recognizable difference between (physicists' talk of "the world" or "Nature itself") and our "*Weltbild* of the future"? Certainly not. For that there is no method of proving such a

difference has, through Immanuel Kant, become common property (*Gemeingut*) of all thinkers.[14]

Kant had shown that there is no way of delineating a "recognizable difference" between the noumenal metaphysically real world with its mind-independent objects, properties and structures and the physicists' "*Weltbild* of the future". The term *Weltbild* therefore has a precise function; it is adopted "as a precaution (*Vorsicht*), to exclude certain illusions from the start"; that is, to exclude the conceit of assuming that any physical theory, even an ideal future one, can ever portray with complete and faithful accuracy a mind-independent reality. Planck would make abundantly clear in subsequent essays that an "unbridgeable chasm" always remains between the phenomenal world characterized by physical theory and the metaphysically real world that is presupposed as the target of physical inquiry. Between the two there is "a constant, unbalanced tension" that gives exact science "an irrational element (it) can never shake off".[15] This can be an attenuated realism at most; physics is about an independent external world; physical theory aspires to be a complete description of such a world; yet the non-philosophically naïve physicist must continue to work in an existential state of tension, possessing the knowledge that such an aspiration can never be fully realized.

Planck allowed that even the contemporary physical *Weltbild* contains some permanent elements. These are universal principles, laws, and constants that always shall be elements of yet more refined and complete future world-images. His examples are the second law of thermodynamics, the principle of conservation of energy, the principle of least action (governing reversible processes), and the physical constants of universal character, independent of reference to any particular body or substance. In the course of work on blackbody radiation in 1900, Planck gave an explicit list. They are h and k, both appearing in Planck's radiation law, the velocity of light *in vacuo* c and Newton's gravitational constant G; together they allow a characterization of physical measurement transcending the conventions of particular experimenters at particular times. These "natural unit constants" provided the possibility of establishing units of length, mass, time, and temperature, that independently of special bodies or substances, necessarily retain their significance (*Bedeutung*) for all times and for all cultures, even extraterrestrial and nonhuman ones.[16]

To this list, Planck would later append relativistic invariants such as the interval ds^2 of space-time. Planck's 1908 polemic elicited a spirited reply from Mach (his *Leitgedanken*) and from Mach's followers, but Planck was adamant. His final retort to Mach fairly drips with sarcasm: "the physicist, if he wants to promote science, has to be a realist, not an economist".[17] Though we shall refer to "Planck's realism", this sense of "realism", whose goal is the construction of a physical *Worldbild* purporting to *define* physical reality, is to be understood primarily in opposition to positivism. And on account of the "unbridgeable chasm" that always remains between theory-image and reality, Planck's realism is more aptly described as a variety of neo-Kantianism. It lacks the "epistemic optimism" needed to bridge the two, a bridge required by the semantics of truth and reference of scientific realism in the contemporary sense.

Einstein and Planck

Einstein will always deny the attribution "realist". The most vivid of these denials is made in a September 25, 1918 letter to Bonn mathematician Eduard Study, responding to Study's 1914 monograph, *Die realistische Weltansicht und die Lehre vom Raume* (*The Realist World Outlook and the Doctrine of Space*). As its title suggests, Study had written a polemic against the opponents of realism ("Idealists, Positivists, Pragmatists") and a defense of the realist *Weltbild* (*Das realistische Weltbild*), that is, of the *hypothesis* of the existence of an external world independent of the knowing subject that is the topic of natural science. To Study, Einstein objected,

I do not feel comfortable and at home in any of the "isms". It always seems to me as though such an ism were strong only so long as it nourishes itself on the weakness of its counter-ism. But if the latter is struck dead and it is alone on an open field, then it proves to be wobbly on its legs. Therefore, *away with the grousing* (**los mit der Stänkerei**)!

"The physical world is real". That is supposed to be the basic hypothesis. What does "hypothesis" mean here? To me, a hypothesis is an expression whose truth is presupposed for the time being, *but whose meaning has to be elevated above any ambiguity.* However, the above statement appears to me to be meaningless in itself, as if one said "The physical world is cock-a-doodle-doo (*ist kikeriki*)". It seems to me that "real" is an in-itself meaningless category (pigeon hole) whose monstrous importance only lies in this, that I can do certain things inside it and not certain other things. . . . I grant you that natural science concerns the "real" and yet I am not a "realist".[18]

On the other hand, earlier that year in the full flush of success following the completion of the general theory of relativity, Einstein embraced Planck's attempt to define the "real" in terms of a striven-for constant *world image* (*Weltbild*) and that the task of the theoretician is to advance such a world image through establishing the permanence of certain of its constituents, and of its most general basic laws. This occurs in the text of Einstein's warm appreciation of Planck at the Berlin Physical Society's celebration of Planck's 60th birthday on April 26, 1918. Einstein did not take this responsibility lightly. When inviting Arnold Sommerfeld to attend the festivities some weeks before, he wrote, "I'll be happy that evening if the gods grant me the gift to speak profoundly, because I am very fond of Planck, and he will certainly be pleased when he sees how much we all care for him and how highly we treasure his life's work".[19] It cannot be a coincidence that the prominent message of Einstein's keynote tribute is an endorsement of Planck's signature philosophical position, the characterization of the task of theoretical physics as the search for a unifying and universal "*Weltbild*".

Do the results (of the theoretical physicist) deserve the proud name "*Weltbild*"? . . . the proud name is well-deserved since the most general laws, upon which the thought-structure [*Gedankengebäude*] of theoretical physics is based, raise the claim of being valid for all natural occurrences . . . Therefore the highest task of the physicist is the search for those most general elementary laws from which the world-image (*Weltbild*) is to be obtained by pure deduction.[20]

Einstein's audience undoubtedly understood that this embrace of Planck's conception of "the highest task of the physicist" is a deliberate public alignment with Planck and so with the latter's passionate polemic against Mach and positivist ideas of physical theory. The theoretical physicist is to be understood as engaged in building up a "physical world image", much of which is an admittedly mental construction. Employing symbols for the metaphysically real, the *Weltbild* is not readily expressible in ordinary language, and it may posit elements that outstrip present capacities for observational test. At the same time it must yield consequences that are possible to observationally confirm, and in addition it must be flexible enough to accommodate new phenomena. Though in a sense a creation of mind, its implications purport to refer to a mind-independent real external world. As it is always incomplete, it can never be supposed in satisfactory agreement with mind-independent states of affairs, and in any case such agreement can never be directly ascertained but at most indirectly inferred. Perhaps its most important functions are to serve as a platform within which further thought may develop, as well as to suggest experiments whereby it can be transcended or amended.[21] It is worth mentioning that in 1933, Einstein allowed the text of this 1918 lecture to appear as a "Prologue" to a collection of Planck's philosophical articles in English.[22]

Some dozen years after 1918, Einstein again expressed fundamental philosophical agreement with Planck's anti-positivist realism, this time with Planck's article "Positivism and the Real External World" (*"Positivismus und reale Aussenwelt"*). The article is the text of Planck's lecture on November 12, 1930 at Harnack-Haus, in the Dahlem district of Berlin. Harnack-Haus was completed in 1929 as an international guest house for the *Kaiser Wilhelm Gesellschaft*; since 1917 Einstein had been titular head of the *Kaiser Wilhelm Institut für Physik*, and in 1929 he had received the Max Planck medal of the German Physical Society from Planck himself. There are reasons to assume that Einstein was in the audience; he would leave for a second visit to Pasadena and Caltech beginning in December.

Planck's theme is the looming dual crisis at the end of 1930, both material (between summer 1929 and early 1932, German unemployment had risen from 1.3 million to 6 million, some 24 percent of the labor force[23]) and "spiritual" (by which Planck certainly included "political": since 1927 street fighting between "reds" (communists) and "browns" (Nazis) had become common in Berlin). Though physics should have claim to be a firm foundation on which to base a modern outlook on the world, Planck took the opportunity to express his disquiet at the confusion and contradiction now active among certain physicists. Mach was long dead, but quantum physicists had resuscitated positivism in physical science, fundamentally posing limits on how far and in what way the human mind is capable of attaining knowledge of the external world, and by challenging the law of causality. Again in opposition to positivism Planck articulates the *Bild* conception of physical theory, and the invariable tension between the presupposition of a real world in the metaphysical sense and the realization that theories are images, never to be known capable of completely grasping its nature.

> Positivism always rigorously maintains that there are no other sources of knowledge except sense perceptions. Now, the two theorems (*Sätze*): (1) *there is a real external world independent of us* and (2) *the real external world is not directly knowable* form together the hinge point (*Angelpunkt*) of the whole of physical science. And yet they stand in a certain opposition to one another. In this way they at the same time disclose the purely irrational element that adheres to physics as to every other science. The result is that a science is never able to fully complete its task. We must accept that as an irrefutable fact, one that cannot be removed, as positivism tries to do, by restricting the task of science at its very start. The work of science therefore poses itself to us as an incessant struggle toward a goal that will never be reached and fundamentally is unattainable. For this goal has a metaphysical character, lying beyond any experience.[24]

Planck's lecture marks not so much an intervention in, but a response to, the debate about the meaning of the Heisenberg uncertainty relations and quantum indeterminism. Whereas Planck deplores the quantum physicists' abandonment of the "law of causality", Einstein's criticisms by 1930 had shifted, as seen in Chapter 4, to questioning whether the Ψ-function of wave mechanics could be considered a complete description of the states of individual systems. The quantum physicists' resort to positivist doctrine, then under vigorous revival by philosophers particularly in Vienna and Germany, had spurred Planck into action. One of these philosophers was Moritz Schlick, Planck's former student and now leader of the Vienna Circle. Indeed, Planck's lecture prompted a response from Schlick retorting that statement (1) above is meaningless.[25]

Two documents show that Einstein was in full accord with Planck's conception of realism while opposed to Schlick's positivism. A handwritten draft note from 1931 in the Einstein Archives was possibly intended as the preface to the published pamphlet of Planck's lecture that appeared on March 9, 1931. For one reason or another Einstein's note was never published. Lauding Planck's article, it states:

> I presume I may add that both Planck's conception of the logical state of affairs as well as his subjective expectation concerning the later development of science corresponds entirely with my own understanding.

Citing the above passage, physicist and historian Gerald Holton observes that from this time forward, "Einstein's and Planck's writings on these matters are often almost indistinguishable from each other".[26]

Of course by 1930 it had become increasingly apparent to Einstein that many of the leading quantum physicists sought to cloak what he considered to be the theory's shortcomings under the mantle of positivist strictures upon the meaningfulness of scientific statements. This in turn led to more and more forceful counters of a realism à la Planck. In the autumn of 1930, Schlick, in 1917 a realist supporter of relativity theory but now a leading logical empiricist, completed a lengthy essay purporting to attain "philosophical clarity" about the standing of the principle of causality in the new physics of quantum mechanics. It appeared in the German scientific weekly *Die Naturwissenschaften* at the beginning of 1931. Upon finishing it, sometime in November 1930, Schlick posted a typescript copy to Einstein. If Schlick had any awareness of Einstein's objections to quantum mechanics, he knew beforehand that Einstein was not disposed to favor the essay's conclusion, that "quantum physics teaches us . . . that within the bounds established by the uncertainty relations the (causality) principle is *bad*, useless or idle, and incapable of fulfillment".[27] But Schlick's essay had transgressed much further, impugning the conception of physical theory guiding both Planck and Einstein. At the beginning, Schlick lampooned the "obstinacy" of the physicist's traditional belief that the explanation of nature required comprehensible models, a rejection of the *Bild* conception of physical theory. At the end, he proclaimed "the human imagination is incapable of conjecturing the world structure revealed to us by patient research".[28] Einstein was quick to disagree with these two claims in particular. In a letter to Schlick of November 28, 1930, two weeks after Planck's lecture, Einstein condemns the constraints of positivism's conception of physical theory. Alluding to the irrational bifurcation between real world and the world of experience that Planck viewed as the hinge point of natural science, Einstein freely admitted to the metaphysical character of the theoretician's realist impulses:

> I tell you straight out: Physics is the attempt at the conceptual construction of a model of the *real world* and of its lawful structure. To be sure, (physics) must present exactly the empirical relations between those sense experiences to which we are open; but only *in this way* is it chained to them. . . . In short, I suffer under the (unsharp) separation of Reality of Experience and Reality of Being.
> . . . You will be astonished by Einstein the "metaphysicist". But every four- and two-legged animal is *de facto* in this sense metaphysicist".[29]

Without acknowledging the scientist's motivational realism, admitted even by descriptivists like Duhem, positivism can give only a misleading account, perhaps only a caricature, of scientific method. This will become an increasingly prominent theme in Einstein's response to quantum mechanics.

Realism as a program

Near the end of Chapter 4, it was seen that after WW II, two previously distinct lines of argument against quantum mechanics had
coalesced into the thesis of "macrorealism". Prior to then, Einstein
wielded a two-tiered critique. First, a parade of heuristic examples of
individual systems purporting to show incompleteness of quantum
mechanical description without violating Heisenberg's strictures on
simultaneous measurement of exact values of canonically conjugate
observables. At the same time, on a methodological/philosophical
level, he repudiated complementarity's positivist inference that since
simultaneous measurements cannot reveal exact values of conjugate
properties, such values cannot be said to exist and accordingly need
not be described. By the late 1940s, Einstein began to emphasize
that even positivists found it difficult to deny that macroscopic objects
possessed "real states" independent of observation. An incompleteness argument could then be marshaled if the quantum mechanical
characterization of the transition from micro- to macro-object, such
as is required by allowing the mass of single particle to increase
to classical scale (e.g., 1 g), hence a transition quantum mechanics
was obliged to describe, did not terminate in a "real state" of the
corresponding macro-object; that is, did not produce the definite
kinematical and dynamical properties that classical particle mechanics takes for granted.

A first step is to throw down the gauntlet to positivism, postulating what in 1953 Einstein termed the "thesis of reality":

> "thesis of reality": *there is such a thing as the "real" state* of a physical
> system existing independently of any measurement or observa-
> tion that in principle can be described by the means of expres-
> sion of physics.[30]

The thesis, representing clearly at least a shift in philosophical
emphasis from the response to Study in 1918, is a fundamental posit
of what in 1949 is termed realism as a "program"; that is, a program
for complete description of "real states":

> The "real" in physics is to be taken as a type of program to
> which we are . . . not forced to cling *a priori*. No one is likely to

give up this program within the realm of the "macroscopic" ...
But the "macroscopic" and the "microscopic" are so interrelated
that it appears impractical (untunlich) to give up this program in
the "microscopic" alone.[31]

Three related points will be mentioned here (see also the discussion
at the end of Chapter 4 and further in Chapter 9). First, realism "as
a program" is widely assumed, though perhaps only implicitly, in
classical physics. Objects or states of objects are regarded as having
definite values of observable properties independently of observation or measurement. Naturally the temperature of a solution may
be altered by the insertion of a thermometer not at equilibrium with
it. Or the value of a magnetic field at a point of a region may be
changed by an encounter with a charged particle whose altered trajectory indicates the strength of the field. But the operative assumption is that the unmeasured values exist and are definite, and that
refined procedures of measurement can reduce the disturbances
involved to negligibly small quantities. Second, the entwined "inter-
relation" between microscopic and macroscopic systems gives reason to think the operative assumption above should not be restricted
to the realm of the macroscopic. Third, adherence to the program is
not compelled a priori; after all, the assumption of the real external
world is a metaphysical assumption that may be consistently denied,
as by solipsism. Such an assumption is not, after all, terribly controversial. Even quantum theoreticians adhere to it "so long as they are
not discussing the foundations of quantum theory".[32]

The thesis of realism is affirmed to be generally applicable, and so
to apply indifferently to all physical systems. Controversy ensues only
when it is claimed that there are "real states" of micro-objects that are
only incompletely described by the Ψ-function. But then the quantum theoretician, as indeed Bohr broadcasts, has had to admit that
the notion of "real state" is not univocal between classical and quantum, and that there is a fundamental difference in kind between the
character and method of the classical physical description of objects,
and the quantum mechanical description of quantum objects. This is
the postulate, even the defining characteristic, of complementarity.
Einstein insists on viewing this postulate as a positivist inference, and
is prepared to be a labeled a "metaphysicist" for doing so:

What does not satisfy me in that theory (quantum mechanics) from the principled standpoint is its advocated (*vertretene*) attitude towards what appears to be the programmatic aim of all physics: the complete description of the real states of affairs possible according to the laws of nature (*die vollständige Beschreibung der naturgesetzlich möglichen realen Sachverhalte*). If the modern physicist of positivist persuasion hears such a formulation, his reaction is that of a pitying smile. He says to himself: "There we have the naked formulation of an empty metaphysical prejudice without content, the conquest of which has been the major epistemological achievement of physicists within the last quarter-century. Has anyone ever perceived a "real state of affairs"? Can anyone say in general what is to be understood by "real state of affairs"? How is it possible that a reasonable person could still believe today that he can refute our principal cognitions by allowing such a bloodless ghost to be conjured?" Patience! The above laconic characterization was not meant to convince anyone; it was merely to indicate the point of view around which the following elementary considerations freely group themselves.[33]

In place of the incoherent EPR "criterion of reality", the "thesis of reality" is an attempt to reinstate the conception of physical theory outlined above, in which the intended meaning of the concepts <physical reality>, <external world>, and <real state of (an individual) system> characterize what it is that physical theory attempts to describe. Realism as a "program" thus pertains to the intended target of description of any physical theory, the real external world as partitioned by physics into systems and subsystems. With the advent of quantum mechanics, positivism reappeared in physical theory in the form of Bohr's complementarity. And to the extent that complementarity enjoins a sharp conceptual separation between the respective realms of classical and quantum, it presents a vulnerable target to the "thesis of reality". Not because the thesis merely affirms what Bohr and quantum orthodoxy deny, the existence of definite states of quantum objects at all times: this is moot and merely begs the question. Rather the intent of the thesis is to highlight that, as a matter of consistency and as a *theoretical requirement of quantum mechanics*, there must be a faithful quantum mechanical characterization of the transition in the limit from the domain of quantum phenomena to macroscopic physics. This quantum mechanics does not do, then or arguably now, and so the "thesis of reality" points out a glaring gap in the case that complementarity can make to be an adequate and definitive philosophical interpretation of quantum mechanics.

Einstein's realism vis-à-vis the quantum physicists is also a dispute over the explanatory goals of physical theory. "Realism as a program" delineates categories of understanding, not necessarily of nature. Yet it supports the characterization of quantum mechanics as incomplete. Realist theories, as Einstein conceived of them, share a common world picture, i.e., they represent a cumulative *understanding* of nature as comprised of relativistic (in the sense of the general theory) deterministic (causal) non-quantum continuous fields in which individual systems are spatiotemporally separable and can be characterized completely by properties having determinate values independent of any act of measurement or observation. The concepts of this realist program of description are not to be assumed either *a priori* valid or as impossible. They are not essential to the practice of science but neither are they to be merely dismissed as excrescences of the metaphysically diseased mind. Like all other concepts they are justifiable only by experience. Occasionally rivals (e.g., "purely algebraic physics") are mentioned.[34] Characteristically, after laying out his vision, Einstein will immediately observe there is absolutely no compelling reason, conceptual or metaphysical, to regard it as correct. Justification for any choice of a physical theory and its attendant concepts and presumed ontology lies principally with the ability of such theories to implement coordination with experience (confirmation) "with advantage" (*mit Vorteil*).[35] The latter phrase is significant, for it is an oblique indication of the requirement that a theory must provide *understanding*,

> a conceptual model (*Konstruction*) for the comprehension of the interpersonal whose authority lies solely in its confirmation (*Bewährung*). This conceptual model refers precisely to the "real" (by definition), and every further question concerning the "nature of the real" appears empty.[36]

Summary

Two distinct components comprise Einstein's realism: 1) metaphysical, affirming the existence of an external mind-independent world; and 2) a motivational or aspirational aspect, affirming the aim of fundamental physical theory to provide a model of this world. They share the common dialectical purpose of opposition to positivist conceptions of physical theory that Einstein perceived to be widely, and largely uncritically, assumed by quantum physicists under the aegis of Bohr's complementarity. Einstein's realism lacks the semantic apparatus of truth and reference of scientific realism. It also lacks the "epistemic optimism" encouraging the scientific realist to commit to claims of "approximate truth" for current theories. On the other hand, following Planck's example, Einstein can endorse claims that certain elements of the current "conceptual model" (*Konstruktion*) may well be permanent fixtures of the future "world image" that gives meaning to the term "physical reality". Such elements are not entities or laws of particular theories but meta-level constraints on theories, values of physical constants, principles of connection (e.g., between entropy and probability) or symmetries that remove the bias of particular observers or frames of reference. With these provisos, Einstein is indeed a "paradigm realist".

Notes

1 Leeds, Steven, "Correspondence Truth and Scientific Realism", *Synthese* v. 157 (2007), pp. 1–21; p. 19.
2 Psillos, Stathis, *Scientific Realism: How Science Tracks Truth*. London and New York: Routledge, 1999; Anjan Chakravartty (2011), "Scientific Realism", in Edward N. Zalta (ed.), *The Stanford Encyclopedia of Philosophy* (Fall 2015 Edition), available at http://plato.stanford.edu/archives/fall2015/entries/scientific-realism/
3 Van Fraassen, Bas, *The Scientific Image*. New York: Oxford University Press, 1980.
4 Maxwell, Grover, "On the Ontological Status of Theoretical Entities", in Herbert Feigl and Grover Maxwell (eds.), *Scientific Explanation, Space, and Time, Minnesota Studies in the Philosophy of Science*, vol. 3. Minneapolis: University of Minnesota Press, 1962, pp. 3–27.
5 Smart, John Jamieson Carswell, *Philosophy and Scientific Realism*. London: Routledge and Kegan Paul, 1963, p. 151.
6 Fine, Kit, "The Question of Realism", *The Philosophers' Imprint* v. 1, no. 1 (2001), p. 25.
7 Physicist-turned-epistemologist Paul Volkmann wrote of "*eine auf dem Boden der Naturwissenschaften erwachsene Erkenntnistheorie*" in his *Erkennistheoretische Grundzüge der*

Naturwissenschaften, Zweite Auflage, 1910. B.G. Teubner: Leipzig, Germany & Berlin, p. 241.

8 Kirchhoff, Gustav, *Vorlesungen über mathematische Physik: Mechanik.* Leipzig, Germany: Teubner, 1876, p. 1.

9 Duhem, Pierre, *The Aim and Structure of Physical Theory.* Second edition (1914). As translated by P. Wiener. Princeton, NJ: Princeton University Press, 1991, p. 54.

10 Mach, Ernst, "Über das Prinzip der Vergleichung in der Physik", in Ernst Mach, *Populär-Wissenschaftliche Vorlesungen.* Leipzig, Germany: Johann Barth, 1896, pp. 251–74; p. 269; translation altered from "On the Principle of Comparison in Physics", in T. J. McCormack (trans.), *Ernst Mach Popular Scientific Lectures.* Chicago: Open Court Co., 1898, pp. 236–58; p. 254.

11 "Autobiographical Remarks", p. 45; p. 47.

12 Planck, Max, "Die Einheit des physikalischen Weltbildes", as reprinted in *Wege zur physikalischen Erkenntnis: Reden und Vorträge.* Zweite Auflage. Leipzig, Germany: S. Hirzel Verlag, 1934, pp. 1–32; pp. 28–9.

13 *Ibid.*, p. 30.

14 *Ibid.*, p. 32.

15 Planck, Max, "Sinn und Grenzen der exakten Wissenschaft", *Die Naturwissenschaften* v. 130 (February 27, 1942), p. 130: "from the standpoint of exact science there always remains an unbridgeable chasm between the phenomenological and the metaphysically real worlds (that) engenders a constant, unbalanced tension. . . . In this bifurcation, which expresses itself in that we view the presupposition of a real world in the absolute sense as inevitably required but, on the other hand, we are never capable of completely grasping its nature, lies an irrational element which exact science can never shake off".

16 Planck, Max, "Über irreversible Strahlungsvorgänge", *Annalen der Physik*, Vierte Folge Bd. 1 (1900), pp. 69–122.

17 Planck, Max, "Zur Machschen Theorie der physikalischen Erkenntnis: Eine Erwiderung", *Physikalische Zeitschrift* Bd. 11, pp. 1186–90 (1911), p. 1188, translation as "On Mach's Theory of Physical Knowledge: A Reply" in Stephen Toulmin (ed.), *Physical Reality: Philosophical Essays in Twentieth-Century Physics.* New York: Harper and Row, 1970, pp. 44–52; p. 46.

18 CPAE 8 (1998) Part B, Doc. 624.

19 Hermann, Armin (ed.), *Albert Einstein/Arnold Sommerfeld Briefwechsel.* Basel/Stuttgart: Schwabe & Co. Verlag, 1969, Doc. 10, p. 48. The letter has the approximate date, "end of February, beginning of March 1918".

20 Einstein, "*Motive des Forschens*" ("Motives of Research"), April 26, 1918. CPAE 7 (2002), Doc. 7.

21 See the insightful review by Harry T. Costello ("H.T.C.") of Ernst Zimmer, *The Revolution in Physics.* New York: Harcourt, Brace and Co., 1936. *The Journal of Philosophy* v. 33 (1936), pp. 527–8.

22 Planck, Max, *Where Is Science Going?* Prologue by Albert Einstein. Translated and edited by J. Murphy. New York: W.W. Norton and Co., 1933.

23 Dimsdale, Nicholas H., Nicholas Horsewood, and Arthur Van Riel, "Unemployment and Real Wages in Weimar Germany", University of Oxford, Discussion Papers in Economic and Social History, Number 56, October 2004.

24 Planck, Max, "Positivismus und reale Aussenwelt" as reprinted in *Wege zur physikalischen Erkenntnis: Reden und Vorträge*. Zweite Auflage. Leipzig: S. Hirzel Verlag, 1934, pp. 208–32; p. 217.

25 Schlick, Moritz, "Positivismus und Realismus", *Erkenntnis* v. 3 (1932), pp. 1–31; English translation in Henk L. Mulder and Barbara F. B. van de Velde-Schlick (eds), *Mortiz Schlick Philosophical Papers*, vol. 2 (1925–1936). Dordrecht: D. Reidel, 1979, pp. 259–84.

26 Physicist and historian Gerald Holton reports the note was "written on or just before April 17, 1931". Holton, "Mach, Einstein, and the Search for Reality", 1968, p. 262.

27 Schlick, Mortiz, "Die Kausalität in der gegenwärtigen Physik", *Die Naturwissenschaften* v. 19 (1931), pp. 145–62; English translation in H. Mulder et al (eds), *Mortiz Schlick Philosophical Papers*, vol. 2 (1925–1936). Dordrecht: D. Reidel, 1979, pp. 176–209; p. 196.

28 Ibid., p. 206.

29 EA 21–603.

30 Einstein, "Einleitende Bemerkungen über Grundbegriffe", in *Louis de Broglie: Physicien et Penseur, collection dirigée par André George*. Paris: Éditions Albin Michel, 1953, pp. 4–14; p. 6.

31 "Replies to Criticisms", p. 674.

32 Einstein, "Einleitende Bemerkungen über Grundbegriffe", pp. 4–15; p. 6.

33 "Replies to Criticisms", p. 667; German, p. 494.

34 See Stachel, John, "Einstein and the Quantum: Fifty Years of Struggle", in R. Colodny (ed.), *From Quasars to Quarks: Philosophical Problems of Modern Physics*. Pittsburgh: University of Pittsburgh Press, 1986, pp. 349–85.

35 Einstein, "Elementare Überlegungen zur Interpretation der Grundlagen der Quanten-Mechanik", in *Scientific Papers presented to Max Born on his Retirement from the Tait Chair of Natural Philosophy in the University of Edinburgh*, Edinburgh: Oliver & Boyd, 1953, pp. 33–40; p. 34.

36 "Replies to Criticisms", p. 680.

Further reading

Fine, Arthur, *The Shaky Game: Einstein Realism and the Quantum Theory*. Second edition. Chicago: University of Chicago Press, 1996.

Howard, Don, "Einstein's Philosophy of Science", in Edward N. Zalta (ed.), *The Stanford Encyclopedia of Philosophy*. Winter 2015 edition, URL = <http://plato.stanford.edu/archives/win2015/entries/einstein-philscience/>.

Scheibe, Erhard, "The Origin of Scientific Realism: Boltzmann, Planck, Einstein", in M. Pauri (ed.), *The Reality of the Unobservable*. Dordrecht: Kluwer Publishers, 1999; reprinted in B. Falkenburg (ed.), *Between Rationalism and Empiricism: Selected Papers in the Philosophy of Physics*. New York: Springer Verlag, 2001, pp. 142–55.

Nine
Philosophy of science – constructivism

Physics is an attempt to grasp existence (*das Seinde*) as something conceptual (*etwas begrifflich*). In this sense one speaks of the "physically real".[1]

Introduction

Responding in 1949 to the quantum orthodoxy that it is meaningless to speak of the definite disintegration time of an individual radioactive atom, Einstein pronounced the positivist attitude of the quantum physicists as scarcely distinct from Berkeley's *esse est percipi*. Impugning subjective idealism as absurd, Einstein proceeded to elucidate how he understood claims of physical theory about unobserved objects and processes:

> "Being" (*Das "Sein"*) is always something intellectually constructed by us (*von uns gedanklich Konstruiertes*), hence free statutes (*frei Gesetztes*) by us (in the logical sense). The justification of such posits (*Setzungen*) does not lie in their derivation from what is given to the senses. Never and nowhere is there a derivation of this kind (in the sense of logical deducibility), not even in the domain of prescientific thought. The justification of the constructs that represent the "real" (*das "Reale"*) for us alone lies in their more perfect, or less imperfect, suitability for making intelligible what is given in sensation (the vague character of this expression is forced here upon me by my striving for brevity).[2]

Brevity has not served Einstein well. It has led scholars to select one or another of these, or other similar, remarks in an attempt to identify a distinctive philosophical tendency. But there are difficulties in placing all in harmony together. How can the ostensibly realist retort to Schlick cited in the previous chapter be reconciled with the apparently idealist nuances of this passage? What is the significance of the fact that Einstein persistently places the term "real" in scare quotes? How is it possible to consider "existence" or "being" as an "intellectual construction"? What is meant by "free statute"?

An identifiable philosophical position, established by an earlier generation of philosopher-scientists, does encompass all these distinct aspects, but it is one familiar today perhaps only to historians of physics. Its incongruity with contemporary philosophical understandings of "realism" or "scientific realism" threatens to guarantee that efforts to assimilate Einstein's views to contemporary positions remain anachronistic. The underpinnings of Einstein's idea of physical theory as a conceptual construction aiming to grasp the real drew from, and built upon two related *fin de siècle* epistemological currents, one in pure mathematics, one in theoretical physics. The former, associated with George Cantor and Richard Dedekind, emphasized the freedom of concept formation in modern mathematics; the latter, stemming above all from Heinrich Hertz, sought to make precise the meaningful sense in which a physical theory could be said to be conceptual framework, an image (Bild) that seeks to "grasp" or "picture" – in short, "construct" – physical reality. Einstein's assessment of the quantum theory as incomplete by 1930 implied it was incapable of being a physical theory of this kind. Even more, that the quantum physicists spurned as antiquated such a conception of physical theory elevated Einstein's opposition to the quantum theory into a clash concerning the very meaning of physical theory.

Epistemological credo

Between 1930–1950, Einstein published two extensive expositions – the latter prefaced by remarks terming it "my epistemological credo" – of the philosophical considerations ostensibly underlying his practice as a theoretical physicist. The Latin term calls for comment.

"Credo" means literally "I believe", while a credo is an avowal of religious conviction. We may therefore presume intent, though tinged with gentle irony, to give expression to a resolute and fundamental faith. The question then is: what belief or set of beliefs could elicit such steadfast commitment from a theoretical physicist cognizant of the imperfect state of physical theory?

Unremarkably, it is a conviction regarding the core character of the scientific enterprise itself. And this is a persuasion that the methods, observational procedures, and theories of natural science have value insofar as they contribute to the overriding goal of finding out how the real external world is; that is, as it exists independently of any mode of perception. As Einstein will demonstrate, much in this declaration called for, and indeed received, clarification and epistemological qualification. But clothed even in vulgar dress, this is a *motivational realism* regarding the nature and purpose of natural science that was then, and remains today, perhaps the implicitly held principal *reason for doing science* of many, even most, scientists. Physicists other than Einstein, Duhem and Planck in particular, had recognized both the allure and purported necessity of this belief as well as the impossibility of ever completely justifying it. Yet quantum mechanics, particularly as portrayed by Bohr, renounced the project of seeking physical reality altogether, refusing to recognize as meaningful statements about what might be true of individual physical systems independently of devices for measurement of particular observables. As concordant positivist and non-realist philosophical currents swept over the physics community, Einstein felt compelled to articulate a new defense based upon his understanding of the development and transformation of physical theory since the turn of the century. A crucial component of this understanding is a retrospective reflection on his experience with the general theory of relativity. This reflection forms the backdrop of his criticism of the new quantum theory, not only because he regarded it as an incomplete description of individual physical systems but also since he considered it a theory of a new kind, surrendering and even dismissing the aim of science to find out how nature really is. Aided and abetted by logical positivism, these views had spread also into philosophy and beyond, into the lay understanding of science. To Einstein a view of science offering only positivism and instrumental control was both repugnant and false. In the face of this clearly perceived threat, he decided to respond.

The two presentations, affirming essentially the same viewpoint, were written some ten years apart during his residence in Princeton, the first in 1936 and the other in 1946. On each occasion, Einstein's German text is published in its entirely, together with accompanying English translation. It may be presumed that Einstein insisted on this unusual format not only on account of his far greater confidence of expression in his native tongue but undoubtedly also because the peculiarly German language nuances of his epistemological views, inevitably coarsened in translation, would be available in the public record. The later and more diffuse presentation appeared in "Autobiographical Notes" (written 1946, published 1949, with facing page translation by philosopher Paul Schlipp) together with scattered remarks in the "Replies to Criticisms" at the end of the same volume. In this second version, he took care to observe that his credo evolved slowly over a lengthy period of time and "does not correspond with the point of view I held in younger years" (1946/1949, 11). This appears correct and in accord with an assessment that articulating a "credo" was itself prompted by the challenge posed by quantum mechanics to Einstein's belief in the aim of physical science and the consequent task of the theoretician.

The 1936 presentation, prominently titled *"Physik und Realität"*, is the more systematic exposition. Thirty-four pages in length, it was published in a journal of engineering, popular science, and science education.[3] Immediately follows a complete English translation ("Physics and Reality") by Swiss-American chemist and engineer Jean Piccard, at the time a world-record-holding high altitude balloonist. The unusual venue was an indirect result of the award of the Franklin Medal to Einstein in March 1935.[4] It would appear that Einstein already had conceived of the idea of writing such an essay. Today it is remembered mostly for a brief reiteration of the incompleteness argument against quantum theory given in the famous EPR paper, coincidentally published in the May 15, 1935 issue of *The Physical Review*.

In early summer 1935 two apparently striking recent results may well have fortified Einstein's resolve to articulate his conception of physical theory: first, the famous EPR paper arguing for the incompleteness of quantum mechanics (see Chapter 4); then, the July 1 issue of *The Physical Review* contained a paper written together with his assistant Nathan Rosen (the "R" of EPR) claiming a solution to "the particle problem" in general relativity.[5] For more than ten years, Einstein had considered the theory of general relativity as essentially an approximation to, and as a source of insight for, a further generalization to a unified theory of fields. The Einstein-Rosen paper of 1935 accordingly sought solutions of the now joint field equations of gravitation and electromagnetism corresponding to the particulate structure of matter and electricity. The difficulty had been to avoid point particles and the singularities associated with them. The Einstein-Rosen paper claimed to find such singularity-free solutions in discrete "bridges" connecting two congruent "sheets" of physical space; the bridges are portions of space representing both neutral and (massless) elementary electrical particles.[6] The paper created little notice at the time, but since its revival in the 1980s, it has been celebrated in both popular and physical literature for originating the exotic concept of "wormholes" in space-time, topological "bridges" between distant regions of space-time that theoretically suggest the possibility of time travel. (Most recently, theorists have appealed to Einstein-Rosen bridges to resolve the so-called "firewall paradox" implying an inconsistency of general relativity with quantum mechanics; see Chapter 11.) Not surprisingly, Einstein invoked both the EPR incompleteness result and the Einstein-Rosen result at the corresponding strategic place in his 1935 essay. As with the Herbert Spencer Lecture in Oxford in June 1933 (discussed in Chapter 10), putative breakthroughs within his own research program encouraged Einstein to have the confidence for sustained philosophical pronouncement about the aim and future of theoretical physics.

   Both versions of the credo appear to have been composed with the same double purpose in mind: first, to articulate, clarify, and justify a "realist" (in the sense outlined in the previous chapter and to be further elucidated below) conception of physical theory; secondly, to show by reference to the history of physics that a reasonable course for future fundamental physical theory lay through a further generalization of field theory beyond the space-time geometry of general relativity, the geometry of a unified theory of fields. This path is presented as an evolutionary next step beyond general relativity and a continuation of the conception of theory presupposed in all previous attempts to find a "basis of physics", i.e., a foundational theory universally valid at all scales, capable of embracing all natural phenomena. In contrast, Einstein casts the quantum theory as a significant and disagreeable departure from this conception. Einstein's epistemological credo is not then a disquisition on sundry epistemological issues of natural science but rather as the title of the 1935 essay bluntly affirms, the expression of a deeply rooted conviction that "physics", or rather physical theory, has a determinate relation to "reality" though the nature of that relation required both articulation and rather elaborate justification.

The structure of the argument is essentially the same in both versions of the credo. Einstein begins with epistemological fundamentals, a doctrine of concepts and of their relation to sense experience. Concepts, he insisted, are "free creations of the human mind" not inductively derivable from sense experience. Anti-inductivism is a recurrent theme in the credo. Next an axiomatic conception of physical theory is outlined, emphasizing first, the indirect relation of the theory's basic concepts and relations to experience; and second, the dual criteria distinguishing a desirable theory: empirical adequacy and "inner perfection", i.e., lack of superfluous or unnecessary elements in the theory's formulation. There follows a chronicle of physical theories that, since 1900, were considered fundamental or universal in scope in the sense of providing explanation, at least in principle, of any physical phenomenon. The search for "the basis of physics" informs the historical narrative: mechanics, thermodynamics, and electrodynamics had been successively proposed; each had been subsequently shown inadequate.

In the 1946 version, an autobiographical interlude interrupts the narrative at this point, discussing first the influence of Planck's discovery of the quantum of energy, the use of Boltzmann's conception of probability as entropy, and the inconsistency (pointed out by Einstein in 1906) in Planck's derivation of his empirically justified radiation formula. The interlude provides occasion to refer to his 1905 work on Brownian motion in criticism of philosophical prejudices of positivist skeptics (Ostwald, Mach) of the reality of atoms: they mistakenly held that the facts themselves "can and should yield scientific knowledge without free conceptual construction".

At this juncture, the common narrative resumes through consideration of the two theories of relativity, highlighting the needed revision in the concepts of time and space while stressing that already in the special theory, despite the unjustified concept of inertial frame, it appeared "unavoidable that physical reality must be described by continuous functions in space". The quantum mechanics of Heisenberg, Born, de Broglie, Schrödinger, and Dirac, described in the 1946 version as "the most successful physical theory of our time", is briefly characterized. Then follows a summary of Einstein's reasons for the verdict that, considered as describing individual systems, that theory is incomplete, a defect that however can be removed in Einstein's opinion by viewing it as an ensemble (purely statistical) theory. The narrative concludes by posing the question of "how the theoretical foundation of the physics of the future will appear" as lying in a choice between quantum mechanics and some kind of generally relativistic unified field theory mentioned above. As indicated above, this is also regarded as a choice between two opposing conceptions of physical theory, and in both the 1935 and 1946 versions of the credo, Einstein supports his decision to pursue the latter path, a choice that "departs most widely from that of contemporary physicists", by invoking what he obviously considered to be significant theoretical results in his pursuit of a unified theory of fields. To be sure, the results appealed to in 1935 and in 1946 come from distinct proposals for such a theory.

The "constructing human mind"

The flat-footedly realist avowal that the scientific enterprise pursues the overriding goal of finding out how the real external world is independently of any mode of perception requires rather extensive epistemological tempering to be at all defensible, and this is the task assigned the two lengthy epistemological presentations. What immediately strikes the reader of the 1936 essay is the phrase "the constructing human mind" in its declaration of aim:

> The goal of the following lines is to show which paths the constructing human mind (der konstruierende Menschengeist) has pursued in order to attain a logically simplest unified conceptual basis of physics.[7]

What is "constructive" about the human mind? It will be apparent this is a different sense of "construction" than that discussed in the

Introduction between "principle" and "constructive" theories since both types of theory are the product of the "constructing human mind". Rather the sense here is close to a variety of epistemological views, broadly termed *constructivism*, holding illusory any claim that there can be knowledge of mind-independent objects as they are in themselves, without significant tincture of mentation, what William James called "the trail of the human serpent". According to constructivism, cognition of extra-mental objects is always to be understood as refracted through human conceptual, perceptual, and even cultural and social prisms. The object becomes an object of cognition (i.e., object as known) only insofar as it is constructed, constituted, made, or produced by the cognizing subject (singular or plural), that is, only insofar as it is recognized within the cognitive framework of the knowing subject or, as Kuhn has it, located within the governing paradigm of the scientific community. For constructivism there is, in principle, always a distinction to be drawn between object itself and the object as cognized, a distinction Kant captured as that between noumenal and empirical reality, and Planck between the physical *Weltbild* of the future and unadorned metaphysical reality. But though Kant is the most influential constructivist, the view itself is much older, going back to ancient Greek mathematics and the geometrical method of proof by construction of figures in the plane. The hurdle that any viable constructivism must cross is to reconcile the premise that constructing objects is an unavoidable condition of having knowledge of them with a claim that such knowledge is nonetheless objective, that the objects known are indeed "real" without lapsing into solipsism or subjective idealism. We saw Einstein alluding to this already in 2) in "The changing face of scientific realism" section of Chapter 8, in his remarks improving upon the quantum physicists and Berkeley's *esse est percipi*.

The 1936 essay's first section, entitled "General Considerations Concerning the Method of Science" (*Allegemeines über die wissenschaft-liche Methode*), takes up this theme by expressing a conviction that thinking (or thought) in general is *by its nature* a constructive activity of mind. Einstein shies away from psychology in distinguishing thinking from other mental activities like free association and "dreaming" by the dominating cognitive role performed by concepts. This presupposes a core assumption about cognition, familiar in the Kantian tradition: a sharp distinction between an active, spontaneous conceptual faculty of mind and a largely receptive sensory part whose content is given as sensations or "complexes of sense impressions", interconnected perceptual data whose external causes lie outside the mind. Cognition requires concepts that have a relative autonomy from sense experience; they are in general "constructs", creative products of mentation or rather of imagination.

Thinking then is primarily discursive, operating with concepts, producing and applying determinate relations between concepts, and coordinating sense experience to the concepts employed in inquiry. Einstein pointed to the character of everyday thought as illustration. Through a mental process to some extent arbitrary and intuitive, from the "totality of sense experiences" certain patterns are selected for attention; to these the mind correlates a concept of bodily object ("the first step towards positing a real external world"); this concept then serves as a sign communicable to others for such patterns. The characteristics or extension of the concept is suggested by experience but by no means determined by experience since thinking is also imaginative, fueled by analogies, resemblances, figurative non-literal language, and simplifying idealizations. The activity of mind is creative in both the formation and the application of concepts. On the other hand, the justification for any concept lies in the totality of occurrences or fulfilled expectations of patterns of sense experience associated with the use of the concept. Application of the concept in new experiential situations continually changes the scope of the anticipated associations.

It is this view of thought and the relations of concepts to sense experience that undergirds Einstein's statement that "all science is only a refinement of everyday thinking", a notion of science differing considerably from the Newtonian or Cartesian tradition of natural philosophy. Yet regarding science as a sharpened continuation of "everyday thinking" was not a new perspective. It already was the core conception of science of American pragmatism, most prominently in theory of inquiry of Peirce and Dewey. Portraying science as a "refinement of everyday thinking" appears to have been the concomitant of a current of British 19th-century natural science whose most representative figures are the non-mathematicians

Faraday and Darwin. T.H. Huxley (to become known as "Darwin's bulldog") expressed this conception in a picturesquely unequivocal manner from the perspective of natural history as early as 1854:

> Science is, I believe, nothing but *trained and organized common sense*, differing from the latter only as a veteran may differ from a raw recruit: and its methods differ from those of common sense only as far as the guardsman's cut and thrust differ from the manner in which a savage wields his club.[8]

Huxley had explicitly in mind what he elsewhere termed the "unconscious logic of common sense" in everyday matters, for example, reasoning by simple analogy and inferences seeking to account for observed effects by appeal to causes known from experience to be competent to produce such effects. Presumably Einstein agreed, but he added to this conception a doctrine of concepts, an element completely foreign to Huxley's notion.

There is a distinctive idealist and anti-inductivist cast to Einstein's view of concepts as constructs that are, by their origin as it were, speculative. Even so, they enable (or should make possible) forecasting and prediction. In the now familiar Popperian sense, concepts are conjectures, and indeed Popper's characterization of scientific method as that of conjectures and refutations arose from his reflection on Einstein's general relativity. Einstein even charted the course of the development of the concepts of physical science along an axis showing accelerating distancing from sense experience. In the "childhood" of physical science, at the birth of classical mechanics, an inductive methodology was appropriated – urged by Newton as a corrective to Descartes' rationalist physics. But in the course of the 19th century, particularly following Maxwell's field theory of electromagnetism, the basic concepts of theory lie further and further from confirming observations. General relativity next showed that a speculative mathematical methodology of theory construction appeared necessary to advance the foundation of physics.

After the completion of general relativity, Einstein would occasionally argue that speculative methodology had always been part of physical science, going back to the early 17th century. In a 1930 essay commemorating the 300th anniversary of Kepler's birth, Einstein identified Kepler as its most notable proponent:

> It seems that human reason has first to construct forms independently (*die menschliche Vernunft der Formen erst selbstständig konstruieren muß*) before we can find them in things. From Kepler's marvelous lifework we particularly beautifully recognize that knowledge cannot spring from mere experience alone but only from the comparison of inventions of the intellect (*Erdachtem*) with the observed.[9]

This high praise of Kepler's attempt to use the five Platonic solids to represent the distance relations between the known planets is further proof that reflection on the experience of completing general relativity convinced Einstein that the only way forward in pursuit of a fundamental physical theory capable of unifying widely disparate realms of phenomena in a common theoretical framework lay in mathematical speculation. This belief resounds in repeated admonitions that concepts (and the conceptual structures into which they enter) are "free creations of the human mind" and in an avowal that scientific knowledge itself arises only through "free conceptual construction".

The adjective "constructive" is then Einstein's generic term for the theoretical physicist's balancing act or continual pivot between the opposing tendencies of rationalism and empiricism. From the standpoint of physics, empiricism requires the distinction of "primary concepts" from "all other notions"; the former alone are directly connected to typical patterns of sense experience. The validity or justification of all physical concepts consists in their connection, however indirect, to the "primary concepts". If not explicitly definitional in nature, they have physical meaning only insofar as they can be linked in various ways, however remotely by deduction, to the primary concepts, and so to experience. In a letter to Solovine in 1952, Einstein clearly portrayed his conception of the "eternally problematic connection between everything intellectual (*alles Gedanklichen*) and what is tangible (sense experience)" in a diagram with four articulated steps (Figure 9.1).

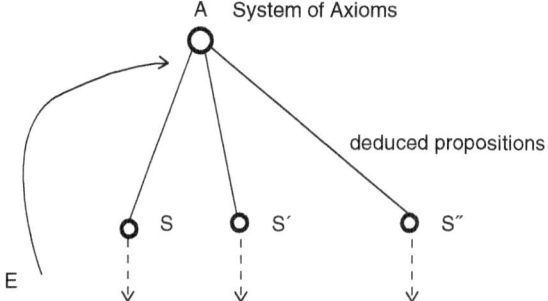

(1) The E's (experiences) are given to us.
(2) A are the axioms from which conclusions are drawn. Psycho-logically, the A's depend on the E's. But there is no logical path from the E's to the A's, rather only an intuitive (psychological) connection, an always "re-turning".
(3) *Logically*, specific propositions S, S′, S″ are deduced from A; these propositions can lay claim to accuracy (*Richtigkeit*).
(4) The S's are brought into relation with the E's (confirmation by experience). Under more exact consideration, this procedure also belongs to the extra-logical (intuitive) sphere because the relations between the concepts appearing in S and the immedi-ate experiences are not logical in nature.[10]

Figure 9.1

The diagram, and accompanying elucidating remarks, are an almost textbook presentation of confirmational holism in the method of hypothetico-deductive testing of theories.

A rationalist moment enters with the axiomatic conceptual struc-tures that are contemporary physical theories. Here speculative abstract concepts or principles may stand at considerable deductive remove from any sense experience, and so are implicated or condemned by observation only indirectly. In this balancing sense of "construc-tive", Einstein fashioned a "constructive" epistemological framework to attempt to justify his conviction about the character of physical theory and the inherently realist aim of science. The constructive framework is not a purely philosophical imposition but stems from, and is traceable to, two distinct currents arising within the previous half-century of exact science. The first comes from pure mathematics but is easy to identify in Einstein's doctrine of concepts by use of Richard Dedekind's locution, "free creations of the human mind". The other arises within physics, but is only implicitly indicated and so must be inferred: this is the *Bild* conception of physical theory first formulated by Heinrich Hertz, reiterated by Ludwig Boltzmann, and extended by Max Planck. These are considered in turn.

Free creations of the human mind

As noted in Chapter 7, in the lecture "Geometry and Experience" at the Berlin Academy in January 1920, Einstein characterized axioms in the modern axiomatic treatment of geometry as "free creations of the human mind" (*"freie Schöpfungen des menschlichen Geistes"*). In his lectures at Princeton in May 1921, he returned to this expression to signal a relative but not complete autonomy of concepts from experience:

> Concepts and conceptual systems are justified only in so far as they provide an overview of the complexity of experience; no other legitimation is possible. For that reason it is my conviction that one of the most corrupting deeds of the philosophers is to have removed certain conceptual foundations of natural science from the accessible control of domains of the empirical expedient to the unassailable heights of necessities of thought (*a priori*). For also if it turns out that concepts cannot be logically or somehow derived from experience but are, in a certain sense free creations of the human mind (*freie Schopfungen des menschlichen Geistes*) then they are just as little independent of the nature of experience as clothes are of the human body. This particularly holds of our concepts of time and space which physicists – compelled by facts – have brought down from the Olympus of the *a priori* in order to repair them and again put them in a usable state.[11]

The accolade "free inventions of the human mind" is applied to both concepts and basic laws of physics in the Herbert Spencer lecture at Oxford in June 1933.[12] In "Physics and Reality", the concept of bodily object is "a free creation of the human (or animal) mind" (*"eine freie Schöpfung des menschlichen [oder tierischen] Geistes"*) and so distinct from the totality of relevant sense impressions.[13] The pattern continues in The Evolution of Physics where "Physical concepts are free creations of the human mind" ("freie Schöpfungen des menschlichen Geistes"),[14] and in "Remarks on Bertrand Russell's Theory of Knowledge": "The concepts arising in our thought and linguistic expressions are all – logically considered – free creations of thought" ("freie Schöpfungen des Denkens").[15] In "Autobiographical Notes" written in 1946, the locution ("eine Schöpfung des Menschen") is extended broadly to conceptual systems in general, i.e., to "the conceptual system together with the syntactic rules comprising the structure of the system";[16] whereas Einstein's last publication affirmed that "logically considered", the concepts of space and time are "free creations of human intelligence ("freie Schöpfungen des menschlichen Intelligenz").[17] Clearly this is a view of concepts deeply rooted in Einstein's epistemology. What is its significance? Against what view or doctrine is it directed?

The Einstein texts are unequivocal in refusing any positivist or empiricist theory of concepts according to which meaning or content is inductively derived or abstracted from sense experience. The rejection extends to the only probabilistic meaning (in terms of measurement outcomes) given by quantum mechanical orthodoxy to the core kinematical or dynamical concepts of "position", "momentum", or "energy". Emphasis is placed instead on the cleft between sense experience and concept, rendering the latter "logically independent" (i.e., not inductively derivable) from the former. As seen above, Einstein also employed the metaphor of "free creation" to underscore a denial that any concepts (archetypically those of space or time) are immutable "necessities of thought". At times the metaphor is accompanied, as in his 1916 obituary of Mach[18] and the quotation from 1921 above, with an admonition to not forget the "earthly origin" of all concepts. In short, the doctrine of "free creation" is a bulwark against positivist phenomenalism, empiricist inductivism and Kantian a priorism. But what positive construal might the doctrine have?

The adjective "free" underscores that concepts are constructed, neither determined by nor derivable from experience nor necessities of thought. "Creation" also connotes freedom in the sense of imaginative pluralism, that distinct concepts may be equally compatible within some domain of experience. The different ways these two aspects are expressed often manifests the language of conventionalism: concepts are "posits" (*Setzungen*), or "creations" (*Schöpfungen*). Logical empiricists and others more recently have interpreted Einstein's remarks about "free creation" as accordingly indicating a strong current of conventionalism in his philosophy of science. A classic attempt along these lines is that of Phillip Frank, physicist, philosopher and Einstein biographer. Frank attempted to characterize Einstein's philosophy, as well as logical positivism as a whole, as resting upon a coherent integration of Machian positivism (on the "extreme empiricist side") with the conventionalism of Henri Poincaré (on the "extreme logical side"). This involves something of a delicate balancing act stressing, with Mach, that concepts must have an interpretation in the realm of observable facts while holding, as did Poincaré, that the meaning of certain general principles of physics or geometry, including terms such as "force", "energy" and "straight line", are established not by experience but through stipulation ("disguised definition"). A section of Frank's Einstein biography, subtitled "The General Laws of Physics Are Free Creations of the Human Mind", is accordingly devoted to Poincaré.[19]

On the other hand, Einstein frequently emphasized, against conventionalism, that that the mind's "liberty of choice" in creatively choosing or constructing concepts "is of a special kind". In the 1936 essay "Physics and Reality", the analogy of a crossword puzzle is invoked to underscore Einstein's conviction that only one choice "really solves the puzzle in all its parts". A conviction that physical reality has the character of a well-formulated puzzle or logical ordering (or in Duhem's sense, a "natural classification")[20] is a matter of faith that transcends all evidence, in short, a metaphysical postulate. In the April 1918 tribute to Planck on his 60th birthday discussed in the previous chapter, Einstein continues the passage quoted there to affirm a deeply anti-conventionalist view of physical theory:

> [T]he highest task of the physicist is the search for those most general elementary laws from which the world-image (*Weltbild*) is to be obtained by pure deduction. No logical path leads to these elementary laws; it is instead just the intuition that rests on an empathic understanding of experience. In this state of methodological uncertainty cone can think that arbitrarily many, in

themselves equally justified systems of theoretical principles were possible, and this opinion is, in principle, certainly correct. But the development of physics has shown that of all the conceivable theoretical constructions a single one has, at any given time, proved itself unconditionally superior to all the others. No one who has really gone deeply into the subject will deny that, in practice, the world of perceptions determines the theoretical system unambiguously, even though no logical path leads from the perceptions to the basic principles of the theory.[21]

With this passage in mind, Frank had to reluctantly recognize that Einstein

> felt that even *Logical Positivism* did not give sufficient credit to the role of imagination in science and did not account for the feeling that the "definitive theory" was hidden somewhere and that all one had to do was to look for it with sufficient intensity.[22]

While in the writings of Poincaré conventions are described as "disguised definitions" the expression "free creations of the human mind" is a well-known locution in Richard Dedekind's 1888 monograph, *Was sind und was sollen die Zahlen?* (*The Nature and Meaning of Numbers*). Maurice Solovine, Einstein's lifelong friend and fellow member in Bern of the so-called Olympia Academy, published a considerable number of Einstein's letters and postcards after Einstein's death from a correspondence extending nearly fifty years (a letter from 1952 was cited above). In the "Introduction", Solovine listed Dedekind's book as among those read and discussed by the Academy. Both in the "Preface" and in section 73 of that work, Dedekind characterized natural numbers as "free creations of the human mind" ("*freie Schöpfungen des menschlichen Geistes*"). By this Dedekind meant not only, as stated in the Preface, that the number concept is independent of the notions or intuitions of space and time, but more importantly (in section 73) that it is *justifiable* to regard the natural numbers as a free creation of the human mind ("*eine freie Schöpfung des menschlichen Geistes*") since as the merely distinguishable elements of a simply infinite system, they are *free from every other content*, in particular, from subjective ideas and extraneous connotations.

In Dedekind's account the natural numbers are mathematical objects whose only properties are completely determined by the totality of truths of arithmetic. Since the latter are infinite in number, they require axiomatic characterization, and so Dedekind's characterization of the number concept as a "free creation of the human mind" goes hand in hand with abstract, structural axioms for simply infinite systems (sometimes called the Dedekind-Peano axioms). Simply infinite systems are unique up to isomorphism which means that the laws and relations of arithmetic obtain for any system of objects satisfying the Dedekind axioms, no matter what designations may be given to the system's individual elements. (For example: Is the first element of the system designated "0" or "1"?) An archetypal example of "creation" is the statement (one of Dedekind's axioms) that every natural number has a successor. Any collection of objects encountered in experience is finite, but it is possible to "abstract" from any collection its finite cardinality, increase it by one and then indefinitely continue the process. The possibility of indefinite iteration undergirds the meaning of the statement that there is no largest natural number. Nothing in experience licenses the induction that given any finite number, there is a larger one. The consequent is a conceptual not an empirical conclusion. The successor function, explicitly in Einstein-Infeld (1938, p. 311; see the discussion below) is then a paradigm "free creation of the human mind".

Dedekind deemed a *structural characterization* of the concept of number in terms of the existence and uniqueness of simply infinite systems to be justification for the claim of "free creation" since it freed the concept of number from any subjective connotations imposed by individual minds while retaining only what must be common to all. This *structuralist* sense of free creation has nothing to do with conventionalism. It pertains not to creation of individual concepts but to an objective, because isomorphic, *system of concepts* capable of expressing all arithmetic truths. An analogue showed up above in Einstein's characterization of the method of physics as the search for "a logically simplest unified conceptual basis of physics".

Bildtheorie

By the early 1850s in Germany the epoch of *Naturphilosophie*, the *a priori* philosophies of nature of German idealism (Fichte, Schelling,

Hegel, and lesser epigoni), had run its course. But by so doing, it had badly tarnished the purport of philosophy to speak at all with authority about empirical natural science. Just as a century before in France, when the materialism of Holbach, La Mettrie, and Diderot developed in reaction to the theological metaphysics of Leibniz, within German philosophical circles the speculative metaphysics of the *Naturphilosophien* antagonistically prompted the crude mechanistic materialism of Vogt, Moleschott, Dühring, and Büchner, largely unknown names today except for the pillory received at the hands of Marx and Engels. On the other hand, in England and France, where *Naturphilosophie* found but few adherents, empiricism and positivism were firmly rooted among both philosophers and scientists. J.S. Mill sought to encapsulate what he saw as the inductive method of the natural sciences within the scope of *A System of Logic* (1838, and many later editions), while in France the positivism of August Comte became dominant, especially in the social sciences.

Meanwhile in Germany, philosophical rejection of mechanistic materialism by the 1860s led to the "Back to Kant" movement that gave rise to the different schools of Neo-Kantianism enduring up until the end of WW I. But as a result of speculative excesses of both idealist and materialist philosophers, the overriding tendency among German-speaking natural scientists after 1850 was to ignore all "school philosophy" as simply irrelevant to science. Almost by default, among natural scientists an austere form of positivism emerged that prized exact description of observed facts while remaining inherently skeptical or dismissive of theory. The most influential form of this positivism was that of physicist Ernst Mach. Mach framed positivism within a gradualist evolutionary account of mind according to which scientific hypotheses were considered merely further developments of primitive and instinctual thought serving as "economical" standins or abbreviations of scientific facts. But positivism proved a too-impoverished view of science for some philosophically minded scientists. Foremost in this group was Hermann von Helmholtz, Germany's greatest natural scientist and one of the few scientists who championed the "Back to Kant" tendency in philosophy. With the one exception of Mach, the most influential philosopher-scientists within a generation were Helmholtz's student Heinrich Hertz and the Viennese physicist Ludwig Boltzmann, who also had studied under Helmholtz. In their scientific and philosophical writings, a "scientific philosophy" autonomously appeared in relative isolation from professional philosophy. By 1900, this renewed "scientific philosophy" was in full flower, including now the prominent French mathematician Henri Poincaré as well as physicist and physical chemist Pierre Duhem. While naturally there were differences, of both substance and emphasis, in the respective philosophical perspectives of these philosopher-scientists, there were also broad similarities in outlook on certain core matters. Again with the exception of Ernst Mach who retained positivism's skepticism of theory, the kinship in viewpoint extended to the general characterization of the aims and value of physical theory.

The appearance of a physical theory of any entirely different kind from mechanics, the Maxwell theory of the electromagnetic field, prompted Heinrich Hertz in the early 1890s to write a highly influential epistemological reflection on what, in general, a physical theory can be, and what are its attainable goals. Hertz's conception was laid out in a lengthy "Introduction" to his Principles of Mechanics (1894), a book in press when Hertz perished from a blood infection on January 1, 1894. Despite the title, Hertz had not written a textbook, but an ambitious attempt to reformulate mechanics by eliminating the concept of force that Hertz regarded as logically obscure. In place of action-at-distance forces, Hertz substituted purely hypothetical networks of hidden masses and hidden motions connected to observable masses and motions in such a way as to simulate the contiguous transmission of action of electromagnetic theory.

Hertz's approach to mechanics did not find widespread acceptance. Far more influential was the Introduction, a philosophical discussion of the nature of physical theory that, as Königsberg physicist-philosopher Paul Volkmann observed in 1901, was "read with equal enthusiasm by philosophers and physicists" and was considered a "new gospel" exercising "great influence . . . in particular on the younger generation".[23] Among that younger generation was Albert Einstein for the Bild conception informs Einstein's view of physical theory throughout his lifetime. From his biographer Philipp Frank, it is known that Einstein read and studied Hertz's Mechanics as a student at the ETH in Zurich, as well as in Bern with his friends Habicht and Solovine, co-members of the "Olympia Academy". Decisive methodological aspects of Einstein's epistemological credo are reflections or implications of Hertz's *Bild* conception of physical theory.

Hertz had been the student of Hermann von Helmholtz, and a rapidly rising star in a growing international community of physics. A renowned experimenter, his generation and detection of radio waves in 1888 gave dramatic confirmation to the 1865 prediction of Maxwell that electromagnetic waves propagate with the speed of light and that light itself is such a wave. It was this expertise with Maxwell's electromagnetic theory that led to Hertz's oft-repeated characterization of that theory as "Maxwell's system of equations". In this way, Hertz swept aside Maxwell's own presentation of his theory via the complicated path of empirical discovery of various electromagnetic phenomena, the theory of which Maxwell sought to heuristically illustrate through construction of elaborate purely mechanical analogies to the electromagnetic ether. Though Maxwell explicitly stated he did not believe in the existence in nature of such hypothetical mechanisms, he nonetheless considered them to be helpful means by which electromagnetic phenomena might be reproduced. Hertz renounced the theoretician's supposed need of heuristic resemblances or visualizable models and promoted a new abstract conception of physical theory in which the sole relation between the premises of theory and entities or processes in nature need be only symbolic.

According to Hertz, the office of physical theory cannot be that of capturing the nature of the physical world lying behind the appearances but rather more modestly that of providing only *Bilder* (images or pictures) or even *Scheinbilder* (apparent images) of nature, i.e., models of unobservable aspects of reality. To be sure, scientific pedagogy traditionally instructed students to imagine or picture unobserved objects and processes (e.g., a sound wave) with "the eye of the mind", i.e., to represent the unobserved through analogy to objects and processes both observed and familiar. Maxwell's mechanical models of the electromagnetic ether were heuristic devices of this kind. But in the new sense, the "picture" or "image" of nature provided by a physical theory is of a considerably more abstract genus, dependent on the theory's mathematical assumptions and intellectual posits. Above all, the *Bild* conception underscored Hertz's epistemological assumption that it is not possible to know whether theoretical representations, the mental images comprising physical theories, are "in conformity with the things themselves", i.e., outside of the conditions and conventions imposed by their mathematical representation. That was Hertz's principal philosophical message.

The *Bild* conception of theory is outlined in two widely quoted paragraphs of the "Introduction", articulating the theme of a double parallelism: that between thought and nature, and between theory and phenomena. The first parallelism is that of hypothetical theoretical structures and unobserved causal mechanisms in nature, and the second that between predictions derivable from the theory (observation statements) and observed causal effects of the hidden mechanisms:

> We form for ourselves apparent mental images or symbols (*innere Scheinbilder oder Symbole*) of external objects; and the form we give them is such that the necessary consequents in thought of the pictures are always the pictures of the necessary consequents in nature of the objects pictured.
>
> The pictures which we speak of are our representations (*Vorstellungen*) of things; they are in conformity with the things in *one* important respect, namely in satisfying the above-mentioned requirement. For our purpose it is not necessary that they should be in conformity with the things in any other respect whatsoever. As a matter of fact, we do not know, nor have we any means of knowing, whether our representations of things are in conformity with the things themselves in any other than this *one* fundamental respect.[24]

A diagram makes Hertz's *Bild* conception of physical theory more readily comprehended (Figure 9.2).

The *innere Scheinbilder* or *Symbole* are hardly synonyms but emphasize different aspects of representational freedom. The term *innere* clearly means "mental", while *Scheinbild* is borrowed from German aesthetic or theatrical discourse; it has no exact English equivalent but connotes something like "appearance" or "apparent image", while also suggesting illusion or mirage. *Scheinbild* roughly connotes an image (*Bild*) suspected of, or understood as having, an illusory (*schein*),

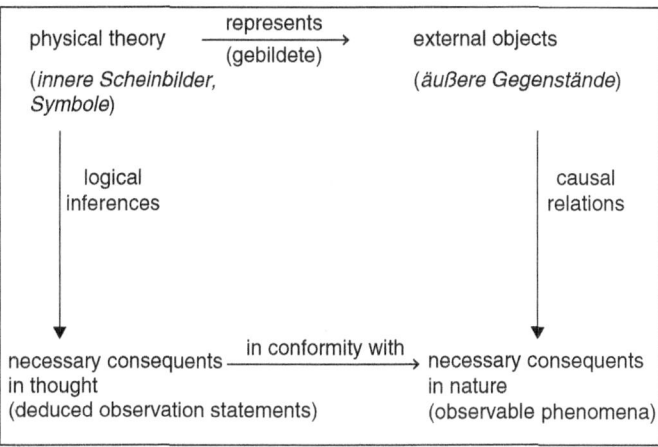

Figure 9.2

or at least not-to-be-taken-as-literal, character. Hertz's readers and followers largely jettisoned the adjectival use of *Schein* while tacitly retaining its non-descriptive nuances, simply referring to physical theories as pictures or images (*Bilder, Abbilder*).

The reference to signs or *Symbole* brings out representational freedom in a different way, signifying that physical theories are stated in terms of algebraic relations between symbols of a calculus or formalism that stand for physical quantities and processes; the term carries the connotations of "conventional" and "abstractness": i.e., as a sign, the particular symbol chosen to stand for some physical quantity or process has no "natural" connection to what it designates. Pierre Duhem, much more familiar to philosophers of science today, followed Hertz in employing precisely this sense of "symbol" when characterizing physical theories as purely symbolic and so neither true nor false.

No positivist, Hertz enjoined that in order to attempt to complete the parallelism between thought and nature, the theorist is free to invoke invisible entities – "clandestine partners (*heimliche Mitspieler*) behind the limits of our senses" (p. 25; p. 30 German original) – in order to formulate a *Bild* of the laws governing physical reality. Thus the main theme of Hertz's "Introduction" is to articulate the sense in which a physical theory may be legitimately said to be a *representation* (*Vorstellung*) of unobservable aspects of the external world. His signature philosophical contribution, drawing upon his formal assessment of Maxwell's theory and upon a contemporary proliferation of theoretical proposals for electrodynamics, is that physical theories are representative of the external world in the sense of being images (*Bilder*) of this world, valuable but not perhaps entirely trustworthy.

Whatever term (*Scheinbild, Bild, Abbild, Symbol*) employed, Hertz's intent was to embed the constructivist transformation in the concept of knowledge initiated by Kant into the very conception of a physical theory. Since, as Hertz emphasized, "we do not know, nor have we any means of knowing" whether physical theories thus constructed are in actual conformity with the mind-independent world, theories can at best project various images of this mind-independent reality, to be compared and evaluated according to antecedently adopted normative criteria. Hertz drew attention to the contexts, conventions, and constraints presupposed by any conceptual representation, observing that different images (*Bilder*) of mechanics are not only possible but also desirable, so long as certain criteria governing representation were satisfied. These are 1) *permissibility* (they must not contradict the laws of our thought – *die Gesetze unseres Denkens*); 2) *correctness*, i.e., empirical correctness; and 3) *appropriateness* (*Zweckmäßigkeit*), by which Hertz meant both simplicity (the theory must contain no superfluous elements) and completeness (the theory omits no known phenomena in its domain). Substituting "consistency" for "permissibility" (which has inflexible Kantian connotations), these criteria are also characteristic guidelines for theory construction according to Einstein.

By 1900 the *Bild* view of physical theory had been taken up by the leading theoreticians of the day. Boltzmann explicitly invoked Hertz's name when lecturing to a gathering of natural scientists in Munich on September 22, 1899, on the recent development of methods of theoretical physics:

> Hertz makes physicists properly aware of something philosophers had no doubt long since stated, namely that no theory can be objective, actually coinciding with nature, but rather each is only a mental picture (*ein geistiges Bild*) of the phenomena, related

to them as is sign to *designatum*. From this it follows that it can never be our task to find an absolutely correct theory but rather an image (*Abbild*) that is as simple as possible and that represents the phenomena as accurately as possible.[25]

Other evidence of Hertz's significant influence is an address of February 8, 1900, by Dutch physicist H.A. Lorentz, of whom Einstein wrote in tribute in 1953, "He meant to me more than all the others I've encountered on life's path".[26] Celebrating the 325th anniversary of the University of Leiden, H.A. Lorentz, then university rector as well as perhaps the world's foremost theoretical physicist, sought to characterize the theoretician's method to his largely lay audience. Though speaking in Dutch, Lorentz used the German term in describing that task as producing "'*innere Scheinbilder*'" ("apparent mental images") of external objects; that is, of positing mathematically characterized physical mechanisms or structures capable of producing the phenomena the theory sought to explain.[27] The expression, as Lorentz expressly noted, originated with the recently deceased Heinrich Hertz.

In the *Bild* conception, physical theories are regarded as axiomatic systems constructed for the purpose of representation of certain aspects of nature. The modern conception of axiomatic system arose not in physics but from developments in 19th-century mathematics, reaching its highest refinement in David Hilbert's influential *Grundlagen der Geometrie* (1899), an axiomatic analysis (not simply presentation) of Euclidean geometry based on five groups of cognate abstract axioms. In this modern sense, one no longer thought of axioms as *a priori* truths, as in Euclid's own treatment of geometry, but simply as hypothetical postulates, chosen for their mutual consistency, independence from one another, and deductive fecundity. From a recently published lecture of Hilbert in 1894, Hilbert reports that his inspiration for this conception of axioms derived from reading Hertz's *Mechanics*.

> Axioms are, as Hertz would say, pictures or symbols (*Bilder oder Symbole*) within our mind such that consequences of the pictures are again pictures of the consequences, that is, what we logically deduce from the pictures again agrees in nature.[28]

Since in Hilbert's conception axioms are not to be regarded as true, the only applicable sense of truth for a proposition of an axiomatized system lay in the requirement of consistency, that the derived consequences of the axioms do not yield a contradiction. Similarly, with Hertz, since *Bilder* concern "images of our own creation not of nature's", the axioms of a physical theory are not to be regarded as true, and so the only meaningful sense of the truth of a physical theory lies in the most complete agreement of deductive consequences of the theory with observable phenomena, i.e., in the theory's empirical adequacy. Along similar lines, Boltzmann, the leading proponent of atomism among physicists, observed at the beginning of his *Lectures on Gas Theory* (1896) that the question "Are atoms real?" was really to ask, "Is the atomistic theory of gases empirically successful?"[29] As only the deductive consequences of an axiomatically formulated theory can be compared with observation, those speculative concepts, postulates and theorems that lie at considerable deductive distance from observation have only an indirect empirical meaning. Although Pierre Duhem drew conspicuous attention to the epistemological and semantic holism associated with the axiomatic conception of physical theory, it was earlier and explicitly recognized by Hertz:

> The statements will be given as facts derived from experience; and experience must be regarded as their proof. It is true, meanwhile, that each separate formula cannot be specially tested by experience, but only the system as a whole. But practically the same holds good for the system of equations of ordinary dynamics.[30]

As seen above, this expression of confirmational holism very well characterizes Einstein's understanding of physical theory.

From *Bildtheorie* to physical world picture (*physikalischen Weltbild*)

Bildtheorie's pluralism — that is, tolerance for and encouragement of construction of empirically equivalent alternative "pictures" of a domain of phenomena — is an aspect Einstein emphatically did not adopt. In recognizing pluralism as a virtue, Hertz had been inspired by alternative theories of electrodynamics, each claiming empirical equivalence to Maxwell's. In the Hertzian conception of physical theory, any *permissible* (i.e., consistent) physical image can be considered *correct* (*richtig*) only if the logical relations between symbols do not contradict observed relations among the phenomena. Selection of one or another of these images was to be made through application of Hertz's third criterion, *appropriateness*, but Hertz allowed considerable latitude in interpreting this principle, admitting that different physicists might apply it differently. In general Hertz deemed theoretical pluralism to encourage novel modes of thought; his reformulation of classical mechanics is a prime illustration of this belief. However, there is little evidence that Einstein endorsed this kind of theoretical pluralism at any stage of his career, and, after 1918, much evidence that he rejected it. Naturally, avowal of belief in theoretical uniqueness brought Einstein closer to a realist outlook on physical theories. But what might realism look like within the *Bild* conception of theory?

As discussed in the previous chapter, an answer was supplied by Max Planck, reluctant discoverer of the quantum of action, in his 1908 Leiden lecture. Planck took the further step in distinguishing natural scientific research by the goal of creating a stable and unified "world picture":

> The constant unity of the world picture (*Weltbild*) . . . is the fixed goal towards which real natural science approaches through all its changes and, in physics, we may affirm with justification that already our contemporary *Weltbild* . . . contains certain contours that no revolution in Nature or Man can obliterate. This constancy, independent of all human and intellectual individuality, is plainly what we call the real(*das Reale*).[31]

The condition of *constancy* mandates use of concepts of universal validity not dependent on particular individuals, circumstances or times, i.e., independent of the peculiarities of particular peoples, periods, and cultures. The condition of *unity* required that the *Weltbild* must, at least in principle, extend all physical phenomena, no matter how disparate or varied, as derivable consequences from a de-anthropomorphized basis of principles, concepts and laws. Planck had taken a sizeable step towards this goal in 1899 in showing that physical constants may be freed from anthropocentric arbitrariness through definition in terms of fundamental physical theories, rather than as interchangeable experimental parameters.

A physical world picture is then a *projected* theoretical representation of the totality of physical phenomena freed from the taint of particular persons, cultures or points of view. As previously discussed, Planck certainly recognized that a completely unchanging picture did not then and even perhaps may never exist. At the same time he argued that it remained a necessary goal of physical theory and even that in referring to "the world", physicists intend just an unchanging "ideal future picture" (*ideale Zufunftsbild*). This usage, according to Planck, gives the physicists' notion of "the real", not the significance of the world as it is, independently of experience or conceptions, but rather that portrayed by an ideal future picture. The former remains merely a *Grenzbegriff*, or limiting concept. The *physika-lischen Weltbild* accordingly retains the ontological modesty of Hertz's constructivist *Bild* conception, namely, that "we do not know, nor have we any means of knowing" whether physical theories as constructed are in actual conformity with the mind-independent world. The world in itself remains an abstract concept, beyond epistemic accessibility and the possibility of ever confirming any ideal future picture as uniquely correct. Planck highlighted this incommensurability in 1908, expressly attributing the point to Kant that there is no method of proving the existence of such a difference between ideal future picture and world in itself:

> For that there is no method of proving such a difference has, through Immanuel Kant, become common property (*Gemeingut*) of all thinkers.[32]

In epistemological writings over four decades, Planck returned to the theme of this incommensurability, portraying it as an irremovable source of tension, even irrationality, within theoretical physics. In the depths of World War II, Planck at age 84 reiterated the message:

> From the standpoint of exact science there always remains an unbridgeable chasm between the phenomenological and the

metaphysically real worlds (that) engenders a constant, unbalanced tension . . . In this bifurcation, which expresses itself in that we view the presupposition of a real world in the absolute sense as inevitably required but, on the other hand, we are never capable of completely grasping its nature, lies an irrational element which exact science can never shake off.[33]

Planck, in 1905–1906, had been the first theoretical physicist of international repute to place the special theory of relativity within his own research agenda. He initiated bringing Einstein to Berlin in 1914. Admiration between Planck and Einstein rapidly became reciprocal. Post general relativity, Einstein surely appreciated that his public endorsement in 1918 of the search for the *Weltbild* as "the highest task of the physicist" would be understood as a deliberate alignment with Planck over Mach and positivist conceptions of physical theory. Essentially from that time forward, Einstein's own search for the yet-to-be-completed world picture, seeking a theoretical structure of greatest possible logical unity capable of comprehending the complete and exhaustive description of all individual events, would be based on a field-theoretic generalization of his theory of gravity.

Reprise: "Physics" and "reality"

For Einstein linking the two terms is a belief in the conception of physical theory as seeking to "grasp" reality, i.e., constructing the picture of a reality law-governed in the manner affirmed in fundamental physical theory. "Grasping" remains a metaphor: The sense in which the theorist speaks of the "physically real" is as a reality "pictured" or "imagined" by the conceptual constructions of a fundamental physical theory. As these conceptual constructions are "free inventions of the human mind", no strong sense of mind independence *can* be involved. From the theoretician's perspective, the supposition of a conceptionless strongly mind-independent reality is strictly speaking meaningless:

> That the posit (*Setzung*) of a real external world is meaningless without that conceptual comprehension (*ohne jene Begreiflichkeit*) is one of the great discoveries of Immanuel Kant.[34]

As seen above, Einstein's credo emphasized a speculative, anti-inductivist – or better, constructivist – view of concepts, axioms, principles, and theories. There is no determinate inductive route from facts of observation to the fundamental concepts and relations of physical theories; the latter are neither inductive generalizations of empirical data, nor based on relative frequencies of outcomes of repeated measurements, nor mere economical compressions of data, nor "a priori necessities of thought". Rather they are hypothetical posits, "free creations of the human mind". With the success of general relativity, Einstein constrained the freedom of the Bild conception with a notable rationalist amendment, the criterion of "logical simplicity" requiring a minimum of primary concepts and relations in the basis of a theory as a determinative factor in its choice. Yet the sine qua non hurdle to be met by any theory remained constructivist: the sole justification (and so "truth") of any theory lies in conformity of its derived consequences with observation.

The abstract character of axiomatic representation of theories permits further and further degrees of deductive separation between the fundamental concepts and relations of a theory and observable confirmation. The result of this distancing is that it is impossible to associate a clear empirical (observational) meaning with each fundamental concept or relation; experimental tests can only pertain to the theory as a whole. An empirical holism in physical theory was an essential component of Hertz's Bild conception particularly emphasized by Duhem. At this juncture looms the problem of underdetermination of theory by empirical evidence, opening the door to theoretical pluralism. Einstein sought to circumvent pluralism, at least in part, by a Planckian-like commitment to the permanence of the core constituents of his physikalischen Weltbild.

Finally, the Bild and Weltbild conceptions of theory give rise to a weakened semantics for physical theories. Reference to unobservable theoretical entities (using Arthur Fine's term[35], see the discussion in Chapter 11), is entheorized, i.e., reference of a theoretical term is not a context-independent primitive and irreducible semantic relation between name and object (like designation) but a function of the term's explanatory role in the setting of the entire theory's success or lack of success in attaining overall empirical adequacy. A prime example of an entheorized term is the concept of space within the general theory of relativity,

According to general relativity, the concept of space detached from any physical content does not exist.[36]

The theory-functional characterization of reference is then another aspect of regarding empirically successful physical theories as Bilder. Just as the meaning of a theoretical concept is given by its occurrences in an axiomatically formulated theoretical system, so a theoretical term has reference only insofar as it remains essential to the empirical success of the theory. Contextual reference cements the truth of referential claims to the constructivist requirement of epistemic access, even though the route from one to other may be highly indirect. It implies that it is not really admissible to state without qualification that e.g., the term "atom" refers to atoms, posited mind-independent particulate entities that are constituents of matter. As entities they may be deemed to exist independent of mind but as "atoms", they are theoretical constructs serving to establish order among sense impressions. To return to Boltzmann's example, the posit of certain entities termed "atoms" in the kinetic theory (possessing the mechanical properties of mass, elasticity, velocity, and mean free path) is justified to the extent that the term serves as a necessary explanatory posit in deductively accounting for the empirical successes of gas theory. This was also the view of Einstein for whom atoms are physically real in the sense that they are posited conceptual objects, the properties of which are not directly perceptible but are assigned hypothetically in the explanation of phenomena.

Similarly, the Bild conception entheorizes truth (see also Chapter 11). The significance of the term "physical reality" is always theory-internal, the conception of reality projected from a fundamental theory satisfying the constraints of empirical adequacy and logical parsimony. As did Planck, Einstein regarded it necessary to formulate the very aim of science as an aspiration to seek a "picture of reality" Weltbild – that ideal limit of physical knowledge to which the human mind may aspire. No reconciliation with the quantum theory was possible so long as it surrendered this quest to represent a mind-independent reality.

Perhaps the clearest expression of what Einstein meant by conjoining "Physics" and "Reality" occurs in the semi-popular book of Einstein and Leopold Infeld, The Evolution of Physics (1938). The book came about, Einstein wrote to Solovine on April 1, 1938, in order "to provide for Mr. Infeld, who was refused a fellowship" at the Institute of Advanced Study. But there was also an ulterior motive. Of his collaboration with Infeld, Einstein reported to Solovine: "We worked out the subject very carefully together, giving particular attention to the epistemological point of view". The declared intent, reminding of Einstein's response to Schlick in 1930, was to counter the "mood of the present", dominated by "the subjective – positivist point of view" that "explains the demand for comprehension of nature as an objective reality as an antiquated prejudice". The credo is essentially stated in the book early on:

> Physical concepts are free creations of the human mind . . . In our endeavor to understand reality we are somewhat like a man trying to understand the mechanism of a closed watch. . . . If he is ingenious he may form some picture of a mechanism which could be responsible for all the things he observes, but he may never be quite sure his picture is the only one which could explain his observations. He will never be able to compare his picture with the real mechanism and he cannot even imagine the possibility or the meaning of such a comparison. But he certainly believes that, as his knowledge increases, his picture of reality will become simpler and simpler and will explain a wider and wider range of his sensuous impressions. He may also believe in the existence of the ideal limit of knowledge and that it is approached by the human mind. He may call this ideal limit the objective truth.[37]

In the book's final subsection, entitled "Physics and Reality" (pp. 310–13), Einstein's view of physical theory is again schematically summarized, while the concept of natural number, as in Dedekind, is denominated one of the "creations of the thinking mind". Four now familiar points are given particular emphasis:

1. Since there is no path from the empirical facts to the setting up of a theory, any theory sought as a basis for physics can only be a product of the "creative scientific imagination". As will be seen in the next chapter, in his pursuit of unified field theory Einstein will seek "the discovery of a logically simple mathematical

condition" in order that relatively little empirical knowledge is needed to set up such a theory.

2.   "The reality created by modern physics", i.e., as projected within contemporary fundamental theory at the time, is very far removed from the more prosaic reality of the mechanical concepts of "force" and "mass" and "inertial frame". The implication is that current theories must and do employ concepts of far greater abstraction than before. For example, the four-dimensionality of space-time, coupled with the symmetry of the metric tensor and the condition of general covariance, comprised the "logically simple mathematical condition" mentioned in 1).

3.   "Without the belief that it is possible to grasp the reality with our theoretical constructions, without the belief in the inner harmony of our world, there could be no science". The entire enterprise of attempting to create a unified fundamental theory is based upon this faith. The demand for comprehension ("to grasp the reality with our theoretical constructions") is the demand for a representation of nature meriting the status of *Weltbild*. Interestingly, this aspiration is "continually strengthened by the increasing obstacles to comprehension". It is fair to say these "obstacles to comprehension" continue to increase in the quantum theory, and the struggle to overcome them continues today in the foundations of quantum mechanics.

4.   The *sine qua non* requirement imposed on any projected fundamental unified theory is that it "succeed in explaining all events by (its) fundamental concepts of reality"; that is, empirical adequacy alone can be the criterion of "truth" of such a theory.

Summary

Near universal acceptance of the quantum theory after 1930 and its apparent alignment with positivism prompted Einstein to articulate and defend an ostensibly realist conception of the aim and meaning of physical theory. Scrutiny of his "epistemological credo", however, shows that in doing so, Einstein drew from a philosophical current, prominent within the previous generation of physical theorists, whose constructivist elements significantly temper the claims of realism. These constructive elements are most significant in understanding Einstein's use of the metaphor that a physical theory is an attempt to "grasp reality". Above all, the metaphor signals a faith that comprehension of nature through theoretical construction is possible. The response to positivism, coupled with a speculative methodology employed in seeking a "theory of the total field", a unified representation of gravitation and electromagnetism, produced extensive epistemological engagement in his last decades. The ensuing philosophy of physical theory combines elements of realism, idealism-constructivism, and as will be seen in the next chapter, rationalism.

Notes

1 "Autobiographical Remarks", p. 81.
2 "Replies to Criticisms", p. 669; German 1951, p. 496.
3 Einstein, "Physik und Realität", *The Journal of the Franklin Institute* v. 221, no. 3 (March 1936), pp. 313–47; English translation, pp. 349–82. Translation reprinted in Einstein, *Ideas and Opinions*. New York: Crown Publishers, 1954, pp. 290–323.
4 In March 1935 the Franklin Institute, a venerable museum of science education in Philadelphia dating to 1824, awarded its Franklin Medal jointly to Einstein and to Sir Ambrose Fleming, the inventor of the vacuum tube. At the ceremony Einstein, not expecting the request, was asked to speak. An embarrassed pause followed. In correspondence with the Institute's director a few days later, Einstein confessed to being "sincerely ashamed" ("*redlich geschämt*") by his silence on that occasion. To make amends, he offered to contribute a "dissertation" to the Franklin Institute's *Journal* that eventually appeared in the March 1936 issue. Reporting the appearance of Einstein's essay, *Time Magazine* (March 16, 1936) tells the story of what happened a year before. "A throng of scientists and dignitaries was assembled to hear what the medalist had to say. Einstein genially informed the chairman that he had nothing to say, that inspiration which he had awaited until the last moment had failed him. The chairman, much more embarrassed than the medalist, conveyed this information to the audience. In atonement, Einstein wrote a 44-page essay entitled 'Physics and Reality'".
5 Einstein and Nathan Rosen, "The Particle Problem in the General Theory of Relativity", *Physical Review*, v. 48 (July 1, 1935), pp. 73–77.
6 In the Einstein-Rosen solutions, electrical charge and mass appear as independent constants of integration.
7 Einstein, "Physik und Realität", *The Journal of the Franklin Institute* v. 221, no. 3 (March 1936), pp. 313–47; p. 318.
8 Huxley, Thomas Henry, "On the Educational Value of the Natural History Sciences", 1854; reprinted in *Collected Papers* v. III ("Science and Education"). New York: D. Appleton and Co., 1897, pp. 38–65; p. 45.
9 Einstein, "Johannes Kepler", originally published 1930; as translated by Sonia Bargmann in *Albert Einstein: Ideas and Opinions. Ideas and Opinions*. New York: Crown Publishers, 1954, pp. 262-66; p. 266.

10 Einstein, letter to Solovine of May 7, 1952, in *Albert Einstein Letters to Solovine*, with an introduction by Maurice Solovine. New York: Philosophical Library, 1987, pp. 138–9.

11 Einstein, *Vier Vorlesungen über Relativitätstheorie* gehalten im Mai 1921 an der Universität Princeton. Braunschweig: Friedrich Vieweg & Sohn, 122; CPAE 7 (2002), Doc 71; translated by Edwin P. Adams, *The Meaning of Relativity*. Fifth edition. Princeton: Princeton University Press, 1956, p. 6.

12 Einstein, "On the Method of Theoretical Physics", *Philosophy of Science* v. 1, no. 2 (April, 1934), pp. 163–9; p. 165.

13 Einstein, "Physik und Realität", *The Journal of the Franklin Institute* v. 221, no. 3 (March, 1936), pp. 313–47; English translation, pp. 349–82. Translation reprinted in Einstein, *Ideas and Opinions*. New York: Crown Publishers, 1954, pp. 290–323; p. 314.

14 Einstein and Leopold Infeld, *The Evolution of Physics: The Growth of Ideas from Early Concepts to Relativity and Quanta*. New York: Simon and Schuster, 1938, p. 33.

15 Einstein, "Bemerkungen zu Bertrand Russells Erkenntnis-Theorie", in Paul A. Schilpp (ed.), *The Philosophy of Bertrand Russell*. Evanston, IL: Northwestern University Press, 1944, pp. 278–91; p. 286.

16 Einstein, "Autobiographical Notes", in Paul A. Schilpp (ed.), *Albert Einstein: Philosopher-Scientist*. Evanston IL: Northwestern University Press, 1949, pp. 2–95; p. 81.

17 Einstein, "Relativität und Raumproblem", in Einstein, *Über die spezielle und die allgemeine Relativitätstheorie. Erweiterte Auflage*. Braunschweig: Friedrich Vieweg & Sohn Verlagsgesellschaft mbH, 1954, pp. 91–109; pp. 95–6. Translation by Robert Lawson as Appendix V of Albert Einstein *Relativity: The Speical and the General Theory*. New York: Bonanza Books, 1961, pp. 135–57; p. 141.

18 "Ernst Mach", *Physikalische Zeitschrift*, v. 17 (1916), pp. 101–104; CPAE 8, Doc. 29; English translation supplement, pp. 141–145.

19 Frank, Phillip, *Einstein: His Life and Times*. New York: Alfred Knopf, 1947; fourth printing January 1953, pp. 40–42.

20 Duhem, Pierre. *The Aim and Structure of Physical Theory*. Second edition. Translated by Philip P. Wiener from orginal 1914 French edition. Princeton: Princeton University Press, 1991, chapter 2.

21 Einstein, *"Motive des Forschens"* ("Motives of Research"), April 26, 1918. CPAE 7 (2002), Doc. 7.

22 Frank, *Einstein: His Life and Times*, p. 218.

23 Volkmann, Paul. "Die gewöhnliche Darstellung der Mechanik und ihre Kritik durch Hertz", *Zeitschrift für den physikalischen und chemischen Unterricht* Bd.14 (1901), 266–93; p. 266.

24 Hertz, Heinrich. *Die Prinzipien der Mechanik in neuem zusammenhänge dargestellt*. Leipzig, Germany: Johann Ambrosius Barth, 1894. *"Einleitung"*, pp. 1–2.

25 Boltzmann, Ludwig, "Über die Entwicklung der Methoden der theoretischen Physik in neuerer Zeit", *Münchener Naturforscherversammlung* (September 22, 1899), as reprinted in Ludwig Boltzmann, *Populäre Schriften, ausgewählt von Engelbert Broda*. Braunschweig/Wiesbaden, Germany: Friedrich Vieweg & Sohn, 1979, pp. 120–49; p. 138.

26 Einstein, Albert, H.A. *Lorentz als Schöpfer und als Persönlichkeit*. Leiden: Mededeling van het Rijksmuseum voor de Geschiedenis der Natuurwetenschappen te Leiden, Netherlands, no. 91. 1953.
27 Lorentz, Hendrik A., "Electromagnetische theorieën van natuurkundige verschijnselen", *Jaarb. Rijksuniv. Leiden, Bijlagen* 1. German translation in *Collected Papers*, vol. 8 (1935), pp. 333–52; p. 337.
28 Hilbert, David, "Die Grundlagen der Geometrie", 1894 lecture in M. Hallett and U. Majer (eds.), *David Hilbert's Lectures on the Foundations of Geometry 1891–1902*. Berlin, Heidelberg, and New York: Springer, 2004, p. 74.
29 Boltzmann, Ludwig, *Vorlesungen über Gastheorie*. I Theil. Leipzig: Ambrosius Barth, 1896, p. 6; as translated by Steven G. Brush, *Ludwig Boltzmann: Lectures on Gas Theory*. Berkeley: University of California Press, 1964, p. 28, translation modified: "The question of the appropriateness (Zweckmässigkeit) of atomistic representations (Anschauungen) is of course ... really whether mere (blossen) differential equations or atomistic views (Ansichten) will one day be established as complete descriptions of the phenomena."
30 Hertz, Heinrich, "On the Fundamental Equations of Electromagnetics for Bodies at Rest", (*Göttinger Nachr.* 19 March 1890; *Wiedemann's Ann.* 40, p. 577) as translated in *Electric Waves*. New York: Dover, 1962, p. 197.
31 Planck, "Die Einheit des physikalischen Weltbildes", as reprinted in *Wege zur physikalischen Erkenntnis: Reden und Vorträge*. Zweite Auflage. Leipzig, Germany: S. Hirzel Verlag, 1934, pp. 1–32; p. 30.
32 Ibid., p. 32.
33 Planck, "Sinn und Grenzen der exakten Wissenschaft", *Die Naturwissenschaften* Bd. 30, Heft 9/10 (27 February 1942), pp. 125–33; p. 130.
34 Einstein, "Physik und Realität", *The Journal of the Franklin Institute* v. 221, no. 3 (March 1936), pp. 313–47; p. 315. Translation in Albert Einstein, *Ideas and Opinions*. New York: Crown Publishers, 1954, pp. 290–323, p. 292.
35 Fine, *The Shaky Game*, pp. 93 ff.
36 Einstein, "On the Generalized Theory of Gravitation", *Scientific American* v. 182, no. 4 (April 1950), p. 13.
37 Einstein and Leopold Infeld, *The Evolution of Physics: The Growth of Ideas from Early Concepts to Relativity and Quanta*. New York: Simon and Schuster, 1938, p. 33.

Further reading

Baird, Davis, Richard I. Hughes, and Albert Nordmann (eds.), *Heinrich Hertz: Classical Physicist, Modern Philosopher*. Boston Studies in the Philosophy of Science. Dordrecht, Netherlands: Kluwer, 1998.
Boltzmann, Ludwig, "On the Development of the Methods of Theoretical Physics in Recent Times", *Populäre Schriften*, essay 14, 1899, translation in B. McGuiness (ed.), *Theoretical Physics and Philosophical Problems*, Dordrecht, Netherlands: Reidel, 1974, pp. 77–100.
Scheibe, Erhard, *Die Philosophie der Physiker*. München: C.H. Beck, 2006.

Ten
Philosophy of science – rationalism

I did not grow up in the Kantian tradition, but came to understand quite late the truly valuable which is to be found in his doctrine, alongside of errors that today are quite obvious. It is contained in the sentence: "The real is not given to us, but rather put to us (*nicht gegeben sondern aufgegeben*) by way of a riddle". This obviously means: There is such a thing as a conceptual construction for the grasping of the interpersonal, the authority of which lies purely in its validation. This conceptual construction refers precisely to the "real" (by definition), and every further question concerning the "nature of the real" appears empty.[1]

Introduction

According to physicist, historian and Einstein scholar Gerald Holton, a significant shift occurred in Einstein's philosophical views as a result of his work on the general theory of relativity. In Holton's influential account, Einstein began his "philosophical pilgrimage . . . starting on the historic ground of positivism", heavily under the influence of Mach.[2] Yet following the completion of general relativity in late 1915, "apostasy" from Mach became more and more apparent. The end point of Einstein's philosophical odyssey lay in his conversion to what Holton termed a "rationalistic realism", namely the conviction

> there exists an external, objective, physical reality which we may hope to grasp – not directly, empirically, or logically or with fullest certainty, but at least by an intuitive leap, one that is only guided by experience of the totality of sensible "facts".[3]

To document this trajectory, Holton pointed to a letter of January 24, 1938, to mathematical physicist Cornelius Lanczos in the course of which Einstein identifies himself as "a believing rationalist":

> Coming from skeptical empiricism of somewhat the kind of Mach's, I was made by the problem of gravitation into a believing rationalist, that is, one who seeks the only trustworthy source of truth in mathematical simplicity.[4]

Lanczos, a Berlin collaborator on unified field theory in 1928–1929, was not only a colleague but a philosophical soulmate. In a 1931 monographic report on what was then the latest Einstein unified field theory (the so-called "distant parallelism theory"), Lanczos pointed out that the distinction between the general theory of relativity and quantum mechanics cannot be captured in a simple opposition of "classical" and "quantum", but signals a far wider divide, a difference "in the whole intellectual orientation which lies at the basis of modern research", namely, that between the "metaphysical-realistic" and "positivistic". He continued:

> Something unique, and perhaps even uniquely lasting, happened with the theory of relativity – we are thinking here of primarily the general theory. It happened that "metaphysical" thinking, or logical-constructive imagination (*Phantasie*), won insight into the mysteries of Nature that arguably would never have been attained along purely empirical paths.[5]

Small wonder Einstein observed in a 1942 letter to Lanczos that he was the only other physicist who held "the same orientation towards physics as I", namely, "belief in the comprehensibility of reality through something logically simple and unified".[6] Einstein's rationalist orientation received full public expression for the first time in the Herbert Spencer lecture, "On the Method of Theoretical Physics", delivered at Oxford on June 10, 1933. Lecturing in English, also for the first time, in no uncertain terms Einstein expressed a conviction that the "right method" of theoretical physics consisted in seeking laws of nature with the simplest mathematical formulation:

> Our experience up to date justifies us in feeling sure that in Nature is actualized the ideal of mathematical simplicity. It is my

conviction that pure mathematical construction enables us to discover the concepts and the laws connecting them which give us the key to the understanding of the phenomena of Nature. Experience can of course guide us in our choice of serviceable mathematical concepts; it cannot possibly be the source from which they are derived; experience of course remains the sole criterion of the serviceability of a mathematical construction for physics, but the truly creative principle resides in mathematics. In a certain sense, therefore, I hold it true that pure thought is competent to comprehend the real, as the ancients dreamed.[7]

As recently as his 1923 "Nobel lecture" a quite different public expression of philosophical orientation had been given. At that time Einstein insisted, in near-textbook positivist fashion, "concepts and distinctions are only admissible to the extent that observable facts can be assigned to them without ambiguity".[8] The striking declaration at Oxford, echoing Platonism in identifying fundamental physical reality with the abstract reality of ideal mathematical structures, was immediately recognized as marking a significant shift of perspective. Appearing a few months later in the second issue of the new journal *Philosophy of Science*, Einstein's lecture was widely discussed and criticized, apparently a shock to many who saw his scattered previous philosophical remarks as rather congenial to logical positivism or empiricism.

Recent scholarship suggests that the Oxford declaration of extreme rationalism has a highly context-sensitive interpretation.[9] The lecture occurred towards the end of a two-year period of intense cultivation of a new physical proposal for which there had been high hopes. Since at least 1919, Einstein had been engaged in the attempt to find a "unified field theory" combining electromagnetism and gravitation, with the goal of representing particles as solutions of the unified field equations set in four-dimensional space-time. The proposal current in 1933 had been developed in collaboration with his assistant Walter Mayer. It featured a four-dimensional metric field "of a new sort" from which Dirac's spinor fields for relativistic fermions, the most recent conquest of quantum theory and a triumph of 20th-century physics, could be regarded as a special case. The Einstein-Mayer "semi-vector fields" seemed "naturally fitted to describe important properties of the electrical elementary particles". In Oxford, Einstein hailed the new concept as found by "adhering to the principle of searching for the mathematically simplest concepts and their connections" (Einstein 1934, p. 168). But even as the lecture proclaimed "in Nature is actualized the ideal of mathematical simplicity", giving unmistakable impression of Einstein's self-assurance in this new theory, his correspondence at that time reveals "waning" confidence in the indicated physical interpretation of semivectors.[10] So perhaps the vainglorious Oxford proclamation should not be taken at face value but rather *cum grano salis*. It certainly falls into an oft-noted pattern of highly enthusiastic statements regarding the bright prospects of each newly minted candidate unified field theory, venting an optimism that invariably proved to be short-lived. In this regard, it is not without interest that Lanczos himself entertained the suspicion that Einstein's claims regarding the extent of Mach's influence on his earlier self "may have been self deception".[11]

Einstein's "believing rationalist" is neither a Platonist, nor really a rationalist at all, at least as that term is understood within the tradition stemming from Descartes, Spinoza, and Leibniz. Notwithstanding well-known railings against Kant's unsustainable doctrine of synthetic *a priori* concepts and judgments, the later Einstein affirmed that while experience alone remained the ultimate arbiter of any theory, a "belief in the comprehensibility of reality through something logically simple and unified" was a methodologically legitimate heuristic, whose general form Kant had thoroughly probed in the "Transcendental Dialectic" of the *Critique of Pure Reason* (A293-704/B249-732). There Kant had given a tempered account of the necessary, although "hypothetical", cognitive role of reason in science as an injunction to seek systematic unity in nature. The message that Einstein came to understand as "the truly valuable in Kant" lay in the doctrine of the regulative principles of reason, according to which the very concept of an order of nature presupposes a decision to seek unity through systematic connections.

What "the problem of gravitation" taught

Einstein began his Oxford address with a provocative challenge:

> If you wish to learn from the theoretical physicist anything about the methods which he uses, I would give you the following piece of advice: Don't listen to his words; examine his achievements.[12]

Puzzling advice from a theoretical physicist lecturing on the method of theoretical physics! But the sage counsel to examine deeds, not words, has been closely followed by a group of recent philosophers and historians of general relativity who have closely scrutinized Einstein's notebooks and papers recording his struggle with the problem of gravitation between 1912–1915.[13] These researchers have determined that during this crucial period, Einstein pursued a "double strategy" in trying to find the field equations of gravitation, oscillating between two distinct and not necessarily complementary tracks: one physical, the other mathematical. Working the physical side, Einstein took the limiting case of Newtonian gravity as the starting point, attempting to recover the limit cases of Newtonian gravity ("weak field approximation") and static gravitational fields in a relativistic theory of gravity. He then sought to determine conservation laws for energy and momentum, and only if finding success there, attempted to make the theory covariant according to the largest possible class of coordinate transformations.

On the other side, the so-called "mathematical strategy" rests on the postulate of a generalized principle of relativity of motion, formally expressed as the requirement that the theory's equations be generally covariant, i.e., that they remain valid under the largest possible ("arbitrary") class of differentiable coordinate transformations. As discussed in Chapter 6, while accelerated reference frames of a generalized principle of relativity are naturally represented by the arbitrary (smooth) curvilinear coordinate systems permitted under the purely formal condition of general covariance, the connection between the two principles is not at all straightforward.[14] However, beginning in 1913 in Zurich, Einstein had discovered through his friend Grossmann the existence of an already formulated mathematical language, the "absolute differential calculus", that seemed ideally suited to a generalized principle of relativity, since the objects of such a calculus – tensors – are invariant under transformation between arbitrary coordinate systems. This ensured that the fundamental mathematical relations between physical quantities represented as tensors, i.e., the field equations, would be by default generally covariant, having the same form in all coordinate systems. So following the mathematical strategy, Einstein started from a generalized principle of relativity of motion. Then under the supposition that the Riemannian metric tensor $g_{\mu\nu}$ represented the "potential" of the gravitational field, he sought (for details, see again Renn and Sauer 2007) a generally covariant differential operator on $g_{\mu\nu}$ coupling the ensuing tensorial gravitational field strengths to the energy-momentum sources of the field. The latter must also be represented by a tensor if the principle of conservation of energy is to be satisfied. Hence the gravitational field equations will be generally covariant tensor equations, in accordance with the generalized principle of relativity. As Einstein already recognized in the so-called "Zurich Notebook" dating from 1912–1913,[15] the starting point for seeking such an operator is naturally found in the Riemann curvature tensor $R_{\mu\nu\sigma\tau}$ since this quantity implicitly contains the complete system of generally covariant differential operators possible in the underlying (pseudo-) Riemannian manifold. However, stymied in part by the "hole argument", in part by difficulties with energy-momentum conservation and his failure to recover the Newtonian limit, two long years passed before Einstein found the right answer, formulating the so-called Einstein tensor on the left-hand side of his field equations from an algebraic combination of differential invariants constructed from the Riemann curvature tensor.

Here it is only necessary to observe that the "mathematical strategy" – the speculative route based upon a generalized principle of relativity and the requirement of general covariance – was motivated essentially for philosophical reasons, by a desire to eliminate from physics *a priori* privileged structures of space-time, in particular global inertial frames of reference. Chapter 6 discussed how Einstein's commitment to a generalized principle of relativity stemmed from a Machian-inspired wish to entirely eliminate privileged frames of reference – inertial systems – from physics. As Einstein later but characteristically observed, such a system, "acts upon all processes but undergoes no reaction", and in this regard the concept of an inertial system "is in principle no better than that of the center of the universe in Aristotelian physics".[16] The concept of a global inertial system, the last vestige of Newtonian absolute space, still survived in the Minkowski space-time of the special theory of relativity, a defect of that theory Einstein had pointed out already in 1907. Philosophical repugnance of a non-dynamical background space thus lay behind Einstein's attraction to a generalized principle of relativity, however imperfectly expressed by the principle of general covariance. In turn, his efforts to implement a philosophically satisfactory remedy led to a gravitational theory according to which there is no coordinate-independent way to distinguish between gravitational and inertial "forces", and so in which there are no global inertial frames, no prior background geometry of space-time. The result, in the general theory of relativity, is that the very metric of space-time becomes a dynamical field quantity, meaning that its properties depend in part on its specific matter content, just like electric and magnetic fields.

While a consensus exists among the above-mentioned scholars regarding the presence of twin physical and mathematical strategies in Einstein's notebooks and papers prior to November 1915, disagreement emerges regarding which strategy ultimately proved to be the successful path leading to the general theory of relativity. Norton (2000) has argued that it was the mathematical strategy; Janssen and Renn (2007) that it was the physical strategy.[17] Einstein himself provided conflicting retrospective assessments. In the immediate aftermath of completing the theory, he sought to defend it by emphasizing its mathematical pedigree in the "absolute differential calculus". On the other hand, against the early and purely mathematically driven speculative attempts of David Hilbert, Hermann Weyl, Arthur Eddington, and Theodor Kaluza to generalize his theory, Einstein insisted on the necessity of physical foundations, tending to downplay the role of mathematical speculation, perhaps in a rhetorical response.[18] But he did not allow this tempered caution to hamper his own research; by 1923 he was explicit about following a speculative methodology based upon mathematical simplicity in pursuit of unified field theory.[19] The result is that during the decade of the 1920s, a cognitive fissure opened between his own pronouncements against the speculative mathematical approach of Weyl and others, and the methodology he himself employed in constructing one failed unified field theory proposal after another. Reflecting back on this decade in 1933, perhaps the opening lines of the Oxford lecture cited above were intentionally self-referential.

By the end of the 1920s, Einstein was quite forthcoming about his conviction that a methodology coupling mathematical speculation with a belief in "the uniformity of natural laws and their accessibility to the speculative intellect" had not only yielded general relativity, but might well do the same in the project of unified field theory.[20] Post-1930, ever more deeply immersed in a paradigm that viewed unified field theory as the necessary completion of the problem of gravitation as well as the fundamental basis of a theory of matter, from which the empirical successes of quantum mechanics might follow as mere statistical approximations, Einstein repeatedly stressed that pure mathematics, not physical insight or physical requirements, had been the winning ticket back in 1912–1915. The message that mathematical simplicity provided the needed key to unlock the door shielding general relativity was reiterated many times in his last years, including to de Broglie on Feburary 15, 1954, just a little more than a year before his death:

> The equations of gravitation were only found on the basis of a purely formal principle (general covariance), that is to say, on the basis of trust in the greatest logical simplicity of laws of nature thinkable.[21]

Scholars who conclude that, to the contrary, the physical strategy was essential to establishing general relativity downplay these later pronouncements as only indicating Einstein had a selective memory. Still other factors help explain why Einstein played the mathematical card as a trump. The fruitfulness of the mathematical strategy was amply illustrated in the example of Hilbert, who certainly pursued a purely mathematical path ("the axiomatic method") that quickly brought him into the race for the generally covariant gravitational field equations in November 1915.[22] Did Einstein switch to the mathematical strategy in late 1915? Whether or not he did, it was only natural for him to justify (perhaps even to himself) the speculative mathematical approach guiding his unified field theory program by a claim that such a methodology had already yielded success. It is readily documented that the further Einstein went down the road of unified field theory, the more he stressed his "belief in the comprehensibility of reality through something logically simple and unified", legitimation for which he found belatedly in Kant.

The "truly valuable in Kant"

Transcendental idealism is most familiar as the thesis that knowledge is only of things as they appear (in Kant's technical sense, and so in accordance with the subjective conditions of the human faculties of understanding and sensibility), that is to say, as mere representations, not of things are in themselves. It is assumed here that, according to Kant, the *Dinge an sich* are not a realm of unknowable objects (a noumenal world) but rather limit concepts, correlates to the cognitive representations of objects.[23] Of course, rather straightforwardly, relativity theory – both special and general – called for distinct modifications of central sections of *The Critique of Pure Reason*.

Recall that the first and by far major portion of *The Critique of Pure Reason*, "The Transcendental Doctrine of Elements", is divided into two parts, a "Transcendental Aesthetic" and a "Transcendental Logic", where the latter is partitioned into two subdivisions: "Transcendental Analytic" and "Transcendental Dialectic". Undoubtedly, the "Transcendental Aesthetic" and the "Transcendental Analytic" are the most widely read and best-known parts of the Critique; the Aesthetic contains the doctrine that space and time are *a priori* forms of sensibility, while the Analytic presents Kant's answer to the question of how synthetic *a priori* judgments are possible: namely, judgments structured by the categories that, as schematized by the forms of intuition, prescribe precise conditions within which all human cognition of objects occurs. Obviously, it is the doctrine of space and time in the Transcendental Aesthetic that most egregiously requires modification in the light of relativity theory; in particular Kant's account of the necessary structure or form of the subjective conditions of human sensibility, understood as mandating a globally Euclidean structure of space and an absolute Newtonian character of time, cannot be sustained.

However, in arguing that synthetic *a priori* judgments can only be established for cognitions within the domain of sensible experience, the Transcendental Analytic already initiated a critique of traditional "dogmatic" metaphysics, wherein *a priori* concepts and judgments transcend the boundaries of possible experience. The goal in the Dialectic is similarly critical though somewhat more indirect. No mere skeptic of metaphysics, Kant wished to show that, although the questions that preoccupy metaphysical inquiry are inevitable, as inherent in the nature of human reason itself, they are nonetheless deceptive, and so always must be understood in the right (i.e., "critical") manner on pain of falling into metaphysical dogmatism.

From this perspective, the central task of the "Transcendental Dialectic" is to complete the account of cognition presented in the "Transcendental Analytic" (pertaining to sensibility and the categories of the understanding) by including the cognitive role of "theoretical reason" in its "critical" employment. In general, Kant viewed reason (both "theoretical" and "practical") as rooted in the human mental capacity to project beyond given experience in order to seek the totality of possible experience, the "totality of all conditions" or indeed the "unconditioned" presupposed by any series of conditions. Since such a totality itself can never be an object of possible experience, it cannot be legitimately considered an object of cognition. Nonetheless, the ideas of theoretical reason that project such a totality are essential to natural science where they have only a regulative, not a constitutive, sense: they give expression to reason's capacity to surpass the confines of experience through the hypothetical adoption of maxims of systematic unity or unity of nature. Kant is quite careful in drawing boundaries around such a use of reason; in particular, the rationality it prescribes is merely projected:

> Systematic unity (as mere idea) is, however, only a projected unity, which one must regard not as given in itself, but only as a problem (*nicht als gegeben, sondern als Problem*).
>
> (A647/B675)

The ideal concepts or principles of this projected unity are the product of the "hypothetical use of reason, on the basis of ideas as problematic concepts", and are "really not constitutive" ("constitutive" in Kant's sense pertains only to objects of experience). Now this, in Einstein's assessment above, is just the core of "the truly valuable in Kant".

From the standpoint of epistemology of science, the purpose in invoking the regulative use of reason is to demonstrate that empirical knowledge presupposes a general framework of unity within which specific empirical claims can be situated. The regulative use of reason, specifying the ideal structure of a completed system of scientific knowledge, provides the context within which specific scientific theories are located. In this way, scientific theorizing requires a transcendental, not merely a logical (methodological, instrumental) use of ideas, articulating the idea of a completely unified system of explanation to which, however, any current state of knowledge of the world only very imperfectly approximates and in fact can never be attained. This is just the distinction Planck attempted to capture by speaking of a "chasm" between the ideal *Weltbild* of future physical theory and the external mind-independent metaphysical reality it purports to portray. The regulative use of reason clearly involves a fundamentally different use (and meaning) of the *a priori* than that in the Transcendental Analytic. There is no *a priori knowledge* but only the recognition that empirical science requires an *assumption* that nature accord with reason's interest in unity. The way, or degree to which, this demand may or may not be satisfied cannot be specified *a priori*. By paying attention to this neglected aspect of Kant's account of the nature of empirical knowledge, Einstein arguably came to see that, despite "errors that today are quite obvious" in Kant's doctrine of necessary *a priori* structures of the Analytic, Kant had shown how ideas transcending experience have an independent and indeed essential role to play in theoretical cognition.

Einstein's Kant

In the course of a celebrated discussion with French luminaries at the Collège de France in April 1922, Einstein pointedly replied to philosopher Leon Brunschvicg's question concerning the bearing of Kant's philosophy upon the theories of relativity:

> As concerns Kant's philosophy, in my opinion, every philosopher has his own Kant.

After remarking that he remained uncertain of Brunschvicg's interpretation of Kant as manifested in the latter's lengthy intervention, Einstein went on to say:

> It seems to me that the most important matter in Kant's philosophy is that one speaks of *a priori* concepts in the construction of science. But here there are two opposing viewpoints: the *apriorism* of Kant, in which certain concepts preexist in our mind, and the conventionalism of Poincaré. These two points of view

agree that science requires, for its construction, arbitrary concepts; with regard to whether these concepts are given *a priori* or are arbitrary conventions, I cannot say.[24]

It is somewhat surprising that in such an august public venue, Einstein chose to sit on the fence, refusing to clearly voice an opinion previously affirmed in correspondence with Schlick and Born among others,[25] that the concepts required for theoretical construction are conventions as well as to state his opposition to any Kantian doctrine holding that "certain concepts preexist in our mind". As is widely known, this dissent emphasized that the concepts of theoretical science ("categories") are not "unalterable"; they are by no means necessary ("conditioned by the nature of the understanding"), but are "(in the logical sense) free conventions". Accordingly, what might be termed the negative moment of Einstein's Kant targets the doctrine of the "Transcendental Analytic" and the "Transcendental Aesthetic" that it presupposes. To the extent that these sections are regarded as comprising the core of Kant's theory of cognition, as logical empiricists such as Schlick and Reichenbach affirmed, there could be but one attitude to Kant in the aftermath of general relativity: utter rejection.[26]

Still, the larger significance of Einstein's remarks on Kant in Paris lies in their implied dissent to pure empiricism or positivism, a criticism contained in the statement that "science requires, for its construction, arbitrary concepts"; here "arbitrary" is an ellipsis for "logically arbitrary" which, in Einstein's somewhat peculiar epistemological vocabulary, has the meaning "not derived from sense experience". That physical theory requires freely posited concepts or "free conceptual construction" is a long-recognized tenet of Einstein's epistemology. And, though not in Paris, it is in an insistence that "all concepts . . . are from the point of view of logic free conventions" that Einstein locates his departure from Kant — that is to say, from the doctrine of *a priori* categories in the Transcendental Analytic. But Einstein's positive appreciation of Kant lies with the Transcendental Dialectic's emphasis not on the constitutive rules of the understanding but on regulative principles, transcendental ideas or concepts of the faculty of reason. In this positive moment, two non-conventionalist components stand out quite distinctly. First there is an abiding concern with "justification" (*Berechtigung*) of the use of theoretical concepts in physics neither analytic nor derived from sense experience. Second is the supreme importance of the idea of systematic unity in Einstein's conception of physical theory. The relationship Einstein affirmed between these two aspects can be brought out in view of our previous discussion of the Transcendental Dialectic.

In response to the new vogue of positivism among philosophers and quantum physicists beginning in the late 1920s, an increasingly prominent feature of Einstein's later writings on the epistemology of science points to the insufficiency of the empiricist thesis that all knowledge rests solely on the deliverances of the senses. Einstein counters empiricism's shortcoming with an emphasis that "reason" or "pure thought" is an uneliminable, albeit "metaphysical" (scare quotes always added by Einstein), factor in cognition. But as Einstein recognized, this presents an epistemological challenge: With what "justification" ("*Berechtigung*") does the physicist employ concepts that are neither analytic nor derived from sense experience? Kant, in the opinion of the late Einstein, at least provided a partly correct statement of the problem:

> The following, however, appears to me to be correct in Kant's statement of the problem: in thinking we use, with a certain "justification" such concepts in thinking, to which there is no access from the material of sense experience, if one considers the matter from the logical standpoint.[27]

He went on to affirm once again that physico-mathematical concepts are "free creations of thought which cannot be inductively derived from sense experience". Now the mantra of concepts as "free creations of the human mind" (see Chapter 9) is one of the most familiar components of Einstein's epistemology of science, one he repeatedly stressed against empiricists, from Russell to Reichenbach. But then if concepts (certain ones, surely, more than others) do lack empirical justification, just what other kind of justification might Einstein have had in mind?

His answer, triangulating between realism and conventionalism, is perhaps most fully stated in the course of an extended discussion covering several pages in the 1949 "Reply to Criticisms". Ostensibly engaging once more in battle against the positivist epistemology of many quantum theorists, Einstein articulated a viewpoint decidedly non-positivist, but also non-conventionalist, by focusing on the justification of one such concept, that of "the real" or "being", whose extension includes the "not observable". Derided by positivists as a metaphysical excrescence on the fabric of science, Einstein proceeded to give reasons for considering the concept not merely convenient, but essential.

> "Being" (*Das "Sein"*) is always something intellectually constructed by us (*von uns gedanklich Konstruiertes*), hence free statutes (*frei Gesetztes*) by us (in the logical sense). The justification of such constructs (*Setzungen*) does not lie in their derivation from what is given to the senses. Never and nowhere is there a derivation of this kind (in the sense of logical deducibility), not even in the domain of prescientific thought. The justification of the constructs that represent the "real" (*das "Reale"*) for us alone lies in their more perfect, or less imperfect, suitability for making intelligible what is given in sensation (the vague character of this expression is forced here upon me by my striving for brevity).[28]

As discussed in Chapter 9, Einstein's striving for brevity should not occlude two important points stated here. First, "being" – in this context a clear reference to the portrayal of physical reality in theoretical physics – is always a "mental construction" (i.e., "conceptual construction for grasping the interpersonal", in the quotation beginning this chapter), freely posited by the theorist. This admission distances Einstein's realism from either a "rationalistic realism" or scientific realism more generally, since the core of any such realism far outstrips the mere avowal of a mind-independent external reality and the intent of scientists to theoretically "grasp" this reality. Rather, any realist interpretation of physical theory worthy of the title must in addition claim *knowledge of the real* (without scare quotes!) holding firm to a conception that the best physical theories are at least approximately true. For such a realist, only *à la façon de parler* might true physical theories be considered "mental constructions", in the rather trivial sense that it takes a theorist to write down a theory, just as it takes a photographer to snap a portrait. But for the scientific realist, the theorist/photographer is irrelevant except for "stylistic" details, while the significance of the ensuing portrait or image is not that it is a "mental construction" at all but that it (approximately) corresponds to, maps, or represents in literal fashion, a mind-independent reality.

Secondly, the passage affirms that the only justification accruing to claims that conceptual constructs represent "being" or the "real" lies in their ability to make our sense experience "intelligible". It cannot be, as conventionalism prototypically maintains, that there is "no fact of the matter" about which concepts are correct, or that justification is the subjective matter of choosing concepts that pragmatically prove most convenient or commodious in managing experience. It is "intelligibility" that matters, and for Einstein, intelligibility required that conceptual constructs of fundamental physical theory manifest to the greatest extent possible the unity of systematic interconnection of an ideal order of nature, an ideal projected by the hypothetical employment of reason. It is just here that Einstein points to "the truly valuable in Kant", that the real is not "given" but "*aufgegeben*" – posed as a problem (for theoretical construction). As appears in the epigram to this chapter, he paraphrases this to mean:

> There is such a thing as a conceptual construction for the grasping of the interpersonal, the authority of which lies purely in its validation. This conceptual construction refers precisely to the "real" (by definition), and every further question concerning the "nature of the real" appears empty.[29]

In place of the term *Bild* Einstein used the expression "conceptual construction" as that enabling a "grasp (of) experiences intellectually" and in this sense, it can be regarded as "knowledge of the real":

> Insofar as physical thinking justifies itself, in the more than once indicated sense, by its ability to grasp experiences intellectually, we regard it as "knowledge of the real".[30]

At this point, he remarks that "the 'real'" in physics "is to be taken as a type of program, to which we are, however, not forced to cling *a priori*". Einstein's last documented discussion of Kant concludes with a reminder that he naturally dissents from the doctrine of *a priori* categories in the "Transcendental Analytic".

> The theoretical attitude here advocated is distinct from that of Kant only by the fact that we do not conceive of the "categories" as unalterable (conditioned by the nature of the understanding) but as (in the logical sense) free conventions. They appear to be *a priori* only insofar as thinking without positing categories and concepts in general would be as impossible as is breathing in a vacuum.[31]

These late philosophical pronouncements stress the role of reason – that is, theoretical speculation based on seeking mathematical simplicity – in creating a unifying conceptual structure within which all physical phenomena find intelligible representation. They are not *ex cathedra* pronouncements but reflect a confluence of two philosophical currents. First, Einstein's assessment that progress in fundamental physical theory required new physical and mathematical ideas and that the most appropriate way of finding them was to use the beacon of mathematical-logical simplicity. Second, pursuit of such a path required a philosophical orientation appropriate to an ever-growing distance between observation and the basic concepts and relations of the theoretical structure, one that is "metaphysical-realistic" and not "positivistic".

Tamed metaphysicist

The evidence shows Einstein's final philosophical vantage point developed as the result of a long preoccupation with what he delicately termed "the present difficulties of physics".[32] These "difficulties" are a reference to the two principal concerns of his last three decades: 1) the critique of the quantum theory as incomplete, and a related attempt to counter positivist conceptions of physical theory; 2) struggles with unified field theory, which epistemologically speaking presented new concerns on account of the tenuous and indirect connection between the fundamental mathematical concepts and possible observable evidence. Dovetailing philosophical commitments emerged from these twin battles. As seen above,

Einstein stressed against positivism the legitimacy of a concept of "the real", a concept that carries the presumption of comprehensibility by the human mind; that is, that nature is intelligible. But a considerably more filled-out picture emerges by examining the several overt expressions of rationalist methodology accompanying his quest for unified field theory, the hypothetical projection of the intelligible-real, framed as a nonlinear, continuum-based field theory whose laws are generally covariant.

Above, it was seen that a famous instance of such a declaration, at Oxford in 1933, may perhaps be tempered by consideration of the context of its occurrence. But there are others that should be considered. A little-known Einstein text of 1929 featured another strident avowal of rationalist aspiration. Appearing in an obscure *Festschrift* for an old Bern teacher, Aurel Stodola, the paper opens by listing the "two ardent desires" of the theoretical physicist. The first is the wish to satisfy the requirement of "completeness" (*Vollständigkeit*), i.e., to contain within one theory "all the relevant phenomena and their connections". But the second points in an orthogonal direction, an aspiration that is "the Promethean element of scientific experience":

> We wish not only to know *how* Nature is (and *how* her processes transpire), but also to attain as far as possible the perhaps utopian and seemingly arrogant goal, to know *why* Nature is *thus and not otherwise*".[33]

This demand, no less than an expression of the principle of sufficient reason, is certainly the ultimate goal of explanation and, if satisfied, would be undeniable proof of the comprehensibility and intelligibility of nature cherished by rationalism. But the full story of Einstein's rationalism requires introducing the "tamed metaphysicist".

Writing for a lay audience in *Scientific American* in 1950, Einstein again returned to endorse the method of mathematical speculation that drove the search for a "generalization" of the theory of gravity, a unified field theory:

> Time and again the passion for understanding has led to the illusion that man is able to comprehend the objective world rationally by pure thought without any empirical foundations – in

short, by metaphysics. I believe that every true theorist is a kind
of tamed metaphysicist, no matter how pure a "positivist" he
may fancy himself to be. The tamed metaphysicist believes that
not all that is logically simple is embodied in experienced reality,
but that the totality of all sensory experience can be "comprehended" on the basis of a conceptual system build on premises
of great simplicity.[34]

Admitting this a "miracle creed", nonetheless Einstein claimed it
"has been borne out to an amazing extent by the development of
science". He made no attempt to justify such a highly controversial
characterization of the history of physics, and perhaps none can be
given. It may only be intended as an indirect reference to the perceived success of previous unification efforts, a narrative of physical
theory related many times elsewhere. The story begins with classical
mechanics as the fundamental basis of all physics, in which the fundamental concepts are of particles and their laws of motion. The next
stage is the Faraday/Maxwell concept of field and the unification of
optics, electricity, and magnetism. A unification between mechanics and field theory is effected through the special and the general
theories of relativity, and points toward a future theory of the "total
field" in which particles and their properties are derived as solutions
to nonlinear field equations. The underlying theme is unity, whereas
the underlying theme of unity is comprehension via logical simplicity. In this way one can view "the grand aim of all science" in the
attempt to "cover the greatest number of empirical facts by logical
deduction from the smallest number of hypotheses or axioms".[35]

Yet in referring to "the *illusion* that man is able to comprehend the
objective world rationally by pure thought", Einstein drops a clue
that theory unification by the method of mathematical speculation
may well be a rationalist pipe dream, that is, of *the illusion of reason*
that comprehension and understanding, in precisely the rationalist
sense of knowing why nature *is thus and not otherwise*, is possible. It
is clear that Einstein recognized the possible deceptive character of
this rationalist faith by the two personae with which he frequently
associated it, Hegel and Don Quixote. Einstein was impressed by
Émile Meyerson's treatment of the *a prioristic* tendency in science in
his 1925 book *La Déduction Relativiste* and by the utility of an analogy

Meyerson employed between the explanatory *pangeometrisim* of the general theory of relativity and Hegel's idealist narrative of the evolution of reason.[36] Einstein returned time and again to the figure of Hegel as exemplifying both the danger and the promise of the rationalist dogma that the human mind can force reality into the mold created by mind, or, in terms relevant to Einstein, that mathematical speculation based on "logical simplicity" can be employed as a methodology of successful theory construction. For example, in the Stodola essay just mentioned, Einstein inserts a footnote after speaking of the theorist's goal of deductively capturing empirical laws as logical necessities:

> Meyerson's comparison with Hegel's goal (*Zielsetzung*) surely has a certain justification; he sharply illuminates the frightening danger here".

And in *The Times* (London) on February 5, 1929, attempting to explain his new "distant parallelism" unified field theory, Einstein noted that the characteristics

> distinguishing the general theory of relativity, and even more so the new third stage of (relativity) theory, the Unitary Field Theory, from other physical theories, are the degree of formal speculation, the slender empirical basis, the boldness in theoretical construction, and finally the fundamental reliance on the uniformity of the secrets of natural law and their accessibility to the speculative intellect. It is this feature which appears as a weakness to physicists who incline towards realism or positivism, but is especially attractive, nay, fascinating, to the speculative mathematical mind. Meyerson, in his brilliant studies on the theory of knowledge, justly draws a comparison of the intellectual attitude of the relativity theoretician with that of Descartes, or even of Hegel, without thereby implying the censure which a physicist would naturally read into this".[37]

Einstein is nonetheless clear that "the speculative mathematical mind" of the theorist must be tamed by sober recognition that the very premise that such a methodology is possible at all may well

be illusory. If so, the grandiose aspirations but utter failure of rationalist understanding, exemplified by Hegel, may then be linked to
the madness afflicting Don Quixote, as in this letter to Max Born,
responding to Born's 1943 attack on the speculative non-empirical
"Hegelian physics" of Eddington and James Jeans:

> I have read with much interest your lecture against Hegelianism
> ("*die Hegelei*"), which with us theoreticians amounts to the Don
> Quixotean element ("*das Don Quijote'sche Element*") or, should I say,
> the seducer? But where this evil or vice is fundamentally miss-
> ing, the hopeless philistine is on the scene.[38]

Here, as elsewhere (see below), Einstein regards theory construction not driven by the rationalist impulse for understanding as no
better than the science of the philistines, i.e., positivists, a jab at the
quantum theorists and their views of quantum mechanics. In point
of fact, a closer reading of the Herbert Spencer lecture shows that
even these hyper-rationalist declarations are tempered by the setting posed by the perennial problem of philosophy and of science,
the "eternal antithesis between Reason and Experience". Phrased in
this way, and confronted with Einstein's late appreciation of "the
truly valuable in Kant" encapsulated in the cryptic catechism *nicht
gegeben, sondern aufgegeben*, Einstein cannot be thought unaware of the
significance of its message in the "Transcendental Dialectic": that
reason in science, as elsewhere, has only a purely *regulative* employment; it poses a mold for an understanding that requires unity. It is
a necessary role because "intelligibility" matters, and for Einstein
intelligibility is the child of reason. The constructs of fundamental physical theory must manifest to the greatest extent possible the
unity of systematic interconnection of *an ideal order of nature*, an ideal
projected by the hypothetical employment of reason. In Einstein this
employment took the form of mathematical speculation guided by
"logical simplicity". Hegel, poster child of rationalism run amuck,
ignored that such a use of reason must be "tamed" or disciplined.
In fact, Kant's most expansive discussion of mathematical knowledge in *The Critique of Pure Reason* occurs in a section entitled "The
Discipline of Pure Reason", where he observed that "discipline" is
to be understood in the punitive sense: the purpose is to *humiliate*

reason's pretensions to knowledge. More might be said about why the regulative role of reason is regarded as a necessary part of the Kantian, and especially neo-Kantian, conceptions of science.[39] But this broadly neo-Kantian aspect of Einstein's methodology of theory construction by mathematical speculation has largely been overlooked.[40] For Einstein's "tamed metaphysicist", philosophy is the "mother" of all inquiry, and without her Don Quixote-like fixation on a "passion for understanding", the sciences are always in danger of sinking into philistinism:

> Philosophy is like a mother who gave birth to and endowed all the other sciences. Therefore, one should not scorn her in her nakedness and poverty, but rather should hope that something of her Don Quixote ideal will live on in her children so that they do not sink into philistinism.[41]

Conclusion

Einstein's "believing rationalist" acknowledged that the summit of comprehension and understanding, in precisely the sense of the *Principle of Sufficient Reason*, the supreme rationalist principle, of knowing *why nature is thus and not otherwise*, could well be an *unavoidable illusion* of reason. The potentially deceptive character of this rationalist faith is exemplified by the two personae with which he frequently associated it, Hegel and Don Quixote. Time and again these figures are invoked to epitomize the dangers, as well as the alluring promise, of the intellectualist conceit that physical reality can be completely comprehended within a unique conceptual structure created by the human mind; for Einstein, a structure erected on the basis of a methodology of mathematical speculation driven by "logical simplicity". The "believing rationalist" holds that this cognitive aspiration for fundamental physical theory cannot be forsworn without opening the door to philistinism, i.e., to an unsophisticated positivist or pragmatic results-driven conception of science. Nonetheless it can and must be "tamed" by what Einstein had come to regard as "the truly valuable in Kant". In choosing to identify this lesson in the cryptic catechism *nicht gegeben, sondern aufgegeben*, he sought to underscore the purely regulative use of reason that is the principal message of the

"Transcendental Dialectic". What conceivable relevance might this have today? Unifiers in contemporary physics can and do have quite divergent assessments of the lasting value of Einstein's own quest for a unified field theory. But insofar as a legacy of unification in physical theory still linked to Einstein survives, the attenuated rationalism behind his own attempts might be better, and more productively, understood.

Notes

1 "Replies to Criticisms", p. 680.
2 Holton, Gerald, "Mach, Einstein, and the Search for Reality", *Daedalus* v. 97 (1968), pp. 636–73; p. 244. Page references are to the reprint in Holton, *Thematic Origins of Scientific Thought: Kepler to Einstein*. Cambridge, MA: Harvard University Press, 1988, pp. 237–77. Holton also identifies non-positivist aspects in the philosophy of the early Einstein but considers these influences subordinate.
3 *Ibid.*, p. 263.
4 *Ibid.*, p. 259.
5 Lanczos, Cornelius, "Die neue Feldtheorie Einsteins", *Ergebnisse der exakten Naturwissenschaften* Bd. 10 (1931), pp. 97–132; p. 99.
6 Einstein to Lanczos, March 21, 1942, (EA 15–294) as cited and translated in Cornelius Lanczos, *Collected Published Papers with Commentaries*. v. IV. Raleigh, NC: North Carolina State University Press, 1998, pp. 2–1526, note 9.
7 Einstein, "On the Method of Theoretical Physics", *Philosophy of Science* v. 1, no. 2 (April 1934), pp. 163–9; p. 167.
8 Einstein, "Fundamental Ideas and Problems of the Theory of Relativity", Lecture delivered to the Nordic Assembly of Naturalists at Gothenburg, Sweden, July 11, 1923. p. 1, available at www.nobelprize.org/nobel_prizes/physics/laureates/1921/einstein-lecture.pdf
9 van Dongen, Jeroen, *Einstein's Unification*. New York: Cambridge University Press, 2010.
10 *Ibid.*, p. 121. Van Dongen documents that early in 1934 Valentin Bargmann, a doctoral student under Pauli in Zurich, soon to be a research assistant to Einstein in Princeton and then professor of mathematics at Princeton, showed that the Einstein-Mayer semivector concept was not more general but essentially equivalent to spinors.
11 Lanczos, Cornelius, *The Einstein Decade 1905–1915*. New York: Academic Press, p. 42.
12 Einstein, "On the Method of Theoretical Physics", *Philosophy of Science* v. 1, no. 2 (April 1934), pp. 163–9; p. 163. Fölsing (1997, p. 673) reports that Einstein read this lecture from a written translation of an original German text. That translation is the one cited; it has the annotation, "reprinted by permission of Oxford University Press". The most widely available (and cited) version of the lecture, in *Ideas and Opinions*, is a translation (by Sonia Bargmann) of the German

text "*Zur Methodik der theoretischen Physik*" published in the 1934 collection *Mein Weltbild*. The two English texts differ in minor detail but mainly in clarity; the *Philosophy of Science* version is more fluid.

13 See Norton, John, "'Nature Is the Realization of the Simplest Conceivable Mathematical Ideas': Einstein and the Canon of Mathematical Simplicity", *Studies in History and Philosophy of Modern Physics* v. 31, (2000), pp. 135–70; Renn, Jürgen and Tilman Sauer, "Pathways Out of Classical Physics: Einstein's Double Strategy in Searching for the Gravitational Field Equations", in Jürgen Renn (ed.), *The Genesis of General Relativity*, Vol. 1, 2007, New York and Berlin: Springer, pp. 113–312.

14 Einstein admitted as much in 1918, *CPAE* 7 (2002), Doc. 4.

15 *CPAE* 4 (1995), Doc. 10.

16 Einstein, letter to George Jaffe of January 19, 1954: "You consider the transition to special relativity as the most essential thought of relativity, not the transition to general relativity. I consider the reverse to be true. I see the most essential thing in the overcoming of the inertial system, a thing that acts upon all processes but undergoes no reaction. This concept is in principle no better than that of the center of the universe in Aristotelian physics". As quoted by John Stachel, *Einstein from 'B' to 'Z'*. Boston, Basel, and Berlin: Birkhäuser, 2002, p. 143.

17 See also Janssen, "'No Success Like Failure . . .': Einstein's Quest for General Relativity 1907–1920, in M. Janssen and C. Lehner (eds.), *The Cambridge Companion to Einstein*. New York: Cambridge University Press, 2014, pp. 167–227.

18 Ryckman, Thomas, *The Reign of Relativity: Philosophy in Physics 1915–1925*. New York: Oxford University Press, 2005.

19 Einstein, "The Theory of the Affine Field", *Nature* v. 112 (1923), pp. 448–9; p. 448: "The search for the mathematical laws which shall correspond to the laws of nature thus resolves itself into the answer to the question: What are the formally most natural conditions that can be imposed upon an affine relation"?

20 The full passage reads: "The characteristics which especially distinguish the General Theory of Relativity and even more the new third stage of the theory, the Unitary Field Theory, from other physical theories, are the degree of formal speculation, the slender empirical basis, the boldness in theoretical construction, and finally the fundamental reliance on the uniformity of natural laws and their accessibility to the speculative intellect" (Einstein, "The New Field Theory: II. The Structure of Space-Time"; *The Times* (London), February 5, 1929, reprinted in *The Observatory* no. 659 [April 1929], pp. 114–18; p. 114.) This is one of several remarks on the epistemological novelty posed by unified field theory, made in the context of the theory based on the concept of "distant parallelism".

21 Einstein to de Broglie, February 15, 1954, in *Annales de la Fondation Louis de Broglie*. Paris: Conservatoire National des Arts et Métiers v. 4, no. 1 (1979), p. 56.

22 See Brading, Katherine A. and Thomas A. Ryckman, "Hilbert's 'Foundations of Physics': Gravitation and Electromagnetism within the Axiomatic Method", *Studies in History and Philosophy of Modern Physics* v. 39 (2008), pp. 102–53.

23 Arguments for this reading of Kant are found in Buchdahl, Gerd, *Metaphysics and the Philosophy of Science: Classical Origins Descartes to Kant*. Oxford, UK: Blackwell, 1969;

Allison, Henry E., *Kant's Transcendental Idealism*. Revised edition. New Haven, CT: Yale University Press, 2004; Bird, Graham, *The Revolutionary Kant: A Commentary on the Critique of Pure Reason*. Chicago: Open Court, 2006.

24 "A propos de la philosophie de Kant, je crois que chaque philosophe a son Kant propre et je ne puis répondre à ce que vous venez de dire, parce que les quelques indications que vous avez données ne me suffisent pas pour savoir comment vous interprétez Kant. Je ne crois pas, pour ma part, que ma théorie concorde sur tous les points avec la pensée de Kant telle qu'elle m'apparaît. Ce qui me paraît le plus important dans la philosophie de Kant, c'est qu'on y parle de concepts a priori pour edifier las science. Or on peut opposer deux points de vue: l'apriorisme de Kant, dans lequel certains concepts preexistent dans notre conscience et le conventionalisme de Poincaré. Ces deux points de vue s'accordent sur ce point que la science a besoin, pour être édifiée, de concepts arbitraires; quant à savoir si ces concepts sont donnés a priori, ou sont des conventions arbitraires, je ne puis rien dire". *Bulletin de la Société Française de Philosophie* v. 17 (1922), p. 101.

25 See, for example, Einstein's letter to Moritz Schlick dated December 14, 1915 (*CPAE* v. 8, Doc. 165) or his undated letter to Max Born from the summer of 1918 (*CPAE* v. 8, Doc. 575).

26 See Ryckman, *The Reign of Relativity*, Chapter Two.

27 "Bemerkungen zu Bertrand Russells Erkenntnis-Theorie", in P.A. Schilpp (ed.), *The Philosophy of Bertrand Russell*. Evanston, IL: Northwestern University Press, 1944, pp. 278–87; p. 285 and p. 287.

28 "Replies to Criticisms", p. 669; German 1951, p. 496.

29 "Replies to Criticisms" (1949), p. 680.

30 *Ibid.*, pp. 673–4.

31 *Ibid.*, p. 674.

32 "*Bemerkungen zu Bertrand Russells Erkenntnis-Theorie*", in P.A. Schilpp (ed.),The Philosophy of Bertrand Russell. Evanston, IL: Northwestern University Press, 1944, pp. 278–87; p. 279.

33 Einstein, "Über den gegenwärtigen Stand der Feld-Theorie", in Ernst Honegger (ed.), *Festschrift Prof. Dr. A. Stodola zum 70. Geburtstag*. Zurich und Leipzig: Orell Füssli Verlag, 1929, pp. 126–32; p. 126.

34 "On the Generalized Theory of Gravitation", *Scientific American* v. 182, no. 4 (April 1950), as reprinted in *Ideas and Opinions*, pp. 341–56; p. 342.

35 Quoted in Barnett, Lincoln, 'The Meaning of Einstein's New Theory', *Life Magazine*, January 9, 1950.

36 In his review of Meyerson's book, Einstein notes the aptness of Meyerson's "very ingenious" comparison of the deductive-constructive character of relativity theory with the systems of Descartes and Hegel because "the human mind wants not only to propose relationships, it wants to *comprehend*" (*il veut comprendre*)". "A propos de La Déduction relativiste de M. É. Meyerson", *Revue philosophique de la France et de l'étranger*, t.105 (1928), pp. 161–66; p. 164.

37 "The New Field Theory: II. The Structure of Space-Time"; reprinted in *The Observatory* no. 659 (April 1929), pp. 114–18; p. 114.

38 Einstein to Born September. 7, 1944, in *Albert Einstein-Max Born Briefwechsel* 1916–1955. München: Nymphenburger Verlagshandlung, 1969, p. 199 (my translation).
39 See Gerd Buchdahl, *Metaphysics and the Philosophy of Science*, 1969; and Susan Neiman, *The Unity of Reason*. New York: Oxford University Press, 1994.
40 A notable exception is Beller (2000).
41 "Die Philosophie gleicht einer Mutter, die alle übrigen Wisssenschaften geboren and ausgestattet hat. Man darf sie in ihrer Nackheit und Armut daher nicht geringschätzen, sondern muss hoffen, dass etwas von ihrem Don-Quichote-Ideal auch in ihren Kindern lebendig bleibe, damit sie nicht in Banausentum verkommen". Letter to Polish writer Bruno Winawer, September 28, 1932 (EA 52–267); as translated in H. Dukas and B. Hoffmann, *Albert Einstein, the Human Side*. Princeton, NJ: Princeton University Press, 1979, p. 106.

Further reading

Beller, Mara, "Kant's Impact on Einstein's Thought", in Don Howard (ed.), *Einstein: The Formative Years 1879–1909* (Einstein Studies v. 8). Basel-Boston-Berlin: Birkhäuser, 2000, pp. 83–106.
Dongen, Jeroen van, *Einstein's Unification*. New York: Cambridge University Press, 2010.
Paty, Michel, *Einstein philosophe: La physique comme pratique philosophique*. Paris: Presses Universitaires de France, 1993.
Ryckman, Thomas, "Early Philosophical Interpretations of General Relativity", in Edward N. Zalta (ed.), *The Stanford Encyclopedia of Philosophy*. Spring 2014 edition, <http://plato.stanford.edu/archives/spr2014/entries/genrel-early/>.

Eleven
Influence and legacy

"If philosophy is interpreted as a quest for the most general and comprehensive knowledge, it obviously becomes the mother of all scientific inquiry".[1]

Assessing influence merely in terms of impact upon the number of other lives, undoubtedly the most influential person of the 20th century is to be found among the terrible trio of dictators, Adolf Hitler, Joseph Stalin, and Mao Tse Tung. *Intellectual* or (to use an unfashionable term) *spiritual* influence is not a quantifiable effect of this kind; it can be gauged only more or less imprecisely and impressionistically. Nonetheless, it can be fairly said that in the initial decades of the 21st century the outlook of nearly everyone has been shaped, even if only in reaction to it, by the work and influence of Albert Einstein. Naturally, the claim is most directly confirmable by the present state of physical theory, where consequences of Einstein's achievements extend to literally every aspect of the current understanding of nature's most basic entities and processes. Nearly the same might be said for his purported failures, ranging from the critique of quantum mechanical completeness to the introduction of the cosmological constant. More broadly considered, spillovers from both achievements and failures have been developed into technologies, those ubiquitous today (lasers, GPS) and those promised for the future (quantum information transmission and processing).

Still, over and above any impact on science or technology, in his lifetime and continuing today, Einstein, like Freud, is one of few thinkers associated with a fateful alteration of human awareness. Whereas Freud pried open the Pandora's box of the unconscious,

Einstein transformed the concepts of space and time, among the basic constituents of thought. To many, then, and still today, this most fundamental revolution is an unwanted dislocation, an unfathomable, even rude, intrusion into a complacent trust that experience of the everyday world has been, and will continue to be, a reliable template for cognition of nature. As a later, and second chapter, quantum mechanics further reinforced the estrangement from the world as it appears introduced by relativity theory. The consequences of these two epistemological ruptures still play out in remarkable ways in the arenas of contemporary physical theory. But Einstein's transformation of the concepts of space and time has ramifications extending to human culture as a whole, as Picasso and Braque seem to have recognized with cubism. [2]

Taking the sage advice to focus on deeds, rather than words, on which achievement should emphasis be placed? An earlier era turned to the special theory of relativity. Positivists pointed to Einstein's identification of Hume and Mach as anti-metaphysical precursors for dismantling absolute time into the relativity of simultaneity. No doubt they also expected that his critique of quantum mechanics would surely fade away with the passage of time. But ever since John S. Bell reconsidered the EPR argument in the mid-1960s, this expectation has proven faulty. Downstream consequences of Einstein's ideas and opinions regarding the incompleteness of quantum mechanics have overturned the once-hegemonic reign of complementarity, prompting a number of novel interpretations of quantum mechanics.

Realist tendencies in the foundations of quantum mechanics take inspiration from Einstein's storied confrontation with Bohr. Such tendencies uphold the existence of an external world independent of the observer, regard physical theory as the description of this mind-independent world, and even adopt a conception of physical state in which quantities have definite values at all times independent of any act of measurement. Quantum entanglement, brought to attention by EPR as an intended criticism of quantum mechanical completeness, is now a primary focus of investigation in the burgeoning field of quantum technology and information theory, not to mention foundational pursuits such as quantum decoherence. Beyond the theories of relativity and quantum mechanics, a case might be made that attention should be paid to accomplishments that Einstein would dismiss as *Gelegenheitsarbeit*, that is, works performed as the occasion arose. Among these must be counted the 1905 hypothesis of the existence of light quanta, the 1917 derivation of Planck's law that introduced the concept of stimulated emission, and the first recognition of quantum statistics in 1924. The expanse of Einstein's influence and legacy is simply too vast for an accurate overview. In lieu of any more comprehensive summation, consideration here is given to just three areas, both physical and philosophical, where his impact is, or should be, readily recognized: 1) the further developments in what he termed his "lifework", general relativity; 2) the pursuit of theoretical unification in fundamental physics; and 3) the innovative understanding of the scope and limits of physical theories that can, with certain provisos, be recognized as a "realism".

100 years of GR

For much of its first half-century, the general theory of relativity stood well outside the mainstream of theoretical physics. Due to its limited contact with the empirical world (until the 1960s there were few observational tests of the theory), to its unusual mathematical demands, and to the fact that theoretical attention shifted to quantum mechanics and its applications in the mid-1920s, many theorists regarded it as a mainly mathematical theory of little physical relevance beyond its small corrections to the much simpler Newtonian theory of gravity. Astronomers proved an exception, particularly after 1929 when Hubble provided convincing evidence that the universe is expanding; his famous graph related distance of galaxies linearly to their speed of recession. Neglect from physicists was only reinforced for several more decades as the center of theoretical gravity turned in the 1930s to the newly discovered complexities of the atomic nucleus, and in the 1950s to the high-energy physics of particle accelerators and the failing theoretical attempts to comprehend an ever-expanding catalogue of particles. As a result, curious graduate students encountered general relativity either on their own or in mathematics departments where it was considered a formal science largely divorced from observations and the rest of physics. Even at Princeton University, next door to the Institute of Advanced Study, the physics department did not offer a full year course on general relativity until John Wheeler did so in the academic year 1952–1953.[3] During this period so few physicists worked on general relativity that one of them (Peter Bergmann) reportedly quipped, "You only had to know what your six best friends were doing and you would know what was happening in general relativity".[4]

It is not as if there was no theoretical work to be done. On account of general covariance alone, core concepts of general relativity required critical scrutiny: Is there a meaningful (i.e., generally covariant) notion of the gravitational field's local energy density? What are "observables" in general relativity? Does general covariance have any physical significance, and if so, what is it? How can general relativity be formulated without violating general covariance in a way that reconciles it with quantum theory? These and similar foundational pursuits, to one or another extent, continue to be explored. On the experimental side, there were other questions to be answered: Would new terrestrial experiments demonstrate the predicted yet still unobserved (until 1960) gravitational redshift? Does gravitational radiation really exist? What possible astrophysical sources might produce detectable gravitational waves?

Fittingly, two surprising observations moved general relativity again to the forefront of theoretical physics. The first was the discovery of enormously energetic quasi-stellar radio sources (quasars) in the early 1960s. The discovery prompted an initial theorem by English mathematician Roger Penrose showing that a sufficiently massive spherically symmetric star, having radiated away enough thermal energy, will gravitationally contract to a physical singularity at Schwarzschild radius $r = 0$.[5] Analyzing the properties of spacetime with new mathematical methods, Penrose in collaboration with Stephen Hawking subsequently proved that physical singularities occurred generically in general relativity and are not, as many had earlier believed and as Einstein surely would have preferred, mere coordinate artifacts of the high degree of symmetry required by the Schwarzschild solution. A second telling observation was the accidental discovery in 1964 of the cosmic microwave background (CMB), a remarkably uniform 2.7 K across the entire sky. The CMB is fossil radiation produced as the universe cooled, permitting radiation to decouple from matter some 300,000 years after an initial violent event. The discovery amply confirmed the hypothesis of a Big Bang universe and ruled out rival static or steady state cosmological models. Cosmologists in 1992 identified tiny fluctuations in this primordial radiation as the seeds of formation of galactic structure in the universe. The study of anisotropies in the CMB continues today to be at the forefront of observational cosmology, a data-driven established branch of physics, assisted by satellites (COBE, WMAP, Planck, Webb) advanced technologies and general relativistic models of an expanding universe.

One hundred years after its birth, there has been a tremendous growth in the understanding of general relativity and of its relation to experiment. The theory continues to pass the most exquisite tests; in May 2011 NASA announced the results of satellite (Gravity Probe B) measurements precisely confirming the frame-dragging precession effects near a rotating mass first pointed out by Einstein in 1913.[6] General relativity is a lively field of experimental and theoretical research in astrophysics, cosmology, and the dynamics of the early universe. It is used in studying stellar structure and the formation of neutron stars, pulsars, and black holes. It is also at the center of three disciplines that were largely speculative or didn't exist fifty years ago: black hole physics, gravitational wave research, and inflationary cosmology. These will be briefly reviewed here.

Singularities and black holes

Near the end of Chapter 6, it was noted that Einstein returned to the Schwarzschild solution in April 1939 with a paper in *Annals of Mathematics* arguing that the so-called Schwarzschild singularities do not exist; in particular, he argued that the total gravitational mass within a given radius of a contracting spherical star cluster would always remain below a certain bound for any realistic system of solar masses. The existence of such a bound would mean that gravitational collapse would cease and the Schwarzschild radius would never be reached. Ironically, within a few months of Einstein's paper, Robert Oppenheimer and his student Hartland Snyder published a paper (appearing on September 1, the day WW II began in Europe) claiming that, according to general relativity (and excluding various possible interrupting processes), a sufficiently massive spherical star, having exhausted its thermonuclear resources, would gravitationally contract indefinitely to its Schwarzschild radius. In doing so, it would "cut itself off from the rest of the universe",[7] producing what is known today as a "black hole" – the term coined only in 1967.[8] As

Oppenheimer and Snyder indicated, a black hole is a body created by the gravitational implosion of a massive star with gravity so strong that it is surrounded by an event horizon, a region across which things can enter but from which nothing, not even light, can escape. Today, from the observed motions of gas and stars, it is generally assumed that supermassive black holes (millions or even billions of times more massive than the Sun) lie at the center of most galaxies.

Einstein never directly responded to the Oppenheimer-Snyder paper. But his aversion to singularities was well-known, often expressed, and well-summarized by the relativist Peter Bergmann, one of his assistants in Princeton in the late 1930s:

> It seems Einstein was always of the opinion that singularities in classical field theory are intolerable . . . because a singular region represents a breakdown of the postulated laws of nature. . . . (O)ne can turn this argument around and say that a theory that involves singularities, and involves them unavoidably, moreover, carries within itself the seeds of its own destruction.[9]

The seeds of destruction sprouted within general relativity in the 1960s as Roger Penrose proved that a singularity exists within every black hole, and then Penrose, Hawking, and Robert Geroch proved theorems demonstrating the existence of singularities in a wide class of solutions to the Einstein field equations of gravitation.[10] It is not entirely certain that Einstein would have been dismayed. His last opinion on the matter seems to suggest that he thought singularities in general relativity might be avoided by either denying the validity of the gravitational field equations in the case of "very high density of field and matter" or, reminding that general relativity presumes an unaesthetic separation of "gravitational field" and "matter", by holding out the hope that in a unified theory eliminating this dualism, no singularities would arise.[11] Today it is widely accepted that general relativity does not coherently extend to the smallest scales and highest energies of the Planck regime, the theoretical meeting place of gravity, quantum mechanics, space and time (10^{-33} cm; 10^{-43} sec.; 10^{19} GeV). One commonly encountered goal of the program of quantum gravity is to demonstrate that at these scales, singularities are theoretically removed by the disappearance of the space-time continuum, replaced perhaps by a discrete structure or a structure from which space and time emerge. In any case, Einstein's antipathy to singularities is well represented by Penrose's cosmic censorship conjecture (1969), affirming that a singularity cannot be "naked" but must be "clothed" behind a horizon so that it remains invisible to observation.[12]

Black hole physics

A black hole described by the Schwarzschild solution is static, but a more realistic scenario of gravitational collapse to a black hole will be rotational, creating a spinning object. Roy Kerr in 1963 found another exact solution to the Einstein field equations describing the space-time geometry surrounding an electrically neutral spinning black hole; Kerr's solution was generalized to a charged configuration by Ted Newman in 1964.[13] Remarkably, the Kerr-Newman description of the space-time surrounding a spinning black hole can be specified by only three numbers: the object's mass, its angular momentum, and its charge. In the early 1970s, Jacob Bekenstein showed that black holes have an entropy proportional to the surface area of their horizon, and though very cold, radiate and so have a temperature; the temperature of a small black hole of only three solar masses was estimated to be less that a millionth of a degree above absolute zero. Hawking and others subsequently developed the "laws of black hole mechanics" in terms of an analogy to thermodynamics.[14] In general relativity, the mass of a black hole is the same quantity as its total energy, so its mass plays the role of thermodynamic energy, while a black hole's surface gravity is analogous to temperature, and the area of its event horizon is analogous to thermodynamic entropy. On the assumption of black hole entropy, Bekenstein in 1972 and Hawking in 1974 formulated an analogue of the second law of thermodynamics showing that black hole entropy will not decrease as matter falls across the event horizon.[15]

Hawking then was able to theoretically characterize the process by which black holes radiate by pointing to a well-known aspect of quantum theory: particle pair production due to quantum energy fluctuations in empty space. According to quantum field theory, pairs of particles and their corresponding antiparticles are constantly created by energy fluctuations; they promptly annihilate each other due to their opposite charges, hence these particles are called *virtual*. According to Hawking, when this process occurs at the horizon of a black hole, it can lead to the radiation of a real particle, and the eventual complete evaporation of the black hole. Consider a pair of low mass virtual particles (say, a neutrino and its hypothetical antiparticle, an antineutrino) created at the horizon. In the Hawking process, one will have energy $-E$, the other energy E. Supposing the one of energy $-E$ falls inwards to the black hole, the other may escape to become a real particle (since its annihilating partner has disappeared); this particle would be observed as Hawking radiation. As the black hole will have to pay the energy debt brought in by the negative energy member of the pair, the outgoing radiation will carry off energy from the black hole. While the process occurs at a stunningly negligible rate, over eons of time (many times longer than the age of the universe), the mass and horizon of a black hole will shrink, leading ultimately to the black hole's evaporation. Of greater theoretical consequence is that Hawking radiation suggests also that all information regarding the quantum state of the particle falling into the black hole is irretrievably lost. That is a problem for quantum mechanics, since despite its probabilistic character, in principle quantum information is never lost: from complete information about how a process ends – its final state – one can always reconstruct how it began – its initial state (this assumes, as quantum mechanics does, that probabilities of all possible processes must always sum to 1).

In sum, if information is truly lost, quantum mechanics is strictly inconsistent with black hole thermodynamics and its basis in general relativity. The so-called black hole information paradox, once thought resolved in the 1990s, has been revived most recently as one of the central problems of theoretical physics.[16]

Cosmological constant and "dark energy"

After the completion of general relativity in 1915, Einstein sought on Machian grounds to find global (cosmological) solutions to the gravitational field equations. As was common at the time, he assumed the universe to be unchanging and eternal at the largest scales. Seeking a static global solution in 1917, Einstein added a "cosmological term" to his field equations to counter the gravitational tendency of matter to clump together. The attempted "fix" was contrived, and couldn't possibly work. Russian mathematician Friedmann would show a few years later that the most natural global solutions of the field equations, without the new term, correspond to a non-static universe either expanding or contracting. After Hubble provided convincing evidence of an expanding universe, Einstein retracted the new term in 1931.[17] Curiously, his final pronouncement was not to condemn it on empirical grounds. Writing in 1945, he deemed it "possible from the point of view of relativity" but "to be rejected from the point of view of logical economy".[18]

Recall that originally Einstein added the term to the left-hand side of the field equations,

$$R_{\mu\nu} - \frac{1}{2} g_{\mu\nu} R + \Lambda g_{\mu\nu} = 8\pi T_{\mu\nu}.$$

Modern practice, however, places the term on the right-hand side,

$$R_{\mu\nu} - \frac{1}{2} g_{\mu\nu} R = 8\pi T_{\mu\nu} - \Lambda g_{\mu\nu}$$

where it represents the possible stress-energy of empty space (the quantum vacuum) thus contributing to the stress-energy tensor. If empty space is (as quantum theory requires) filled with vacuum energy, it will exert a repulsive "negative pressure", a force indistinguishable from Einstein's cosmological term.[19] In short, if quantum theory is correct, the cosmological term is naturally included on the right-hand side of the field equations; moreover, the term represents an energy density of empty space with a constant value. It is a remarkable property of the vacuum that its energy density is not diluted with the expansion of space.

In its re-emergence, the cosmological constant has become a central thorn in the side of fundamental physical theory; Susskind recently termed it "the greatest enigma of theoretical physics".[20] According to quantum theory, vacuum energy arises from energy fluctuations rooted in the Heisenberg uncertainty relations. But if all theoretically possible contributions to vacuum energy stored in all known fields and particles are summed up (admittedly, these contributions are not theoretically well understood), an impossibly large value of vacuum energy density results. Beginning in the 1970s, the idea of *supersymmetry* was invoked to skirt this result. According to supersymmetry, every boson has a fermionic partner and vice versa; supposing fermionic and bosonic contributions to the vacuum energy to be of different signs, their respective contributions would exactly cancel. A significant difficulty with this putative solution is that supersymmetry has yet to be observed. Nonetheless for some decades many theoretical physicists believed that the value of the cosmological term, corresponding to the energy density of the vacuum, had to be zero; somehow the different contributions from all particles and fields would effectively sum to 0.

Then in 1998, two rival teams of astronomers revealed that observations of redshifts of light from the most distant supernovae pointed to an accelerating expansion of the universe; a completely unexpected finding, the two teams shared the Nobel Prize in Physics in 2011.[21] Their finding suggests some kind of energy is fueling the increasing rate of expansion of the universe; this so-called "dark energy" is consistent with a very small, but positive, value of the cosmological constant.[22] Moreover, taking the CMB to be a measure of the total energy in the universe after subtracting the total amount of matter (both familiar and "dark") contained in galaxies and galaxy clusters (about 27 percent of the former), the remaining 73 percent of all that is believed to exist in the universe resides in the constant energy density of empty space, energy of the vacuum. Depending on how the various theoretical contributions to the vacuum are assessed, the observed value of the cosmological constant is outrageously smaller than the theoretically predicted value, anywhere from approximately 10^{-120}–10^{-60} times less. With good reason, this mismatch has been called "the largest failure in physics, ever".[23] From a completely different direction, attempts to argue that the observed value of the cosmological constant must lie within anthropic bounds remain ill-defined and highly speculative (see §2 below).

Inflationary cosmology and gravitational waves

Since the 1930s most cosmological models of an expanding universe have employed a particular class of general relativistic models, so-called Friedmann-Lemaître-Robertson-Walker (FLRW) metrics.

FLRW metrics are standardly taken as the background model for the universe at the largest scales; such universes are exactly spatially homogeneous (with no center, and no distinguishing features) and isotropic (at any point all directions are equivalent). More realistic refinements are introduced by considering small perturbations of FLRW universes (see below). However the large-scale isotropy of the CMB, together with the cosmological principle (if isotropic for us, then isotropic for any observer), gave rise to the following questions: How can parts of the universe separated by great distances be so observationally similar? How might such an extraordinary degree of spatial homogeneity have been produced? In standard Big Bang models, the distances between regions observed similar today are so large that the regions could not have been in causal contact in the past. Any similarities observed between distantly separated regions would appear to be due to very special initial conditions in the early universe, a highly unsatisfactory situation for physical theory.

Furthermore, the universe is observed to be nearly flat at the largest scales; this means that the mean matter-energy density of the universe is incredibly close to the "critical" value at which the spatial geometry of the universe is Euclidean. But in general relativity, curved space is generic and flatness is a largely unstable exception away from which the universe will almost certainly evolve. As with Einstein's cosmological constant, a fine balance is required to maintain flatness. As the mean density is related to the rate of expansion, if too large, it would have overwhelmed expansion resulting in a closed space with finite volume and no edge wherein parallel lines would be observed to converge. If the mean density is too little, expansion would have overwhelmed it and the universe would be infinite and open: parallel lines would be observed to diverge. The third possibility, that the large-scale geometry of the universe is Euclidean, corresponds to observation but appears to require another special initial condition, a highly fine-tuned flatness in the very early universe.

In 1980, particle physicist Alan Guth, working on another problem, posited a hypothetical vacuum energy in the very early universe driving a cosmic spatial expansion that would be both enormous and extremely rapid (an exponential function of time).[24] Any mechanism of this kind itself requires a bit of fine-tuning (it must not end too soon; it must end when the universe is almost flat) but by rapidly stretching everything out and diluting space, inflation can in principle address the above two fine-tuning quandaries, widely called the "horizon" and "flatness" problems. While inflation is not a theory (there are literally hundreds of possible models of the mechanism producing inflation), it is widely accepted today that an era of inflation did occur in the very early universe. And it is widely believed that the observed value of the cosmological constant, accounting for the accelerating rate of expansion of the universe discovered in 1998, is a weakened relic of the much larger vacuum energy-producing inflation in the very early universe and then effectively switched off.

Traces of an inflationary epoch should show up in the CMB as slight density perturbations consistent with those observed (about 1 part in 100,000). These perturbations are theoretically produced by quantum fluctuations in an inflating sea of vacuum energy, with a distinct spectrum of nonuniformities produced by the different hypothetical mechanisms. The perturbations can be decomposed into scalar and tensor fluctuations; the former are energy density fluctuations that are scale-invariant (hence independent of the space-time metric). The latter arise from perturbations of the FLRW metric and its coupled energy-momentum tensor and are thought to give rise to gravitational radiation that cause a particular polarization pattern in the CMB. In March 2014 it was widely and dramatically reported that such patterns were observed in the CMB by the BICEP2 team, comprising the first direct observational evidence of inflation.[25] If confirmed, the finding would have been an extraordinary achievement on two fronts: strong evidence of cosmological inflation, together with confirmation of the existence of gravitational waves. However, further analysis showed that the observed patterns were most likely produced by CMB interactions with intervening cosmic dust, and both claims were withdrawn within a year.

The hypothetical existence of gravity waves goes back to 1916. Exploiting an analogy to the way that a system of moving charges emits electromagnetic waves, Einstein predicted the existence of gravitational radiation.[26] To be sure, Einstein had second and even third thoughts about the reality of gravitational waves, famously denying their existence in a 1937 paper with Nathan Rosen that, however, contained an error negating its conclusion.[27] The reality of gravitational waves as a legitimate implication of general relativity was not fully accepted by the (very small) community of general relativists until the 1960s after systematic analysis of full nonlinear general relativity by Hermann Bondi and Roger Penrose. They found that changing gravitational fields created by moving masses create waves that propagate at the speed of light, slightly stretching or shrinking the distance between objects lying in their path. Unlike electromagnetic waves, gravitational waves are not vibrations in space-time but of space-time geometry. On account of its extreme weakness, gravitational radiation has been for one hundred years the last significant prediction of the general theory of relativity yet to be observed.

It has long been thought that a characteristic place to look for gravitation waves is a binary star system, two stars orbiting around a common center of mass like a spinning dumbbell, the paradigm of a quadrupole system. Indirect evidence for the existence of gravitational waves appeared in the discovery of the first binary pulsar system (PSR B1913 + 16) by radio astronomers Russell Hulse and Joseph Taylor in 1974; this is a pair of compact neutron stars orbiting a common center very rapidly (period of 59 milliseconds) and losing orbital energy. For their discovery, Hulse and Taylor were awarded the Nobel Prize in 1993. Theoretically, the diminishing orbital energy is radiated away as gravitational waves and observations over three decades have precisely confirmed the rate predicted by general relativity at which the system's orbit is shrinking. However, this did not comprise a direct observation of gravitational waves.

On September 14, 2015, gravitational waves were detected by the advanced LIGO (Laser Interferometer Gravitational-Wave Observatory) scientific collaboration operated by Caltech and MIT. The signal, detected nearly simultaneously at LIGO observatories in Hanford, Washington, and in Livingston, Louisiana, had the expected sharp "chirp" signature of a violent event, identified as consistent with a computational model of the inspiraling and coalescence of two "surprisingly massive" black holes approximately 1.3 billion light years from Earth.[28] Gravitational wave detection is not just another confirmed prediction of general relativity, for it opens an entirely new observational window into the

early universe. Prior to 2015, all astronomical and cosmological observations were based on the electromagnetic spectrum from infrared to
ultraviolet. It may well be that further observations of the CMB will
not be sensitive enough to identify unmistakable evidence of inflation
produced by primordial gravitational waves. But since these are slight
ripples of space-time produced during inflation, it may be possible
to "look behind" the CMB. Their direct detection would furnish an
entirely new band of observation, theoretically extending back in time
earlier than the formation of the CMB, produced when the universe
first became transparent to light. It is widely hoped that gravitational
waves will provide direct observational confirmation of an inflationary
epoch in the very early universe, and possibly a means of choosing
between the current plethora of inflationary models.

The dream of unification: A theory of everything

By contemporary lights, the unified field theory program occupying
Einstein for three decades was a hugely premature, bound-to-fail
search for the cosmic world order. It produced no new physics but
only a trail of equations too unwieldy or complex to be solved. While
Einstein worked in Berlin in the 1920s, before quantum mechanics
had monopolized the attention of theorists, the unification enterprise could be viewed as a longshot pursuit of limited inherent interest. But with the discovery of the physics of the nucleus in the early
1930s, tolerance quickly ebbed for a program content to grapple
purely formally with a unification of general relativity and classical
electromagnetism. Several generations of physicists, schooled in the
results-oriented pragmatism of particle physics, dismissed the later
Einstein as a stubborn old man incapable of coming to terms with
quantum mechanics while choosing to remain willfully ignorant
of the exciting new physics of the atomic nucleus. By 1935, Robert
Oppenheimer could aptly sum up the view of a large segment of
the theoretical community in referring to Einstein (and his dogged
pursuit of the unified theory program) as "cuckoo".[29]

Freeman Dyson at the Institute of Advanced Study since 1948
relates a story with a similarly harsh assessment.[30] On first arriving
in Princeton, the young English mathematician and quantum field
theorist was most eager to meet Einstein. He thought to prepare

beforehand by reading Einstein's most recent papers on unified field theory. Finding them to be "junk", Dyson in embarrassment skipped his appointed meeting with Einstein, then carefully avoided making contact for the next seven years until Einstein's death in 1955. Much later Dyson admitted to being under the influence of a certain arrogance inflicting the theorists who had just developed quantum electrodynamics, at the time the cutting edge of physical theory. Yet as late as the 1990s, particle physicist Abraham Pais – Einstein's scientific biographer – could remark in a TV documentary:

> If Einstein had stopped doing physics in the year 1925 and had gone fishing, he would be just as beloved, just as great. It would not have made a damn bit of difference.[31]

That Pais's verdict represents the prejudice of his generation will be confirmed by even the slightest glance at the contemporary literature on foundations of quantum mechanics. But postwar theorists took an understandably dim view of seeking to unify fundamental laws through a speculative method relying on such a vague criterion as mathematical simplicity while ignoring the new frontier of high-energy experiments. The furthest reach of quantum theory through most of the 1960s produced a largely instrumentalist or descriptivist conception of physical theory, as the attempt to provide an adequate taxonomy for the proliferating data of elementary particles emerging in particle colliders. It was easy then to say of Einstein's speculative method that physical theory is *not like that* – an ironic judgment since, by 1980 or so, much of fundamental physical theory had *become like that*. So rapid and prevalent was the new trend that already at the end of the 1980s, Harvard particle physicist Howard Georgi would satirize theorists engaged in pursuit of elegant geometrical unifications of all fundamental interactions, including gravity. According to Georgi, such theorists, led by the sirens of philosophical principles and mathematical elegance while ignoring the real physics of quantum mechanical interactions, suffered from an "Einstein Complex", that is,

> a desire to work on difficult and irrelevant questions just because Einstein did.[30]

The most prominent example of the contemporary unification program, in 1989 and today, is string theory, the favored candidate for a unified understanding of the basic laws of the universe. String theory (really superstring theory since it requires supersymmetry) is not so much a theory but a so-far-incomplete mathematical framework that makes no testable predictions. It poses questions of considerable philosophical interest about the scientific standing of theorizing that may only allow evaluation through methods of so-called non-empirical confirmation.[31] When string theory is coupled with inflationary cosmology (and in particular, with so-called eternal inflation), quite another set of philosophical questions arises.[32] Unless some principled reason is found to severely constrain the enormous number of string theories believed possible (admittedly a vague notion, but the range between 10^{100}–10^{500} is widely bruited about[33]), each string theory may correspond to a bubble universe produced in the unending ("eternal") process of inflation. An overabundance of different and distinct universes, each with its own particular values of the fundamental constants of physics, is said to "populate" the landscape of solutions of possible string theories. The result is the so-called "inflationary multiverse".

How does any of this bear on Einstein's legacy of unification? Recall Einstein's seeming declaration of rationalist faith when characterizing the task of the physical theorist:

> We wish not only to know *how* Nature is (and *how* her processes transpire), but also to attain as far as possible the perhaps utopian and seemingly arrogant goal, to know *why* Nature *is thus and not otherwise.*[34]

No less than an avowal of the principle of sufficient reason, it asserts the theorist is ultimately satisfied only when attaining the very apex of cognitive understanding. Admittedly, immediately following the quoted passage, Einstein quickly refers to this aspiration as "the Promethean element of scientific experience", implying perhaps a rather unpleasant penalty for hubris. And as seen in Chapter 10, on various occasions Einstein took the opportunity to distance himself from any straightforward endorsement of this kind. As a "tamed metaphysicist" this rationalist aim is deemed a largely motivational and regulative injunction. Nevertheless, it might be argued that Einstein only articulated the most ambitious goal of physical theory since the rise of modern science, whether held consciously or unconsciously, realistically or only as a matter of aspirational faith. On the other hand with the inflationary multiverse, the hope to rationally understand the universe on the basis of a unified theory constrained only by fundamental principles and mathematical simplicity is radically changed, if not surrendered entirely. Consider the attempt to explain the values of physical constants and parameters that today must be put in "by hand" (i.e., with their measured values) in the standard model and in other areas of fundamental physics (such as the observed value of the cosmological constant). Under the mantra of "anything that can happen will happen an infinite number of times",[35] proponents of the inflationary multiverse seek to account for the particular values of constants and parameters we observe as a selection effect: e.g., the fact that the cosmological constant has the tiny value it is observed to have is simply a reflection of our existence: if it were otherwise, we would not be here to measure it. The implications of a sea change of this magnitude upon the very notion of what constitutes an adequate explanation in fundamental physics are currently under debate. It is certainly a "historic fork in the road" and, if taken, a "radical change" from the style of theoretical physics that Einstein exemplified.[36]

Einstein's realism and NOA

Readers of Arthur Fine's *The Shaky Game: Einstein, Realism and the Quantum Theory* (second edition, 1996) will know that the view of science Fine recommends to philosophers as the Natural Ontological Attitude (NOA) is closely fashioned along the lines of a sympathetic portrayal of Einstein's realism. NOA is a minimalist stance towards science. Both scientific realism and antirealism are distinguished from NOA by what they respectively add to the "core position" that characterizes NOA, namely

> (to) accept the results of scientific investigations as "true", on a par with more homely truths,[37]

i.e., statements of purported fact in ordinary discourse. The scare quotes around the term "true" receive comment below; the phrase "on a par" is elliptical for something like *on a par with respect to the normally assumed trust that a statement is made in good faith, and on the basis of the kind of evidence that warrants its utterance*. In placing statements of science on a par with those based on the evidence of the senses, NOA invokes the spirit of Einstein's remark that "the whole of science is nothing more than a refinement of everyday thinking".[38] Neither scientific realism nor antirealism, what then is NOA? To Fine it is a "nonrealism", others claim to find that "NOA is a thoroughly realist view".[39] Considering its pedigree, it seems more appropriate to view NOA as a realism, though not a thoroughly realist one. NOA's realism is metaphysically deflated, taking its lead from a close analysis of Einstein's responses to quantum mechanics. That analysis convincingly argued that Einstein's metaphysically realist remarks are "not to be taken at face value".[40]

Following Fine's reconstruction, the core of Einstein's realism is found not in any particular thesis or doctrine but in two critical moments operating in conjunction; together they enable "Einstein to use the vocabulary of metaphysical realism" but also "to pull its metaphysical sting".[41]

The first moment is one of *entheorizing*. To "entheorize" is to specify a "family of constraints" governing what can be considered a properly realist theory. This is just the "programmatic" aspect of realism set in opposition to the perceived positivist irrealism and instrumentalism of quantum mechanics. In this book, these constraints are termed principles; they include observer-independence and property definiteness (the properties of physical objects have definite values, whether or not they are observed), causality in the sense that the laws of physics should permit univocal prediction of the state of a physical system from exact knowledge of the system at a given time, and a principle of separation, that physical systems are individuated by spatial separation so that an intervention made on one system has no influence on another spatially separate from it.

The list can be lengthened to include representation in space-time (the theory must describe physical phenomena in space and time), general covariance (in the sense that no fixed background space or space-time may appear in a theory), a monist principle that fields are primary (particles and their laws of motion are derivable from the laws of the field), and a principle of unification (that it is conceptually possible to comprehend all physical phenomena and interactions in a single but logically simple mathematical scheme). By suggesting a second-order set of requirements a theory should satisfy in order to be considered a viable representation of physical reality, the pivotal step of entheorizing has more to do with favored contours of understanding and intelligibility, what Toulmin once called "ideals of natural order",[42] rather than attributes of nature itself. It supports a characterization of quantum mechanics as an incomplete theory, and the attempt to portray its empirically confirmed predictions as those of a statistical averaging over the solutions of some underlying deterministic field theory. Characteristically, having laid out such a vision, Einstein will observe there is absolutely no compelling reason, conceptual or metaphysical, why such a world picture must be correct. This underscores the programmatic aspects. No *a priori* reason can be given that the posed list of constraints is correct or unique, or that any such list can determine a unique resultant theory; certainly more than one theory may satisfy all of a given list, and perhaps do so at the same time.

Nonetheless, there is a strong motivational reason to adhere to this program for realism. It lies in a complex conviction that the independently existing physical world is a highly complex riddle to be solved, that a solution exists, that without such a system of constraints, the number of possible theories is simply too vast to ever allow progress towards it, and finally that general relativity suggests a particular set of such constraints. Analogous to Planck's striving for permanent features of the physical *Weltbild*, Einstein's programmatic realism entertains theories subordinated to ideal regulative criteria based largely on the success of general relativity, in the hope of articulating a more unified image of nature consistent with that success.

The second dialectical moment, operating in tandem with the first, is one of *meaning avoidance* or *deflection*. Deflection signals Einstein's refusal to engage on the traditional philosophical ground of realism, turning attention instead to whether a theory meets the necessary criterion of empirical success that alone merits appellation of the theory as "true", with the appropriate scare quotes for emphasis.

The deflecting move is found above all in Einstein's rejection of the realist label, and of -isms more generally. The refusal is above all a rejection of realism's account of a physical theory's truth or approximate truth as the theory's correspondence to, or isomorphism with, an external world structure that underlies and produces the phenomena of observation. Hand in hand with rejection of correspondence truth is a dismissal of the realist semantics of reference, that theoretical terms in true physical theories pick out precise structures in the world that are strongly similar to, if not exactly characterized by, the mathematical structure of the referring terms. In place of the usual realist semantics for truth and reference, deflection retains the use of the term "true" but when applying it to physical theories always appends scare quotes (as in the "core position") to signal that its sole meaning can only be an aim of complete agreement of theory with the totality of observation, in short, empirical adequacy. Justification for this conception of the adequacy of physical theory is always qualified, but appears to be based on Einstein's experience (above all, with general relativity) of the ability of such theories to obtain coordination with experience (confirmation) "with advantage" (mit Vorteil).[43] The latter phrase is significant; "advantage" signals the programmatic entheorizing agenda, an oblique indication that a theory be

> a conceptual model (Konstruktion) for the comprehension of the interpersonal whose authority lies solely in its confirmation (Bewährung). This conceptual model refers precisely to the "real" (by definition), and every further question concerning the "nature of the real" appears empty.[44]

This passage, minus the deflecting bit about sole authority, well illustrates the sense in which Einstein is indeed the "paradigm realist". With it, it is a prototypical manifestation of Einstein's realism: not a metaphysical doctrine but a motivational impulse necessary to engage in science at all, accompanied by a fervent belief akin to a religious faith in the ability of thought to comprehend what is:

> I have found no better expression than "religious" for confidence in the rational nature of reality insofar as it is accessible to

human reason. Wherever this feeling is absent, science degener-
ates into uninspired empiricism.[45]

Although Einstein's realism is a godparent of NOA, it is also
clear that there are aspects of NOA lacking analogues or precursors there. The entheorizing and deflecting moves are projected
by NOA into a broad and encompassing picture of science and its
multifarious practices. It is a view of science broadly pragmatic,
in the sense of Dewey. Indeed, like Dewey, NOA highlights "the
social in science"; once that is taken into account "no further special framework is required for understanding scientific practice . . .
nothing beyond the common framework of everyday pragmatism".[46] Diametrically opposed to everyday pragmatism, however,
is the Einstein who called attention to "the Promethean element of
scientific experience" and the one who could declare "our experience hitherto justifies us in believing that nature is the realization of the simplest mathematical ideas", a proclamation as far
from refined everyday thinking, and as metaphysical, as can be. Of
course the term "our experience hitherto" refers here not to everyday sensory experience but to the retrospective assessment that the
mathematical strategy of general covariance produced the winning
ticket for his biggest success, the general theory of relativity (see
Chapter 10).[47]

It is Einstein, the "tamed metaphysicist" or "enlightened rationalist"[48] lurking behind the purple prose of the 1933 Herbert
Spencer Lecture. Even as "enlightened", it is a rationalism that cannot be deconstructed in quite the same way as his realism, i.e., as
stemming from "the pre-rational springs of human behavior".[49]
The philosophical passion for theoretical unity is an attitude integral to nearly all of Einstein's many achievements. From his later
failures, one can, as did several generations of physicists, dismiss
this passion as simply tilting at windmills. On the other side of
the coin, if their utterances are taken at face value, many in the
current group of theorists seeking unity of the laws of nature lack
Einstein's intellectual humility in conceding that a methodology
of mathematical speculation may simply be rationalist illusion. In
something of the spirit of NOA's "letting science speak for itself",
Einstein's life and science exhibit a profound role for philosophy in
understanding the practice of science. Whether or not this leads to

an anti-naturalist picture of science, it reveals, as much as anything, that science is also a practice of faith (and hope) that the world is ultimately comprehensible in human terms. Getting the full picture matters. Einstein remains "the paradigm realist" from whom contemporary philosophy of science may still learn.

Notes

1 Einstein, "Physics, Philosophy, and Scientific Progress", 1950 speech delivered in English to the International Congress of Surgeons, Cleveland, Ohio. Text reprinted in *Physics Today* (June 2005), pp. 46–8; p. 46.
2 See Miller, Arthur I., *Einstein, Picasso: Space, Time and the Beauty That Causes Havoc*. New York: Basic Books, 2001.
3 Wheeler, John A., "Mentor and Sounding Board", in John Brockman (ed.), *My Einstein*. New York: Vintage Books (Random House), 2007, pp. 27–37; p. 33.
4 Reported in Pais, Abraham, *"Subtle Is the Lord ..."The Science and the Life of Albert Einstein.* New York: Oxford University Press, 1982, p. 268.
5 Roger Penrose, "Gravitational Collapse and Space-Time Singularities", *Physical Review Letters*, v. 14 (no. 3), January 18, 1965, pp. 57–9.
6 Gravity Probe B, in an orbit of 640 km, confirmed that the rotating Earth drags a tiny amount of space-time around with it. This is the so-called Lense-Thirring effect, after the two Austrian physicists who in 1918, with Einstein's assistance, carried out the general relativistic calculation. See C.W.F. Everitt et al, "Gravity Probe B: Final Results of a Space Experiment to Test General Relativity", *Physical Review Letters*, v. 106, 221101 (2011), 3 June 2011.
7 Oppenheimer's characterization in 1958, cited by Werner Israel, "Dark Stars: The Evolution of an Idea", in Stephen Hawking and Werner Israel (eds.), *300 Years of Gravitation*. Cambridge and New York: Cambridge University Press, 1987, pp. 199–276; p. 230.
8 By John A. Wheeler in the Sigma Xi-Phi Beta Kappa Annual Lecture, American Association for the Advancement of Science, New York, December 29, 1967; see "Our Universe: The Known and the Unknown", *American Scientist* v. 56, no. 1 (1968), pp. 1–20; p. 8.
9 Bergmann, Peter, "Open Discussion, following papers of S. Hawking and W.G. Unruh", in H. Woolf (ed.), *Some Strangeness in the Proportion: A Centennial Symposium to Celebrate the Achievements of Albert Einstein*. Reading, MA: Addison-Wesley Publishing Co., 1980, p. 156.
10 More specifically, they showed that space-time is not singularity-free if the following conditions all obtain: 1) the Einstein field equations; 2) a positive energy condition (gravity must act on everything as an attractive force, a condition violated by the cosmological constant); 3) strong causality (time travel is impossible); and 4) there must be a certain amount of matter in the universe. For details, see Earman, John, *Bangs, Crunches, Whimpers, and Shrieks: Singularities and Acausalities in Relativistic Spacetimes*. New York: Oxford University Press, 1995, Chapter 2.

11 "On the 'Cosmologic Problem' ", Appendix added to the second (1945) edition of *The Meaning of Relativity*, Princeton, NJ: Princeton University Press, Fifth edition, 1956, pp. 109–32; p. 129 and note, p. 124.

12 See Clarke, Christopher J.S., "The Cosmic Censorship Hypothesis", in George Ellis, Antonio Lanza, and John Miller (eds.), *The Renaissance of General Relativity and Cosmology*. Cambridge, UK: Cambridge University Press, 1993, pp. 86–99.

13 See Melia, Fulvio, *Cracking Einstein's Code: Relativity and the Birth of Black Hole Physics*. Chicago and London: University of Chicago Press, 2009.

14 Bekenstein, Jacob D., "Black-hole Thermodynamics", *Physics Today* (January 1980), pp. 24-31; Bardeen, John M., Brandon Carter, and Stephen Hawking, "The Four Laws of Black Hole Mechanics", *Communications in Mathematical Physics* v. 31 (1973), pp. 161–70.

15 Ibid., Jacob Bekenstein (1980).

16 For a readable account, see Matt Strassler, "Black Hole Information Paradox: An Introduction", https://profmattstrassler.com/articles-and-posts/relativity-space-astronomy-and-cosmology/black-holes/black-hole-information-paradox-an-introduction/. A recent speculative idea of Leonard Susskind and Juan Maldecena proposes to resolve the paradox while preserving consistency between general relativity and quantum mechanics has the moniker "ER = EPR", wherein EPR correlations between two "entangled" black holes are established via an Einstein-Rosen bridge mechanism. See Tom Siegfried, "A new 'Einstein' equation suggests wormholes hold key to quantum gravity", *Science News*, August 17, 2016: https://www.sciencenews.org/blog/context/new-einstein-equation-wormholes-quantum-gravity

17 The original (1922, 1924) papers of Alexander Friedmann, together with Einstein's appended notes, have been translated from the original German in Jeremy Bernstein and Gerald Feinberg (eds.), *Cosmological Constants: Papers in Modern Cosmology*. New York: Columbia University Press, 1986, pp. 49–67. This volume also contains a reprint of Hubble's 1929 paper, "A Relation Between Distance and Radial Velocity Among Extra-galactic Nebulae", *Proceedings of the National Academy of Sciences* v. 15, no. 3 (March 15, 1929), pp. 168–73. Einstein's 1931 retraction appears in "Zum kosmologischen Problem der allgemeinen Relativitätstheorie", *Sitzungsberichte der Preußischen Akademie der Wissenschaften, Phys-Math. Klasse* (1931), pp. 235–7.

18 "On the 'Cosmologic Problem'", Appendix added to the second (1945) edition of *The Meaning of Relativity*, Princeton, NJ: Princeton University Press, 1956, p. 127.

19 Recall (Chapter 6, note 56) that $\Lambda = \dfrac{8\pi G\rho}{c^2}$.

Space-time curvature not determined solely by mass-energy densities ρ but also by pressure, so $T_{\mu\nu}$ can have the form $\rho + 3P$ where P is the pressure. Cosmological (FRWL) models assume however a simple form of matter as a cosmological fluid with three non-interacting components radiation, vacuum energy however and pressure-less dust – since in ordinary astrophysical objects, pressure is negligible (P ≪ ρ). However, in empty space (the vacuum), consistency with the energy conservation requires pressure to be of the same magnitude as the negative density ($P_V = -\rho_V$), and here the stress-energy tensor takes the form $\rho_V + 3P_V = -2\rho_V$.

20 Susskind, Leonard, *The Cosmic Landscape: String Theory and the Illusion of Cosmic Design.* New York and Boston: Little, Brown and Co., 2005, p. 11.

21 See "The Nobel Prize in Physics 2011, Saul Perlmutter, Brian P. Schmidt, and Adam G. Riess" at www.nobelprize.org/nobel_prizes/physics/laureates/2011/

22 $\Lambda = 8\pi G\rho_\Lambda$, where ρ_Λ is the energy density of the dark energy.

23 Ohanian, Hans C., *Einstein's Mistakes: The Human Failings of Genius.* New York: W.W. Norton and Co., 2008, p. 254.

24 Guth, Alan, *The Inflationary Universe: The Quest for a New Theory of Cosmic Origins.* Reading, MA and New York: Addison-Wesley Pub. Co., 1997.

25 See "Space Ripples Reveal Big Bang's Smoking Gun", *The New York Times*, March 18, 2014; BICEPS 2 abbreviates "Background Imaging of Cosmic Extragalactic Polarization", second experiment, at the South Pole.

26 Solving the weak field gravitational equations in linearized approximation in 1916, Einstein discovered the solutions to have the property of being transverse waves of spatial strain traveling at the speed of light. The paper contains several mistakes.

27 See Kennefick, Daniel, *Traveling at the Speed of Thought: Einstein and the Quest for Gravitational Waves.* Princeton, NJ: Princeton University Press, 2007.

28 As the orbiting distance decreases, the orbital speeds increase, causing the frequency of the gravitational waves to increase until the moment of coalescence. See "Laser Interferometer Gravitational-Wave Observatory (LIGO) Scientific Collaboration" at www.ligo.org.

29 Dongen, Jeroen Van, *Einstein's Unification.* Cambridge, UK and New York: Cambridge University Press, 2010, p. 186.

30 As related in Smolin, Lee, "In Search of Einstein", in John Brockman (ed.), *My Einstein: Essays by Twenty-Four of the World's Leading Thinkers on the Man, His Work, and His Legacy.* New York: Vintage/Random House, 2006, pp. 109–20; pp. 110–11.

31 Cited by Landsman, Nicholas Pieter (Klass), "When Champions Meet: Rethinking the Bohr-Einstein Debate", *Studies in History and Philosophy of Modern Physics* v. 37 (2006), pp. 212–42; p. 213.

32 Georgi, Howard, "Effective Quantum Field Theories", in Paul Davies (ed.), *The New Physics.* New York: Cambridge University Press, 1989, pp. 446–57; p. 456.

33 See Dawid, Richard, *String Theory and the Scientific Method.* New York: Cambridge University Press, 2013.

34 See Susskind, Leonard, *The Cosmic Landscape: String Theory and the Illusion of Intelligent Design.* New York: Little, Brown and Co., 2006.

35 Weinberg, Steven, "Living in the Multiverse", in Bernard Carr (ed.), *Universe or Multiverse?* New York: Cambridge University Press, 2007, pp. 29–42; p. 31.

36 "Über den gegenwärtigen Stand der Feld-Theorie", in *Festschrift Prof. Dr. A. Stodola zum 70. Geburtstag.* Zurich und Leipzig, Germany: Orell Füssli Verlag, 1929, pp. 126–32, p. 126.

37 Guth, Alan, "Quantum Fluctuations in Cosmology and How They Lead to a Multiverse", arXiv:1312.7340v1 [hep-th] December 27, 2013.
38 "We now find ourselves at a historic fork in the road we travel to understand the laws of nature. If the multiuniverse idea is correct, the style of fundamental physics will be radically changed". Steven Weinberg, in an interview with Alan Lightman on July 28, 2011; quoted in Lightman, *The Accidental Universe: The World You Thought You Knew*. New York: Vintage/Random House, 2013, p. 5.
39 Fine, Arthur, *The Shaky Game: Einstein, Realism and the Quantum Theory*. Second edition. Chicago: University of Chicago Press, 1996, p. 128.
40 Ibid., p. 176: "NOA also is trusting in so far as it regards the sciences themselves as a refinement of common human practices; as Einstein remarked, 'The whole of science is nothing more than a refinement of everyday thinking'". That quote is taken from the Sonia Bargmann translation of "Physics and Reality", 1936, in Albert Einstein, *Ideas and Opinions*. New York: Crown Publishers, 1954, p. 290.
41 Musgrave, Alan, "NOA's Ark: Fine for Realism", *Philosophical Quarterly* v. 39 (1989), pp. 383–98; p. 383.
42 Fine, *The Shaky Game*, p. 111.
43 Ibid., p. 106.
44 Toulmin, Steven, *Human Understanding: The Collective Use and Evolution of Concepts*, vol. 1. Princeton, NJ: Princeton University Press, 1972.
45 "Elementare Überlegungen zur Interpretation der Grundlagen der Quanten-Mechanik", in *Scientific Papers Presented to Max Born*. Edinburgh: Oliver & Boyd, 1953, pp. 33–40; p. 34.
46 "Reply to Criticisms", in Paul A. Schilpp (ed.) *Albert Einstein: Philosopher-Scientist*. Evanston, IL: Northwestern University Press, 1949, p. 680.
47 Einstein, letter to M. Solovine, January 1, 1951, in *Albert Einstein: Letters to Solovine*. New York: Philosophical Library, 1987, p. 119.
48 Fine, *The Shaky Game*, p. 188.
49 cf. Barker, Peter, "Einstein's Later Philosophy of Science", in P. Barker and C. Shugart (eds.), *After Einstein: Proceedings of the Einstein Centennial Celebration of Memphis State University*. Memphis, TN: Memphis State University Press, 1981, pp. 133–45; p. 144, n. 13.
50 Bachelard, Gaston, "The Philosophic Dialectic of the Concepts of Relativity", in Paul A. Schilpp (ed.), *Albert Einstein: Philosopher-Scientist*. Evanston, IL: Northwestern University Press, 1949, pp. 565–80; p. 580.
51 Fine, *The Shaky Game*, p. 110.

Glossary

Big Bang term facetiously coined by Cambridge physicist Fred Hoyle around 1950 to designate an initial violent event at the beginning of the universe.

blackbody idealized object assumed to be in perfect equilibrium with electromagnetic radiation. The spectrum of its emitted radiation depends only on the temperature and is independent of the object's material composition.

Boltzmann's principle name given by Einstein to Boltzmann's result holding that the entropy of a thermodynamic (macroscopic) state is directly proportional to the probability of occurrence of that state.

Bose-Einstein condensate unusual state of matter at very low temperatures in which all the constituent atoms or particles have the same energy.

Bose-Einstein quantum statistics quantum statistics obeyed by particles of integer spin (bosons). Bosons are indistinguishable, differing only by differing energies.

Brownian motion a consequence of the kinetic theory of fluids, the random motion of small but observable granules due to thermal motions of the fluid's molecules.

cosmological constant term added to Einstein's field equations representing a repulsive pressure in the vacuum of space.

cosmological principle at very large distance scales ($> 3 \times 10^6$ light years, approximately 3×10^{24} meters), the universe is homogeneous and appears the same in all directions from any point.

curvature in the general theory of relativity, a property of space-time derivable from the metric tensor determining whether straight lines initially parallel remain so when extended indefinitely.

De Sitter's universe a 1917 solution to the matter-free Einstein field equations (supplemented by the cosmological constant) describing a universe infinite in extent, and expanding.

determinism a property of laws of nature, such that with exact knowledge of the initial conditions of a system, the laws precisely predict the state of the system at any other time.

Einstein field equations the gravitational field equations of general relativity. On one side is a tensor expression for the curvature of space-time; on the other, a tensor expression for matter-energy sources of curvature. In all, there are ten equations to be solved for the ten independent components of the metric tensor $g_{\mu\nu}$.

Einstein universe a solution to his field equations derived by Einstein in 1917 describing a spatially closed static universe with a uniform distribution of matter. By closing the universe spatially, the problem of specifying boundary conditions was circumvented, but to render the universe stable against gravitational collapse, the cosmological constant was required. The solution was later shown to be non-static.

energy principle in Newtonian physics and in special relativity, the total energy of any isolated system is constant in time. In general relativity, the conservation of energy holds rigorously only for time-independent space-time geometries.

entropy a measure of the disorder of a system, inversely related to the amount of structure of the system. The second law of thermodynamics holds that the total entropy of any closed system cannot decrease with time.

EPR acronym for the 1935 Einstein-Rosen-Podolsky paper arguing that quantum mechanics is either incomplete or it allows intervention on one part of a separated system to influence the other distant part.

equipartition theorem the prediction of the kinetic theory that at thermal equilibrium, the energy available to a system will be distributed equally among all possible degrees of freedom (varieties of motion) of the system.

event a physical occurrence, or possible occurrence, with a definite spatio-temporal location relative to other occurrences. Events can be considered points of space-time, with the proviso that events are occurrences, not mathematical points.

fluctuation theory a statistical theory developed by Einstein in 1902–1904 showing that systems at thermal equilibrium will fluctuate around the average values considered by thermodynamics.

frame of reference a system of coordinates that an observer constructs to locate events and determine spatial and temporal relations between them.

hole argument a faulty argument persuading Einstein for two years (1913–1915) that generally covariant field equations were incompatible with the requirement that a field theory be deterministic.

inertial frame a frame of reference (reference system or observer plus coordinate system) in which a body on which no accelerative forces act maintains its state of motion. The laws of Newtonian mechanics require the existence of such homogenous frames, as do the laws in special relativity. In the general theory of relativity, space-time curvature allows only local inertial frames, a limited region surrounding a freely falling test body in which the laws of special relativity obtain.

kinetic theory a classical theory of the (Newtonian) motions of atoms or molecules that averages over vast numbers of such particles to explain physical properties of gases, liquids, or solids.

length contraction a core prediction of special relativity that bodies in motion, with respect to the observer measuring them, are shortened in the direction of their motion by a factor that depends upon v^2/c^2, the ratio of the square of the body's velocity to the square of the speed of light in a vacuum.

light principle the principle that the velocity of light in a vacuum is a constant and independent of the velocity of the emitter of light.

linear in mathematics, linear relations between two variables x and y are described by an equation of the form $y = ax + b$ for constants a and b. Physically, a linear relation between two quantities means that the change of one is directly proportional to the change of the other.

Mach's principle a principle adopted by Einstein in recognition of Mach's criticism of Newton's invocation of absolute space and time in explaining the effects of inertia. Mach's proposal that inertial effects result from accelerations with respect to distant cosmic masses (and not absolute space) was used by Einstein to argue for the relativity of acceleration, i.e., a principle of general relativity. In Einstein's hands, the principle states that inertio-gravitational effects are fully determined by matter-energy. Einstein adopted a closed universe cosmology in an attempt to satisfy Mach's principle, but de Sitter's empty universe model clearly violates it.

metric tensor the fundamental mathematical quantity of Einstein's field equations, a tensor $g_{\mu\nu}$ that is a 4×4 matrix with 10 independent components. Its primary role lies in the determination of the space-time interval between two nearby points, $ds^2 = g_{\mu\nu}dx^\mu dx^\nu$ (see Box 6.2). Other quantities pertaining to curvature of space-time are derivable from it.

nonlinear a mathematical relationship between two variables that does not plot as a straight line. Physically, a nonlinear relation between two quantities can mean that a small change in one can result in a disproportionate change in the other. The Einstein field equations are nonlinear partial differential equations; the nonlinearity arises essentially from the fact that the gravitational field carries energy and can act as its own source.

opalescence light passing through a very dense gas or mixture of two fluids is strongly scattered (deviated in all directions) giving off a luminous glow. The scattering has a maximum when the gas or fluid is near its critical point (the temperature of a phase transition).

principle of equivalence a heuristic assisting Einstein in going from the special to the general theory of relativity. In weak form, it asserts the equivalence of inertial and gravitational mass. Since a body's inertial mass is the parameter for its resistance to acceleration by a given force, while a body's gravitational mass is determined by the strength of a given gravitational field, acceleration (in a gravity-free region) can in special circumstances simulate freefall in a gravitational field. In mature form, the principle affirms that the effects of inertia and the effects of gravitation cannot be distinguished in any invariant way.

principle of general covariance narrowly construed, the Einstein field equations, and every other physically meaningful mathematical statement within the general theory of relativity, must retain the same form in all systems of coordinates; more expansively, it asserts that the only space-time quantities that can appear in general relativity are the metric tensor $g_{\mu\nu}$ and quantities derivable from it.

principle of relativity a statement that the laws of physics should be independent of the motion of an observer who performs experiments testing the laws. For the case of uniform motions, the principle was first enunciated by Galileo, recognized by Newton as a corollary to his laws of motion and elevated to an axiom by Einstein with special relativity in 1905. Despite its name, the general theory of relativity does not extend the principle to non-uniform motions. To Einstein, the principle of general relativity was best expressed by the principle of general covariance.

principle of superposition a quantum principle, arising from the linearity of the Schrödinger equation. In quantum mechanics, states are described by wave-like ψ-functions of which sums can be formed. Time evolution by the Schrödinger equation preserves these sums, but measurement always reveals a system to be in a definite quantum state (a definite eigenstate of the observable measured), not a sum of them. This is the so-called collapse of the wave function upon measurement.

relativity of simultaneity Einstein's 1905 assertion that the physical meaning of the relative time of occurrence of two distant events is established only by resting clocks at both events synchronized with each other via light signaling. Since according to the principle of relativity, resting and uniformly moving clocks must be considered as equivalent, the relative time of occurrence of the events will differ for differently moving inertial observers. These differences become physically significant only when the observer's motion is some significant fraction of the speed of light.

Riemannian geometry the geometry of curved spaces (more generally, manifolds) developed by Bernard Riemann (1821–1866) as a generalization of the theory of curved surfaces of Carl Friedrich Gauss (1777–1855). Riemannian manifolds include spaces with variable curvature (varying smoothly from point to point) and with more than three dimensions. Riemann represented the relation between coordinates and distances in curved spaces with the fundamental, or "metric" tensor. Around 1912, Einstein recognized Riemannian geometry as especially suitable for a relativistic theory of gravitation. The geometry of general relativity is "pseudo-Riemannian" since it is necessary to consider time distinctly from the three dimensions of space.

singularity a place where the equations of general relativity break down. According to the modern definition, a particle encountering a singularity in general relativity has no future (its trajectory is not extendible).

space-time the set of all events past, current, and future.

specific heat the heat capacity characteristic of every substance, given by a number that is the ratio of thermal energy (heat) supplied to a unit mass of the substance to its consequent rise in temperature by one degree centigrade.

statistical mechanics the application of statistical methods to the microscopic constituents of a system in order to predict the system's macroscopic behavior. Classically, the microconstituents obey Newtonian laws of motion.

supersymmetry a hypothetical symmetry of elementary particles at very high energies in the early universe; the simplest supersymmetric theories postulate a fermion partner for every boson, and vice versa.

symmetry an invariance of a system under transformation of the laws governing the system's behavior. In special relativity, the symmetry is Lorentz invariance: inertial systems are invariant under Lorentz transformation. In the general theory of relativity, since inertial frames obtain only locally, there are no non-trivial space-time symmetries.

thermodynamics the science of heat and other forms of energy developed in the 19th century, based upon two laws considered absolute. The first law states that heat is a form of energy, and that the energy of a closed system can only be converted from one form to another, but not lost. The second pertains to the flow of heat and the availability of energy to do work. In one form, the second law states the entropy (roughly, measure of disorganization) of a closed system almost always increases. To these two laws are now added a definitional zeroth law (two systems in thermodynamic equilibrium with a third system are in thermodynamic equilibrium with each other) and a third law (due to Walther Nernst) that the entropy of a system approaches a constant minimum as the temperature approaches absolute zero (on the Kelvin scale; $-273.15\ °C$).

time dilation as explained in special relativity, the slowing rate of a clock accompanying a moving body in an amount inversely proportional to the speed of the body.

unified field theory in present parlance, a quantum theory comprehending all known fundamental interactions (gravitational, electro-weak, and strong). The leading contemporary proposal is based upon superstring theory, the basic objects of which are tiny (10^{-35} m) one-dimensional strings with quantum properties. Einstein's conception was rather different and much more restricted: a classical, not quantum, unification of gravitation and electromagnetism in a single mathematical framework, stemming from a generalization of the space-time geometry of general relativity.

wave-particle duality according to quantum mechanics, light and matter have both particle and wave properties, but (standardly) the determination of one by a given experiment precludes determination of the other. In 1909 Einstein first proposed that the theory of light of the future would be an amalgam of the corpuscular and wave theories of light.

Index

Page numbers referring to Notes are designated by "n" followed by note number.